ML

Product Design
and Development

07

Product Design and Development

Third Edition

Karl T. Ulrich
The Wharton School
University of Pennsylvania

Steven D. Eppinger
Massachusetts Institute of Technology

Boston   Burr Ridge, IL   Dubuque, IA   Madison, WI   New York   San Francisco   St. Louis
Bangkok   Bogotá   Caracas   Kuala Lumpur   Lisbon   London   Madrid   Mexico City
Milan   Montreal   New Delhi   Santiago   Seoul   Singapore   Sydney   Taipei   Toronto

The McGraw·Hill Companies

PRODUCT DESIGN AND DEVELOPMENT
International Edition 2003

Exclusive rights by McGraw-Hill Education (Asia), for manufacture and export. This book cannot be re-exported from the country to which it is sold by McGraw-Hill. The International Edition is not available in North America.

10 09 08 07 06 05 04 03 02 01
20 09 08 07 06 05 04 03
CTF   SLP

Library of Congress Cataloging-in-Publication Data
Ulrich, Karl T.
  Product design and development / Karl T. Ulrich, Steven D. Eppinger.—3rd ed.
   p. cm. – (McGraw-Hill/Irwin series in marketing)
  Includes bibliographical references and index.
  ISBN 007-247146-8
   1. Industrial management. 2. Production management. 3. Industrial engineering. 4. New products—Management. I. Eppinger, Steven D. II. Title. III. Series.
HD31.U47 2004
658.5'752—dc21   2003051030

When ordering this title, use ISBN 007-123273-7

Printed in Singapore

www.mhhe.com

To the professionals who shared their experiences with us and to the product development teams we hope will benefit from those experiences.

About the Authors

Karl T. Ulrich   *University of Pennsylvania*
is an associate professor at the Wharton School at the University of Pennsylvania and
also holds a secondary appointment in the Department of Mechanical Engineering and
Applied Mechanics. He received the S.B., S.M., and Sc.D. degrees in Mechanical Engineering from MIT. Professor Ulrich has led the development efforts for many products,
including medical devices and sporting goods, and is a founder of two technology-based
companies. As a result of this work, he has received 18 patents. His current research focuses on coordination of design, manufacturing, and marketing decisions in the product
development process.

Steven D. Eppinger   *Massachusetts Institute of Technology*
is the General Motors LFM Professor of Management Science at the Massachusetts Institute of Technology Sloan School of Management and also holds a joint appointment in
the Engineering Systems Division. He is co-director of the Leaders for Manufacturing
Program and the System Design and Management Program, two of MIT's interdisciplinary masters programs. Professor Eppinger received the S.B., S.M., and Sc.D. degrees in
Mechanical Engineering from MIT. He specializes in the analysis of complex product development efforts and works closely with the automobile, electronics, and aerospace industries. His current research is aimed at the creation of improved product development
processes and project management techniques.

Preface

This book contains material developed for use in the interdisciplinary courses on product development that we teach. Participants in these courses include graduate students in engineering, industrial design students, and MBA students. While we aimed the book at interdisciplinary graduate-level audiences such as this, many faculty teaching graduate and undergraduate courses in engineering design have also found the material useful. *Product Design and Development* is also for practicing professionals. Indeed, we could not avoid writing for a professional audience because most of our students are themselves professionals who have worked either in product development or in closely related functions.

This book blends the perspectives of marketing, design, and manufacturing into a single approach to product development. As a result, we provide students of all kinds with an appreciation for the realities of industrial practice and for the complex and essential roles played by the various members of product development teams. For industrial practitioners, in particular, we provide a set of product development methods that can be put into immediate practice on development projects.

A debate currently rages in the academic community as to whether design should be taught primarily by establishing a foundation of theory or by engaging students in loosely supervised practice. For the broader activity of product design and development, we reject both approaches when taken to their extremes. Theory without practice is ineffective because there are many nuances, exceptions, and subtleties to be learned in practical settings and because some necessary tasks simply lack sufficient theoretical underpinnings. Practice without guidance can too easily result in frustration and fails to exploit the knowledge that successful product development professionals and researchers have accumulated over time. Product development, in this respect, is like sailing: proficiency is gained through practice, but some theory of how sails work and some instruction in the mechanics (and even tricks) of operating the boat help tremendously.

We attempt to strike a balance between theory and practice through our emphasis on methods. The methods we present are typically step-by-step procedures for completing tasks, but rarely embody a clean and concise theory. In some cases, the methods are supported in part by a long tradition of research and practice, as in the chapter on product development economics. In other cases, the methods are a distillation of relatively recent and *ad hoc* techniques, as in the chapter on design for manufacturing. In all cases, the methods provide a concrete approach to solving a product development problem. In our experience, product development is best learned by applying structured methods to ongoing project work in either industrial or academic settings. Therefore, we intend this book to be used as a guide to completing development tasks either in the context of a course project or in industrial practice.

An industrial example or case study illustrates every method in the book. We chose to use different products as the examples for each chapter rather than carrying the same example through the entire book. We provide this variety because we think it makes the book more interesting and because we hope to illustrate that the methods can be applied to a wide range of products, from bowling equipment to syringes.

We designed the book to be extremely modular—it consists of 16 independent chapters. Each chapter presents a development method for a specific portion of the product development process. The primary benefit of the modular approach is that each chapter can be used independently of the rest of the book. This way, faculty, students, and practitioners can easily access only the material they find most useful.

This third edition of the book adds two new chapters: 13, Robust Design; and 14, Patents and Intellectural Property. Based on information gathered from users of the second edition, we discovered that these two were the most important topics for which existing educational materials were inadequate. In addition to developing these two new chapters, we made minor revisions throughout the rest of the book, including updated examples and data, expanded explanations, and new insights from recent research and innovations in practice.

To supplement this textbook, we have developed a web site on the Internet. This is intended to be a resource for instructors, students, and practitioners. We will keep the site current with additional references, examples, and links to available resources related to the product development topics in each chapter. Please make use of this information via the Internet at www.ulrich-eppinger.net.

The application of structured methods to product development also facilitates the study and improvement of development processes. We hope, in fact, that readers will use the ideas in this book as seeds for the creation of their own development methods, uniquely suited to their personalities, talents, and company environments. We encourage readers to share their experiences with us and to provide suggestions for improving this material. Please write to us with your ideas and comments at ulrich@wharton.upenn.edu and eppinger@mit.edu.

Acknowledgments

Hundreds of people contributed to this book in large and small ways. We are grateful to the many industrial practitioners who provided data, examples, and insights. We appreciate the assistance we have received from numerous academic colleagues, research assistants, and support staff, from our sponsors, and from the McGraw-Hill team. Indeed we could not have completed this project without the cooperation and collaboration of many professionals, colleagues, and friends. Thank you all.

Financial support for some of the development of this textbook has come from the Alfred P. Sloan Foundation, the MIT Leaders for Manufacturing Program, the Gordon Book Fund, and from the MIT Center for Innovation in Product Development.

Many industrial practitioners helped us in gathering data and developing examples. We would particularly like to acknowledge the following: Richard Ahern (Herbst LaZar Bell), Liz Altman (Motorola), Lindsay Anderson (Boeing), Terri Anderson (Random Lengths Publications), Mario Belsanti (Castronics), Mike Benjamin (emPower Corporation), Scott Beutler (Motorola), Bill Burton (Magnesium Alloy Corp.), Michael Carter (General Motors), Pat Casey (Motorola), Victor Cheung (Group Four Design), David Cutherell (Design Edge), Tim Davis (Ford Motor), Tom Davis (Hewlett-Packard), John Elter (Xerox), George Favaloro (Compaq Computer), David Fitzpatrick (Boeing), Marc Filerman (Hewlett-Packard), Gregg Geiger (American Ceramic Society), Anthony Giordano (Fitch), David Gordon (Cumberland Foundry), Kamala Grasso (Bose), Matt Haggerty (Product Genesis), Rick Harkey (Meier Metal), Matthew Hern (IDEO Product Development), Alan Huffenus (Brainin Advance Industries), Art Janzen and the Enterprise Design Group (Eastman Kodak), Randy Jezowski (Ramco), Carol Keller (Polaroid), Edward Kreuzer (Brainin Advance Industries), David Lauzun (Chrysler), Peter Lawrence (Corporate Design Foundation), Brian Lee (Intelligent Automation Systems), David Levy (TH Inc.), Albert Lucchetti (Cumberland Foundry), Paul Martin (Hewlett-Packard), Doug Miller (Rollerblade), Leo Montagna (Lee Plastics), Al Nagle (Motorola), John Nicklaus (SGL Carbon), Hossain Nivi (Ford Motor), Paolo Pascarella (FIAT), E. Timothy Pawl (Pawl Inventioneering), Amy Potts (Potts Design), Earl Powell (Design Management Institute), Jason Ruble (Polymerland), Virginia Runkle (General Motors), Nader Sabbaghian (McKinsey), David Shea (General Motors), Wei-Ming Shen (Ucar Carbon), Leon Soren (Motorola), Paul Staelin (emPower Corporation), Michael Stephens (AMF Bowling), Scott Stropkay (IDEO), Larry Sullivan (Barker Lumber), Malcom Taylor (Foster-Miller), Brian Vogel (Product Genesis), David Webb (Stanley Tools), Bob Weisshappel (Motorola), Dan Williams (Motorola), and Mark Winter (Specialized Bicycle Components).

We have received tremendous assistance from our colleagues who have offered frequent encouragement and support for our somewhat unusual approach to teaching and research, some of which is reflected in this book. We are especially indebted to the MIT Leaders for Manufacturing (LFM) Program, the MIT System Design and Management

(SDM) Program, and the MIT Center for Innovation in Product Development (CIPD), which are exemplary partnerships involving major manufacturing firms and MIT's engineering and management schools. We have benefited from collaboration with the faculty and staff associated with these programs, especially Gabriel Bitran, Kent Bowen, Don Clausing, Ed Crawley, Tom Eagar, Charlie Fine, Woodie Flowers, Steve Graves, John Hauser, Rebecca Henderson, Maurice Holmes, Paul Lagace, Tom Magnanti, Kevin Otto, Don Rosenfield, Warren Seering, Shoji Shiba, Anna Thornton, Jim Utterback, Eric von Hippel, Dave Wallace, and Dan Whitney. Most importantly, LFM and CIPD partner companies have provided us with unparalleled access to industrial projects and research problems in product development and manufacturing.

Several faculty members have helped us by reviewing chapters and providing feedback from their in-class trials in teaching with this material. We are particularly grateful to these reviewers and "beta testers": Alice Agogino (University of California, Berkeley), Don Brown (University of Utah), Steve Brown (MIT), Charles Burnette (University of the Arts), Gary Cadenhead (University of Texas at Austin), Roger Calantone (University of Michigan), Cho Lik Chan (University of Arizona), Kim Clark (Harvard Business School), Morris Cohen (University of Pennsylvania), Michael Duffey (George Washington University), William Durfee (University of Minnesota), Josh Eliashberg (University of Pennsylvania), David Ellison (University of Pennsylvania), Woodie Flowers (MIT), Daniel Frey (MIT), Gary Gabriele (Rensselaer Polytechnic Institute), Abbie Griffin (University of Chicago), Marc Harrison (Rhode Island School of Design), Rebecca Henderson (MIT), Tim Hight (Santa Clara University), Mike Houston (University of Minnesota), Marco Iansiti (Harvard Business School), Kos Ishii (Ohio State University), R. T. Johnson (Wichita State University), Viswanathan Krishnan (University of Texas at Austin), Yuyi Lin (University of Missouri), Richard Locke (MIT), Bill Lovejoy (University of Michigan), Jeff Meldman (MIT), Farrokh Mistree (Georgia Institute of Technology), Wanda Orlikowski (MIT), Robert Pelke (University of Michigan), Warren Seering (MIT), Paul Sheng (University of California-Berkeley), Robert Smith (University of Washington), Carl Sorensen (Brigham Young University), Mark Steiner (University of Louisville), Chuck Turtle (University of Rhode Island), Marcie Tyre (MIT), Dan Whitney (MIT), Kristin Wood (University of Texas at Austin), and Khim-Teck Yeo (Nanyang Technological University).

Several industrial practitioners and training experts have also assisted us by reviewing and commenting on draft chapters: Wesley Allen (CWA Engineering), Geoffrey Boothroyd (Boothroyd Dewhurst), Gary Burchill (Center for Quality of Management), Eugene Cafarelli (Center for Concept Development), James Carter (United Technologies), David Cutherell (DTM), Gerard Furbershaw (Lunar Design), Jack Harkins (Roche Harkins), Gerhard Jünemann (Daimler Benz), David Meeker (Digital Equipment), Ulrike Närger (Daimler Benz), B. Joseph Pine II (Strategic Horizons), William Townsend (Barrett Technology), Brian Vogel (Product Genesis), and John Wesner (AT&T Bell Laboratories).

We also wish to acknowledge the several hundred students in the classes in which we have tested these teaching materials over the past few years. These students have been in several teaching programs at MIT, Helsinki University of Technology, Rhode Island School of Design, STOA (Italy), University of Pennsylvania, and Nanyang Technological University (Singapore). Many students provided constructive comments for improving the structure and delivery of the material finally contained here. Also, our experiences in observing the students' use of these methods in product development projects have greatly helped us to refine the material.

Several MIT students served as research assistants to help investigate many of the development methods, examples, and data contained in this book. The students are credited in each individual chapter and we also list them here: Paul Brody (Chapter 10), Tom Foody (Chapter 15), Amy Greenlief (Chapter 12), Christopher Hession (Chapter 3), Eric Howlett (Chapter 7), Tom Pimmler (Chapter 11 Appendices), Stephen Raab (Chapter 16), Harrison Roberts (Chapter 11 Appendices), Jonathan Sterrett (Chapter 4), and Gavin Zau (Chapter 6).

Other MIT students have also contributed by assisting with data collection and by offering comments and stimulating criticisms related to some of the chapters: Tom Abell, E. Yung Cha, Steve Daleiden, Russell Epstein, Matthew Fein, Brad Forry, Mike Frauens, Ben Goss, Daniel Hommes, Bill Liteplo, Habs Moy, Robert Northrop, Leslie Prince Rudolph, Vikas Sharma, and Ranjini Srikantiah. We also appreciate the cheerful and able assistance of the MIT support staff, Cara Barber, Anna Piccolo, Kristin Rocheleau, and Kathy Sullivan.

The staff throughout the Irwin/McGraw-Hill organization has been superb. We are particularly grateful for the enduring efforts of our sponsoring editor Barrett Koger. We also appreciate the efforts of developmental editor Scott Becker, project manager Susanne Riedell, copy editor Carole Schwager, photo editor Kathy Shive, photographer Stuart Cohen, and designer Kami Carter.

Finally, we thank our families for their love and support. Our parents provided much encouragement. Nancy, Julie, Lauren, Drew, Jamie, and Nathan have shown endless patience during the long months of this product development project.

Karl T. Ulrich
Steven D. Eppinger

Brief Contents

Contents

CHAPTER ONE

Introduction

Clockwise from top left: Photo by Stuart Cohen; Copyright 2002 Hewlett-Packard Company. Reproduced with permission; Courtesy of Boeing; Courtesy of Volkswagen of America; Courtesy of Rollerblade, Inc.

EXHIBIT 1-1
Examples of engineered, discrete, physical products (clockwise from top left): Stanley Tools Jobmaster Screwdriver, Hewlett-Packard DeskJet Printer, Boeing 777 Airplane, Volkswagen New Beetle, and Rollerblade In-Line Skate.

The economic success of manufacturing firms depends on their ability to identify the needs of customers and to quickly create products that meet these needs and can be produced at low cost. Achieving these goals is not solely a marketing problem, nor is it solely a design problem or a manufacturing problem; it is a product development problem involving all of these functions. This book provides a collection of methods intended to enhance the abilities of cross-functional teams to work together to develop products.

A *product* is something sold by an enterprise to its customers. *Product development* is the set of activities beginning with the perception of a market opportunity and ending in the production, sale, and delivery of a product. Although much of the material in this book is useful in the development of any product, we explicitly focus on products that are engineered, discrete, and physical. Exhibit 1-1 displays several examples of products from this category. Because we focus on engineered products, the book applies better to the development of power tools and computer peripherals than to magazines or sweaters. Our focus on discrete goods makes the book less applicable to the development of products such as gasoline, nylon, and paper. Because of the focus on physical products, we do not emphasize the specific issues involved in developing services or software. Even with these restrictions, the methods presented apply well to a broad range of products, including, for example, consumer electronics, sports equipment, scientific instruments, machine tools, and medical devices.

The goal of this book is to present in a clear and detailed way a set of product development methods aimed at bringing together the marketing, design, and manufacturing functions of the enterprise. In this introductory chapter we describe some aspects of the industrial practice of product development and provide a road map of the book.

Characteristics of Successful Product Development

From the perspective of the investors in a for-profit enterprise, successful product development results in products that can be produced and sold profitably, yet profitability is often difficult to assess quickly and directly. Five more specific dimensions, all of which ultimately relate to profit, are commonly used to assess the performance of a product development effort:

- *Product quality:* How good is the product resulting from the development effort? Does it satisfy customer needs? Is it robust and reliable? Product quality is ultimately reflected in market share and the price that customers are willing to pay.
- *Product cost:* What is the manufacturing cost of the product? This cost includes spending on capital equipment and tooling as well as the incremental cost of producing each unit of the product. Product cost determines how much profit accrues to the firm for a particular sales volume and a particular sales price.
- *Development time:* How quickly did the team complete the product development effort? Development time determines how responsive the firm can be to competitive forces and to technological developments, as well as how quickly the firm receives the economic returns from the team's efforts.
- *Development cost:* How much did the firm have to spend to develop the product? Development cost is usually a significant fraction of the investment required to achieve the profits.

- *Development capability:* Are the team and the firm better able to develop future products as a result of their experience with a product development project? Development capability is an asset the firm can use to develop products more effectively and economically in the future.

High performance along these five dimensions should ultimately lead to economic success; however, other performance criteria are also important. These criteria arise from interests of other stakeholders in the enterprise, including the members of the development team, other employees, and the community in which the product is manufactured. Members of the development team may be interested in creating an inherently exciting product. Members of the community in which the product is manufactured may be concerned about the degree to which the product creates jobs. Both production workers and users of the product hold the development team accountable to high safety standards, whether or not these standards can be justified on the strict basis of profitability. Other individuals, who may have no direct connection to the firm or the product, may demand that the product make ecologically sound use of resources and create minimal dangerous waste products.

Who Designs and Develops Products?

Product development is an interdisciplinary activity requiring contributions from nearly all the functions of a firm; however, three functions are almost always central to a product development project:

- *Marketing:* The marketing function mediates the interactions between the firm and its customers. Marketing often facilitates the identification of product opportunities, the definition of market segments, and the identification of customer needs. Marketing also typically arranges for communication between the firm and its customers, sets target prices, and oversees the launch and promotion of the product.
- *Design:* The design function plays the lead role in defining the physical form of the product to best meet customer needs. In this context, the design function includes engineering design (mechanical, electrical, software, etc.) and industrial design (aesthetics, ergonomics, user interfaces).
- *Manufacturing:* The manufacturing function is primarily responsible for designing and operating the production system in order to produce the product. Broadly defined, the manufacturing function also often includes purchasing, distribution, and installation. This collection of activities is sometimes called the *supply chain.*

Different individuals within these functions often have specific disciplinary training in areas such as market research, mechanical engineering, electrical engineering, materials science, or manufacturing operations. Several other functions, including finance and sales, are frequently involved on a part-time basis in the development of a new product. Beyond these broad functional categories, the specific composition of a development team depends on the particular characteristics of the product.

Few products are developed by a single individual. The collection of individuals developing a product forms the *project team.* This team usually has a single team leader, who could be drawn from any of the functions of the firm. The team can be thought of as

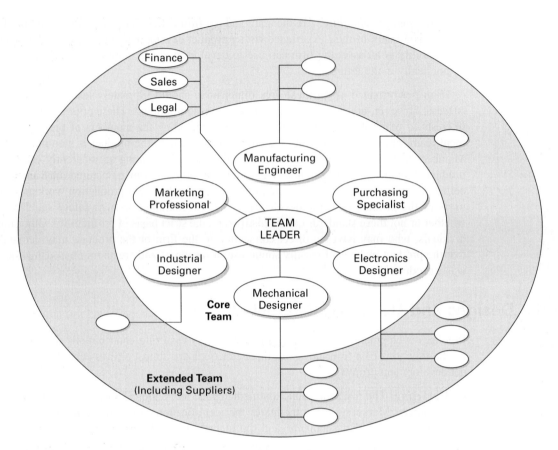

EXHIBIT 1-2 The composition of a product development team for an electromechanical product of modest complexity.

consisting of a *core team* and an *extended team*. In order to work together effectively, the core team usually remains small enough to meet in a conference room, while the extended team may consist of dozens, hundreds, or even thousands of other members. (Even though the term *team* is inappropriate for a group of thousands, the word is often used in this context to emphasize that the group must work toward a common goal.) In most cases, a team within the firm will be supported by individuals or teams at partner companies, suppliers, and consulting firms. Sometimes, as is the case for the development of a new airplane, the number of external team members may be even greater than that of the team within the company whose name will appear on the final product. The composition of a team for the development of an electromechanical product of modest complexity is shown in Exhibit 1-2.

Throughout this book we assume that the team is situated within a firm. In fact, a for-profit manufacturing company is the most common institutional setting for product development, but other settings are possible. Product development teams sometimes work within consulting firms, universities, government agencies, and nonprofit organizations.

	Stanley Tools Jobmaster Screwdriver	Rollerblade In-Line Skate	Hewlett-Packard DeskJet Printer	Volkswagen New Beetle Automobile	Boeing 777 Airplane
Annual production volume	100,000 units/year	100,000 units/year	4 million units/year	100,000 units/year	50 units/year
Sales lifetime	40 years	3 years	2 years	6 years	30 years
Sales price	$3	$200	$300	$17,000	$130 million
Number of unique parts (part numbers)	3 parts	35 parts	200 parts	10,000 parts	130,000 parts
Development time	1 year	2 years	1.5 years	3.5 years	4.5 years
Internal development team (peak size)	3 people	5 people	100 people	800 people	6,800 people
External development team (peak size)	3 people	10 people	75 people	800 people	10,000 people
Development cost	$150,000	$750,000	$50 million	$400 million	$3 billion
Production investment	$150,000	$1 million	$25 million	$500 million	$3 billion

EXHIBIT 1-3   Attributes of five products and their associated development efforts. All figures are approximate, based on publicly available information and company sources.

Duration and Cost of Product Development

Most people without experience in product development are astounded by how much time and money are required to develop a new product. The reality is that very few products can be developed in less than 1 year, many require 3 to 5 years, and some take as long as 10 years. Exhibit 1-1 shows five engineered, discrete products. Exhibit 1-3 is a table showing the approximate scale of the associated product development efforts along with some distinguishing characteristics of the products.

The cost of product development is roughly proportional to the number of people on the project team and to the duration of the project. In addition to expenses for development effort, a firm will almost always have to make some investment in the tooling and equipment required for production. This expense is often as large as the rest of the product development budget; however, it is sometimes useful to think of these expenditures as part of the *fixed costs* of production. For reference purposes, this production investment is listed in Exhibit 1-3 along with the development expenditures.

The Challenges of Product Development

Developing great products is hard. Few companies are highly successful more than half the time. These odds present a significant challenge for a product development team. Some of the characteristics that make product development challenging are:

- *Trade-offs:* An airplane can be made lighter, but this action will probably increase manufacturing cost. One of the most difficult aspects of product development is recognizing, understanding, and managing such trade-offs in a way that maximizes the success of the product.
- *Dynamics:* Technologies improve, customer preferences evolve, competitors introduce new products, and the macroeconomic environment shifts. Decision making in an environment of constant change is a formidable task.
- *Details:* The choice between using screws or snap-fits on the enclosure of a computer can have economic implications of millions of dollars. Developing a product of even modest complexity may require thousands of such decisions.
- *Time pressure:* Any one of these difficulties would be easily manageable by itself given plenty of time, but product development decisions must usually be made quickly and without complete information.
- *Economics:* Developing, producing, and marketing a new product requires a large investment. To earn a reasonable return on this investment, the resulting product must be both appealing to customers and relatively inexpensive to produce.

For many people, product development is interesting precisely because it is challenging. For others, several intrinsic attributes also contribute to its appeal:

- *Creation:* The product development process begins with an idea and ends with the production of a physical artifact. When viewed both in its entirety and at the level of individual activities, the product development process is intensely creative.
- *Satisfaction of societal and individual needs:* All products are aimed at satisfying needs of some kind. Individuals interested in developing new products can almost always find institutional settings in which they can develop products satisfying what they consider to be important needs.
- *Team diversity:* Successful development requires many different skills and talents. As a result, development teams involve people with a wide range of different training, experience, perspectives, and personalities.
- *Team spirit:* Product development teams are often highly motivated, cooperative groups. The team members may be colocated so they can focus their collective energy on creating the product. This situation can result in lasting camaraderie among team members.

Approach of This Book

We focus on product development activities that benefit from the participation of all the core functions of the firm. For our purposes, we define the core functions as marketing, design, and manufacturing. We expect that team members have competence in one or more specific disciplines such as mechanical engineering, electrical engineering, industrial design, market research, or manufacturing operations. For this reason, we do not discuss, for example, how to perform a stress analysis or to create a conjoint survey. These are disciplinary skills we expect someone on the development team to possess. The integrative methods in this book are intended to facilitate problem solving and decision making among people with different disciplinary perspectives.

Structured Methods

The book consists of methods for completing development activities. The methods are structured, which means we generally provide a step-by-step approach and often provide templates for the key information systems used by the team. We believe structured methods are valuable for three reasons: First, they make the decision process explicit, allowing everyone on the team to understand the decision rationale and reducing the possibility of moving forward with unsupported decisions. Second, by acting as "checklists" of the key steps in a development activity they ensure that important issues are not forgotten. Third, structured methods are largely self-documenting; in the process of executing the method, the team creates a record of the decision-making process for future reference and for educating newcomers.

Although the methods are structured, they are not intended to be applied blindly. The methods are a starting point for continuous improvement. Teams should adapt and modify the approaches to meet their own needs and to reflect the unique character of their institutional environment.

Industrial Examples

Each remaining chapter is built around an example drawn from industrial practice. The major examples include the following: a line of bowling equipment, a digital copier, a cordless screwdriver, a mountain bike suspension fork, a power nailer, a dose-metering syringe, an electric scooter, a computer printer, a cellular telephone, an automobile engine, a trackball for a notebook computer, a seat belt system, a coffee-cup insulator, a digital photo printer, and a microfilm cartridge. In most cases we use as examples the simplest products we have access to that illustrate the important aspects of the methods. When a screwdriver illustrates an idea as well as a jet engine, we use the screwdriver. However, every method in this book has been used successfully in industrial practice by hundreds of people on both large and small projects.

Although built around examples, the chapters are not intended to be historically accurate case studies. We use the examples as a way to illustrate development methods, and in doing so we recast some historical details in a way that improves the presentation of the material. We also disguise much of the quantitative information in the examples, especially financial data.

Organizational Realities

We deliberately chose to present the methods with the assumption that the development team operates in an organizational environment conducive to success. In reality, some organizations exhibit characteristics that lead to dysfunctional product development teams. These characteristics include:

- *Lack of empowerment of the team:* General managers or functional managers may engage in continual intervention in the details of a development project without a full understanding of the basis for the team's decisions.

- *Functional allegiances transcending project goals:* Representatives of marketing, design, or manufacturing may influence decisions in order to increase the political standing of themselves or their functions without regard for the overall success of the product.
- *Inadequate resources:* A team may be unable to complete development tasks effectively because of a lack of staff, a mismatch of skills, or a lack of money, equipment, or tools.
- *Lack of cross-functional representation on the project team:* Key development decisions may be made without involvement of marketing, design, manufacturing, or other critical functions.

While most organizations exhibit one or more of these characteristics to some degree, the significant presence of these problems can be so stifling that sound development methods are rendered ineffective. While recognizing the importance of basic organizational issues, we assume, for clarity of explanation, that the development team operates in an environment in which the most restrictive organizational barriers have been removed.

Road Map of the Book

We divide the product development process into six phases, as shown in Exhibit 1-4. (These phases are described in more detail in Chapter 2, Development Processes and Organizations.) This book describes the concept development phase in its entirety and the remaining phases less completely, because we do not provide methods for the more focused development activities that occur later in the process. Each of the remaining chapters in this book can be read, understood, and applied independently.

- Chapter 2, Development Processes and Organizations, presents a generic product development process and shows how variants of this process are used in different industrial situations. The chapter also discusses the way individuals are organized into groups in order to undertake product development projects.
- Chapter 3, Product Planning, presents a method for deciding which products to develop. The output of this method is a mission statement for a particular project.
- Chapters 4 through 8, Identifying Customer Needs, Product Specifications, Concept Generation, Concept Selection, and Concept Testing, present the key activities of the concept development phase. These methods guide a team from a mission statement through a selected product concept.
- Chapter 9, Product Architecture, discusses the implications of product architecture on product change, product variety, component standardization, product performance, manufacturing cost, and project management; it then presents a method for establishing the architecture of a product.
- Chapter 10, Industrial Design, discusses the role of the industrial designer and how human interaction issues, including aesthetics and ergonomics, are treated in product development.
- Chapter 11, Design for Manufacturing, discusses techniques used to reduce manufacturing cost. These techniques are primarily applied during the system-level and detail-design phases of the process.

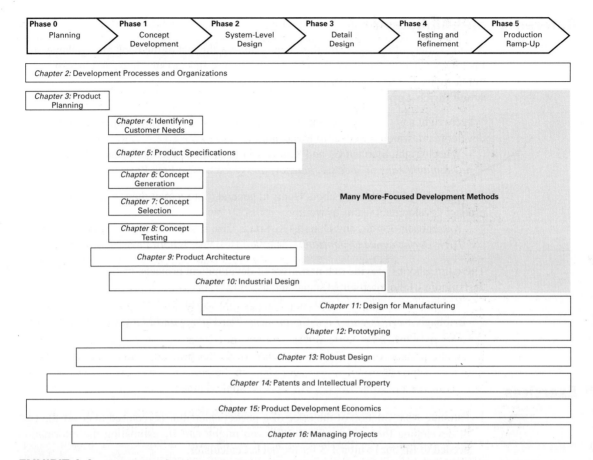

Phase 0	Phase 1	Phase 2	Phase 3	Phase 4	Phase 5
Planning	Concept Development	System-Level Design	Detail Design	Testing and Refinement	Production Ramp-Up

Chapter 2: Development Processes and Organizations

Chapter 3: Product Planning

Chapter 4: Identifying Customer Needs

Chapter 5: Product Specifications

Chapter 6: Concept Generation

Chapter 7: Concept Selection

Chapter 8: Concept Testing

Chapter 9: Product Architecture

Chapter 10: Industrial Design

Many More-Focused Development Methods

Chapter 11: Design for Manufacturing

Chapter 12: Prototyping

Chapter 13: Robust Design

Chapter 14: Patents and Intellectual Property

Chapter 15: Product Development Economics

Chapter 16: Managing Projects

EXHIBIT 1-4   The product development process. The diagram shows where each of the integrative methods presented in the remaining chapters is most applicable.

- Chapter 12, Prototyping, presents a method to ensure that prototyping efforts, which occur throughout the process, are applied effectively.
- Chapter 13, Robust Design, explains methods for choosing values of design variables to ensure reliable and consistent performance.
- Chapter 14, Patents and Intellectual Property, presents an approach to creating a patent application and discusses the role of intellectual property in product development.
- Chapter 15, Product Development Economics, describes a method for understanding the influence of internal and external factors on the economic value of a project.
- Chapter 16, Managing Projects, presents some fundamental concepts for understanding and representing interacting project tasks, along with a method for planning and executing a development project.

References and Bibliography

A wide variety of resources for this chapter and for the rest of the book are available on the Internet. These resources include data, templates, links to suppliers, and lists of publications. Current resources may be accessed via
www.ulrich-eppinger.net

Wheelwright and Clark devote much of their book to the very early stages of product development, which we cover in less detail.

Wheelwright, Stephen C., and Kim B. Clark, *Revolutionizing Product Development: Quantum Leaps in Speed, Efficiency, and Quality,* The Free Press, New York, 1992.

Katzenbach and Smith write about teams in general, but most of their insights apply to product development teams as well.

Katzenbach, Jon R., and Douglas K. Smith, *The Wisdom of Teams: Creating the High-Performance Organization,* Harvard Business School Press, Boston, 1993.

These three books provide rich narratives of development projects, including fascinating descriptions of the intertwined social and technical processes.

Kidder, Tracy, *The Soul of a New Machine,* Avon Books, New York, 1981.

Sabbagh, Karl, *Twenty-First-Century Jet: The Making and Marketing of the Boeing 777,* Scribner, New York, 1996.

Walton, Mary, *Car: A Drama of the American Workplace,* Norton, New York, 1997.

Exercises

1. Estimate what fraction of the price of a pocket calculator is required to cover the cost of developing the product. To do this you might start by estimating the information needed to fill out Exhibit 1-3 for the pocket calculator.
2. Create a set of scatter charts by plotting each of the rows in Exhibit 1-3 against the development cost row. For each one, explain why there is or is not any correlation. (For example, you would first plot "annual production volume" versus "development cost" and explain why there seems to be no correlation. Then repeat for each of the remaining rows.)

Thought Question

1. Each of the chapters listed in Exhibit 1-4 presents a method for a portion of the product development process. For each one, consider what types of skills and expertise might be required. Can you make an argument for staffing the development team from start to finish with individuals possessing all of these skills and areas of expertise?

CHAPTER TWO

Development Processes and Organizations

Courtesy of AMF Bowling Worldwide

EXHIBIT 2-1

A ball return, one of AMF Bowling's products.

The Capital Equipment Division of AMF Bowling is the leading manufacturer of bowling equipment, including pin spotters, ball returns, and scoring equipment. An AMF ball return product is shown in Exhibit 2-1. The general manager of the division asked the engineering manager to establish a well-defined product development process and to propose a product development organization that would allow AMF to compete effectively over the next decade. Some of the questions AMF faced were:

- Is there a standard development process that will work for every company?
- What role do experts from different functional areas play in the development process?
- What milestones can be used to divide the overall development process into phases?
- Should the development organization be divided into groups corresponding to projects or to development functions?

This chapter helps to answer these and related questions by presenting a generic development process and showing how this process can be adapted to meet the needs of particular industrial situations. We highlight the activities and contributions of different functions of the company during each phase of the development process. The chapter also explains what constitutes a product development organization and discusses why different types of organizations are appropriate for different settings.

A Generic Development Process

A process is a sequence of steps that transforms a set of inputs into a set of outputs. Most people are familiar with the idea of physical processes, such as those used to bake a cake or to assemble an automobile. A *product development process* is the sequence of steps or activities which an enterprise employs to conceive, design, and commercialize a product. Many of these steps and activities are intellectual and organizational rather than physical. Some organizations define and follow a precise and detailed development process, while others may not even be able to describe their processes. Furthermore, every organization employs a process at least slightly different from that of every other organization. In fact, the same enterprise may follow different processes for each of several different types of development projects.

A well-defined development process is useful for the following reasons:

- *Quality assurance:* A development process specifies the phases a development project will pass through and the checkpoints along the way. When these phases and checkpoints are chosen wisely, following the development process is one way of assuring the quality of the resulting product.
- *Coordination:* A clearly articulated development process acts as a master plan which defines the roles of each of the players on the development team. This plan informs the members of the team when their contributions will be needed and with whom they will need to exchange information and materials.
- *Planning:* A development process contains natural milestones corresponding to the completion of each phase. The timing of these milestones anchors the schedule of the overall development project.

- *Management:* A development process is a benchmark for assessing the performance of an ongoing development effort. By comparing the actual events to the established process, a manager can identify possible problem areas.
- *Improvement:* The careful documentation of an organization's development process often helps to identify opportunities for improvement.

The generic product development process consists of six phases, as illustrated in Exhibit 2-2. The process begins with a planning phase, which is the link to advanced research and technology development activities. The output of the planning phase is the project's mission statement, which is the input required to begin the concept development phase and which serves as a guide to the development team. The conclusion of the product development process is the product launch, at which time the product becomes available for purchase in the marketplace.

One way to think about the development process is as the initial creation of a wide set of alternative product concepts and then the subsequent narrowing of alternatives and increasing specification of the product until the product can be reliably and repeatably produced by the production system. Note that most of the phases of development are defined in terms of the state of the product, although the production process and marketing plans, among other tangible outputs, are also evolving as development progresses.

Another way to think about the development process is as an information-processing system. The process begins with inputs such as the corporate objectives and the capabilities of available technologies, product platforms, and production systems. Various activities process the development information, formulating specifications, concepts, and design details. The process concludes when all the information required to support production and sales has been created and communicated.

A third way to think about the development process is as a risk management system. In the early phases of product development, various risks are identified and prioritized. As the process progresses, risks are reduced as the key uncertainties are eliminated and the functions of the product are validated. When the process is completed, the team should have substantial confidence that the product will work correctly and be well received by the market.

Exhibit 2-2 also identifies the key activities and responsibilities of the different functions of the organization during each development phase. Because of their continuous involvement in the process, we choose to articulate the roles of marketing, design, and manufacturing. Representatives from other functions, such as research, finance, field service, and sales, also play key roles at particular points in the process.

The six phases of the generic development process are:

0. *Planning:* The planning activity is often referred to as "phase zero" since it precedes the project approval and launch of the actual product development process. This phase begins with corporate strategy and includes assessment of technology developments and market objectives. The output of the planning phase is the project mission statement, which specifies the target market for the product, business goals, key assumptions, and constraints. Chapter 3, Product Planning, presents a discussion of this planning process.

1. *Concept development:* In the concept development phase, the needs of the target market are identified, alternative product concepts are generated and evaluated, and one

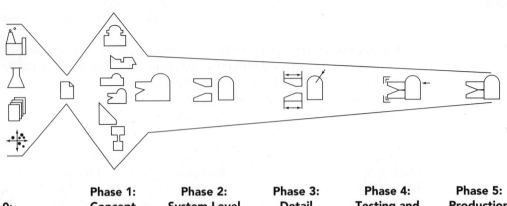

Phase 0: Planning	Phase 1: Concept Development	Phase 2: System-Level Design	Phase 3: Detail Design	Phase 4: Testing and Refinement	Phase 5: Production Ramp-Up
Marketing • Articulate market opportunity. • Define market segments.	• Collect customer needs. • Identify lead users. • Identify competitive products.	• Develop plan for product options and extended product family. • Set target sales price point(s).	• Develop marketing plan.	• Develop promotion and launch materials. • Facilitate field testing.	• Place early production with key customers.
Design • Consider product platform and architecture. • Assess new technologies.	• Investigate feasibility of product concepts. • Develop industrial design concepts. • Build and test experimental prototypes.	• Generate alternative product architectures. • Define major subsystems and interfaces. • Refine industrial design.	• Define part geometry. • Choose materials. • Assign tolerances. • Complete industrial design control documentation.	• Reliability testing. • Life testing. • Performance testing. • Obtain regulatory approvals. • Implement design changes.	• Evaluate early production output.
Manufacturing • Identify production constraints. • Set supply chain strategy.	• Estimate manufacturing cost. • Assess production feasibility.	• Identify suppliers for key components. • Perform make-buy analysis. • Define final assembly scheme. • Set target costs.	• Define piece-part production processes. • Design tooling. • Define quality assurance processes. • Begin procurement of long-lead tooling.	• Facilitate supplier ramp-up. • Refine fabrication and assembly processes. • Train work force. • Refine quality assurance processes.	• Begin operation of entire production system.
Other Functions • Research: Demonstrate available technologies. • Finance: Provide planning goals. • General Management: Allocate project resources.	• Finance: Facilitate economic analysis. • Legal: Investigate patent issues.	• Finance: Facilitate make-buy analysis. • Service: Identify service issues.		• Sales: Develop sales plan.	

EXHIBIT 2-2   The generic product development process. Six phases are shown, including the tasks and responsibilities of the key functions of the organization for each phase.

or more concepts are selected for further development and testing. A concept is a description of the form, function, and features of a product and is usually accompanied by a set of specifications, an analysis of competitive products, and an economic justification of the project. This book presents several detailed methods for the concept development phase (Chapters 4–8). We expand this phase into each of its constitutive activities in the next section.

2. *System-level design:* The system-level design phase includes the definition of the product architecture and the decomposition of the product into subsystems and components. The final assembly scheme for the production system is usually defined during this phase as well. The output of this phase usually includes a geometric layout of the product, a functional specification of each of the product's subsystems, and a preliminary process flow diagram for the final assembly process. Chapter 9, Product Architecture, discusses some of the important activities of system-level design.

3. *Detail design:* The detail design phase includes the complete specification of the geometry, materials, and tolerances of all of the unique parts in the product and the identification of all of the standard parts to be purchased from suppliers. A process plan is established and tooling is designed for each part to be fabricated within the production system. The output of this phase is the *control documentation* for the product—the drawings or computer files describing the geometry of each part and its production tooling, the specifications of the purchased parts, and the process plans for the fabrication and assembly of the product. Two critical issues addressed in the detail design phase are production cost and robust performance. These issues are discussed respectively in Chapter 11, Design for Manufacturing, and Chapter 13, Robust Design.

4. *Testing and refinement:* The testing and refinement phase involves the construction and evaluation of multiple preproduction versions of the product. Early *(alpha)* prototypes are usually built with *production-intent* parts—parts with the same geometry and material properties as intended for the production version of the product but not necessarily fabricated with the actual processes to be used in production. Alpha prototypes are tested to determine whether the product will work as designed and whether the product satisfies the key customer needs. Later *(beta)* prototypes are usually built with parts supplied by the intended production processes but may not be assembled using the intended final assembly process. Beta prototypes are extensively evaluated internally and are also typically tested by customers in their own use environment. The goal for the beta prototypes is usually to answer questions about performance and reliability in order to identify necessary engineering changes for the final product. Chapter 12, Prototyping, presents a thorough discussion of the nature and use of prototypes.

5. Production ramp-up: In the production ramp-up phase, the product is made using the intended production system. The purpose of the ramp-up is to train the work force and to work out any remaining problems in the production processes. Products produced during production ramp-up are sometimes supplied to preferred customers and are carefully evaluated to identify any remaining flaws. The transition from production ramp-up to ongoing production is usually gradual. At some point in this transition, the product is *launched* and becomes available for widespread distribution.

Concept Development: The Front-End Process

Because the concept development phase of the development process demands perhaps more coordination among functions than any other, many of the integrative development methods presented in this book are concentrated here. In this section we expand the concept development phase into what we call the *front-end process*. The front-end process generally contains many interrelated activities, ordered roughly as shown in Exhibit 2-3.

Rarely does the entire process proceed in purely sequential fashion, completing each activity before beginning the next. In practice, the front-end activities may be overlapped in time and iteration is often necessary. The dashed arrows in Exhibit 2-3 reflect the uncertain nature of progress in product development. At almost any stage, new information may become available or results learned which can cause the team to step back to repeat an earlier activity before proceeding. This repetition of nominally complete activities is known as development *iteration*.

The concept development process includes the following activities:

- *Identifying customer needs:* The goal of this activity is to understand customers' needs and to effectively communicate them to the development team. The output of this step is a set of carefully constructed customer need statements, organized in a hierarchical list, with importance weightings for many or all of the needs. A method for this activity is presented in Chapter 4, Identifying Customer Needs.

- *Establishing target specifications:* Specifications provide a precise description of what a product has to do. They are the translation of the customer needs into technical terms. Targets for the specifications are set early in the process and represent the hopes of the development team. Later these specifications are refined to be consistent with the constraints imposed by the team's choice of a product concept. The output of this stage is a list of target specifications. Each specification consists of a metric, and marginal and ideal values for that metric. A method for the specification activity is given in Chapter 5, Product Specifications.

- *Concept generation:* The goal of concept generation is to thoroughly explore the space of product concepts that may address the customer needs. Concept generation includes a mix of external search, creative problem solving within the team, and systematic exploration of the various solution fragments the team generates. The result of

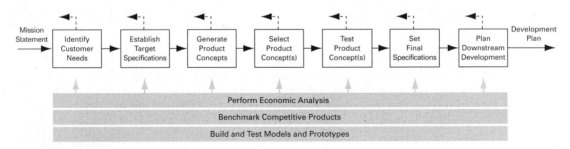

EXHIBIT 2-3   The many front-end activities comprising the concept development phase.

this activity is usually a set of 10 to 20 concepts, each typically represented by a sketch and brief descriptive text. Chapter 6, Concept Generation, describes this activity in detail.

- *Concept selection:* Concept selection is the activity in which various product concepts are analyzed and sequentially eliminated to identify the most promising concept(s). The process usually requires several iterations and may initiate additional concept generation and refinement. A method for this activity is described in Chapter 7, Concept Selection.

- *Concept testing:* One or more concepts are then tested to verify that the customer needs have been met, assess the market potential of the product, and identify any shortcomings which must be remedied during further development. If the customer response is poor, the development project may be terminated or some earlier activities may be repeated as necessary. Chapter 8, Concept Testing, explains a method for this activity.

- *Setting final specifications:* The target specifications set earlier in the process are revisited after a concept has been selected and tested. At this point, the team must commit to specific values of the metrics reflecting the constraints inherent in the product concept, limitations identified through technical modeling, and trade-offs between cost and performance. Chapter 5, Product Specifications, explains the details of this activity.

- *Project planning:* In this final activity of concept development, the team creates a detailed development schedule, devises a strategy to minimize development time, and identifies the resources required to complete the project. The major results of the front-end activities can be usefully captured in a *contract book* which contains the mission statement, the customer needs, the details of the selected concept, the product specifications, the economic analysis of the product, the development schedule, the project staffing, and the budget. The contract book serves to document the agreement (contract) between the team and the senior management of the enterprise. A project planning method is presented in Chapter 16, Managing Projects.

- *Economic analysis:* The team, often with the support of a financial analyst, builds an economic model for the new product. This model is used to justify continuation of the overall development program and to resolve specific trade-offs among, for example, development costs and manufacturing costs. Economic analysis is shown as one of the ongoing activities in the concept development phase. An early economic analysis will almost always be performed before the project even begins, and this analysis is updated as more information becomes available. A method for this activity is presented in Chapter 15, Product Development Economics.

- *Benchmarking of competitive products:* An understanding of competitive products is critical to successful positioning of a new product and can provide a rich source of ideas for the product and production process design. Competitive *benchmarking* is performed in support of many of the front-end activities. Various aspects of competitive benchmarking are presented in Chapters 4–8.

- *Modeling and prototyping:* Every stage of the concept development process involves various forms of models and prototypes. These may include, among others: early "proof-of-concept" models, which help the development team to demonstrate feasibility; "form-only" models, which can be shown to customers to evaluate ergonomics and style; spreadsheet models of technical trade-offs; and experimental test models, which can be used to set design parameters for robust performance. Methods for modeling, prototyping, and testing are discussed throughout the book, including in Chapters 4–6, 8, 10, 12, and 13.

Adapting the Generic Product Development Process

The development process described by Exhibits 2-2 and 2-3 is generic, and particular processes will differ in accordance with a firm's unique context. The generic process is most like the process used in a *market-pull* situation: a firm begins product development with a market opportunity and then uses whatever available technologies are required to satisfy the market need (i.e., the market "pulls" the development decisions). In addition to the market-pull process outlined in Exhibits 2-2 and 2-3, several variants are common and correspond to the following: *technology-push* products, *platform* products, *process-intensive* products, *customized* products, *high-risk* products, *quick-build* products, and *complex systems*. Each of these situations is described below. The characteristics of these situations and the resulting deviations from the generic process are summarized in Exhibit 2-4.

Technology-Push Products

In developing technology-push products, the firm begins with a new proprietary technology and looks for an appropriate market in which to apply this technology (that is, the technology "pushes" development). Gore-Tex, an expanded Teflon sheet manufactured by W. L. Gore Associates, is a striking example of technology push. The company has developed dozens of products incorporating Gore-Tex, including artificial veins for vascular surgery, insulation for high-performance electric cables, fabric for outerwear, dental floss, and liners for bagpipe bags.

Many successful technology-push products involve basic materials or basic process technologies. This may be because basic materials and processes are deployed in thousands of applications, and there is therefore a high likelihood that new and unusual characteristics of materials and processes can be matched with an appropriate application.

The generic product development process can be used with minor modifications for technology-push products. The technology-push process begins with the planning phase, in which the given technology is matched with a market opportunity. Once this matching has occurred, the remainder of the generic development process can be followed. The team includes an assumption in the mission statement that the particular technology will be embodied in the product concepts considered by the team. Although many extremely successful products have arisen from technology-push development, this approach can be perilous. The product is unlikely to succeed unless (1) the assumed technology offers a clear competitive advantage in meeting customer needs, and (2) suitable alternative technologies are unavailable or very difficult for competitors to utilize. Project risk can possibly be minimized by simultaneously considering the merit of a broader set of concepts which do not necessarily incorporate the new technology. In this way the team verifies that the product concept embodying the new technology is superior to the alternatives.

Process Type	Description	Distinct Features	Examples
Generic (Market-Pull) Products	The team begins with a market opportunity and selects appropriate technologies to meet customer needs.	Process generally includes distinct planning, concept development, system-level design, detail design, testing and refinement, and production ramp-up phases.	Sporting goods, furniture, tools.
Technology-Push Products	The team begins with a new technology, then finds an appropriate market.	Planning phase involves matching technology and market. Concept development assumes a given technology.	Gore-Tex rainwear, Tyvek envelopes.
Platform Products	The team assumes that the new product will be built around an established technological subsystem.	Concept development assumes a proven technology platform.	Consumer electronics, computers, printers.
Process-Intensive Products	Characteristics of the product are highly constrained by the production process.	Either an existing production process must be specified from the start, or both product and process must be developed together from the start.	Snack foods, breakfast cereals, chemicals, semiconductors.
Customized Products	New products are slight variations of existing configurations.	Similarity of projects allows for a streamlined and highly structured development process.	Motors, switches, batteries, containers.
High-Risk Products	Technical or market uncertainties create high risks of failure.	Risks are identified early and tracked throughout the process. Analysis and testing activities take place as early as possible.	Pharmaceuticals, space systems.
Quick-Build Products	Rapid modeling and prototyping enables many design-build-test cycles.	Detail design and testing phases are repeated a number of times until the product is completed or time/budget runs out.	Software, cellular phones.
Complex Systems	System must be decomposed into several subsystems and many components.	Subsystems and components are developed by many teams working in parallel, followed by system integration and validation.	Airplanes, jet engines, automobiles.

EXHIBIT 2-4   Summary of variants of generic product development process.

Platform Products

A platform product is built around a preexisting technological subsystem (a technology *platform*). Examples of such platforms include the tape transport mechanism in the Sony Walkman, the Apple Macintosh operating system, and the instant film used in Polaroid cameras. Huge investments were made in developing these platforms, and therefore every attempt is made to incorporate them into several different products. In some sense, platform products are very similar to technology-push products in that the team begins the development effort with an assumption that the product concept will embody a particular technology. The primary difference is that a technology platform has already demonstrated its usefulness in the marketplace in meeting customer needs. The firm can in many cases assume that the technology will also be useful in related markets. Products built on technology platforms are much simpler to develop than if the technology were developed from scratch. For this reason, and because of the possible sharing of costs across several products, a firm may be able to offer a platform product in markets that could not justify the development of a unique technology.

Process-Intensive Products

Examples of process-intensive products include semiconductors, foods, chemicals, and paper. For these products, the production process places strict constraints on the properties of the product, so that the product design cannot be separated, even at the concept phase, from the production process design. In many cases, process-intensive products are produced in very high volumes and are bulk, as opposed to discrete, goods.

In some situations, a new product and new process are developed simultaneously. For example, creating a new shape of breakfast cereal or snack food will require both product and process development activities. In other cases, a specific existing process for making the product is chosen in advance, and the product design is constrained by the capabilities of this process. This might be true of a new paper product to be made in a particular paper mill or a new semiconductor device to be made in an existing wafer fabrication facility.

Customized Products

Examples of customized products include switches, motors, batteries, and containers. Customized products are slight variations of standard configurations and are typically developed in response to a specific order by a customer. Development of customized products consists primarily of setting values of design variables such as physical dimensions and materials. When a customer requests a new product, the firm executes a structured design and development process to create the product to meet the customer's needs. Such firms typically have created a highly detailed development process involving a well-defined sequence of steps with a structured flow of information (analogous to a production process). For customized products, the generic process is augmented with a detailed description of the specific information-processing activities required within each of the phases. Such development processes may consist of hundreds of carefully defined activities.

High-Risk Products

The product development process addresses many types of risk. These include technical risk (Will the product function properly?), market risk (Will customers like what the team develops?), and budget and schedule risk (Can the team complete the project on time and within budget?). High-risk products are those that entail unusually large uncertainties related to the technology or market so that there is substantial technical or market risk. The generic product development process is modified to face high-risk situations by taking steps to address the largest risks in the early stages of product development. This usually requires completing some design and test activities earlier in the process. For example, when there is great uncertainty regarding customer acceptance of a new product, concept testing using renderings or user-interface prototypes may be done very early in the process in order to reduce the market uncertainty and risk. If there is high uncertainty related to technical performance of the product, it makes sense to build working models of the key features and to test these earlier in the process. Multiple solution paths may be explored in parallel to ensure that one of the solutions succeeds. Design reviews must assess levels of risk on a regular basis, with the expectation that risks are being reduced over time and not being postponed.

Quick-Build Products

For the development of some products, such as software and many electronics products, building and testing prototype models has become such a rapid process that the design-build-test cycle can be repeated many times. In fact, teams can take advantage of rapid iteration to achieve a more flexible and responsive product development process, sometimes called a *spiral product development process.* Following concept development in this process, the system-level design phase entails decomposition of the product into high-, medium-, and low-priority features. This is followed by several cycles of design, build, integrate, and test activities, beginning with the highest-priority items. This process takes advantage of the fast prototyping cycle by using the result of each cycle to learn how to modify the priorities for the next cycle. Customers may even be involved in the testing process after one or more cycles. When time or budget runs out, usually all of the high- and medium-priority features have been incorporated into the evolving product, and the low-priority features may be omitted until the next product generation.

Complex Systems

Larger-scale products such as automobiles and airplanes are complex systems comprised of many interacting subsystems and components. When developing complex systems, modifications to the generic product development process address a number of system-level issues. The concept development phase considers the architecture of the entire system, and multiple architectures may be considered as competing concepts for the overall system. The system-level design phase becomes critical. During this phase, the system is decomposed into subsystems and these further into many components. Teams are assigned to develop each component. Additional teams are assigned the special challenge of integrating components into the subsystems and these into the overall system.

Detail design of the components is a highly parallel process in which the many development teams work at once, usually separately. Managing the network of interactions across the components and subsystems is the task of system engineering specialists of many kinds. The testing and refinement phase includes not only system integration, but also extensive testing and validation at all levels.

Product Development Process Flows

The product development process generally follows a structured flow of activity and information flow. This allows us to draw *process flow diagrams* illustrating the process, as shown in Exhibit 2-5. The generic process flow diagram depicts the process used to develop market-pull, technology-push, platform, process-intensive, customized, and high-risk products. Each product development phase (or stage) is followed by a review (or gate) to confirm that the phase is completed and to determine whether the project proceeds. Quick-build products enable a spiral product development process whereby detail design, prototyping, and test activities are repeated a number of times. The process flow diagram for development of complex systems shows the decomposition into parallel stages of work on the many subsystems and components. Once the product development process has been established within an organization, a process flow diagram is used to explain the process to everyone on the team.

The AMF Development Process

AMF Bowling is a market-pull enterprise. AMF generally drives its development process with a market need and seeks out whatever technology is required to meet that need. Its competitive advantage arises from strong marketing channels, strong brand recognition, and a large installed base of equipment, not from any single proprietary technology. For this reason, the technology-push approach would not be appropriate. AMF products are assembled from components fabricated with relatively conventional processes such as molding, casting, and machining. So the AMF product is clearly not process intensive in the way a food product or a chemical is. Bowling equipment is rarely customized for a particular customer; most of the product development at AMF is aimed at new models of products, rather than at the customization of existing models. For this reason, the customization approach is also inappropriate.

AMF chose to establish a development process similar to the generic process. The process proposed by the AMF engineering manager is illustrated in Exhibit 2-6. The representation of the development process used by AMF is a hybrid of those used in Exhibits 2-2 and 2-5, in that it shows the individual activities in the development process as well as the roles of the different development functions in those activities. Note that AMF defines the key functions in product development as marketing, engineering/design, manufacturing, quality assurance, purchasing, and customer service. Also note that there are three major milestones in the process: the project approval, the beginning of tooling fabrication, and the production release. Each of these milestones follows a major review.

Although AMF established a standard process, its managers realized that this process would not necessarily be suitable in its entirety for all AMF products. For example, a few of AMF's new products are based on technology platforms. When platform products are developed, the team assumes the use of an existing technology platform during concept development. Nevertheless, the standard development process is the baseline from which a particular project plan begins.

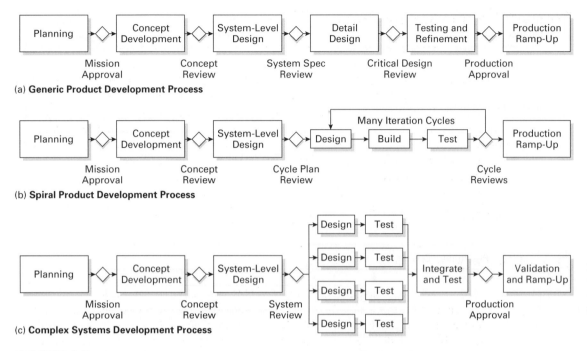

EXHIBIT 2-5 Process flow diagrams for three product development processes.

Product Development Organizations

In addition to crafting an effective development process, successful firms must organize their product development staffs effectively. In this section, we describe several types of organizations used for product development and offer guidelines for choosing among these options.

Organizations Are Formed by Establishing Links among Individuals

A product development organization is the scheme by which individual designers and developers are linked together into groups. The links among individuals may be formal or informal and include, among others, these types:

- *Reporting relationships:* Reporting relationships give rise to the classic notion of *supervisor* and *subordinate*. These are the formal links most frequently shown on an organization chart.

- *Financial arrangements:* Individuals are linked by being part of the same financial entity, such as that defined by a particular budget category or profit-and-loss statement.

- *Physical layout:* Links are created between individuals when they share the same office, floor, building, or site. These links are often informal, arising from spontaneous encounters while at work.

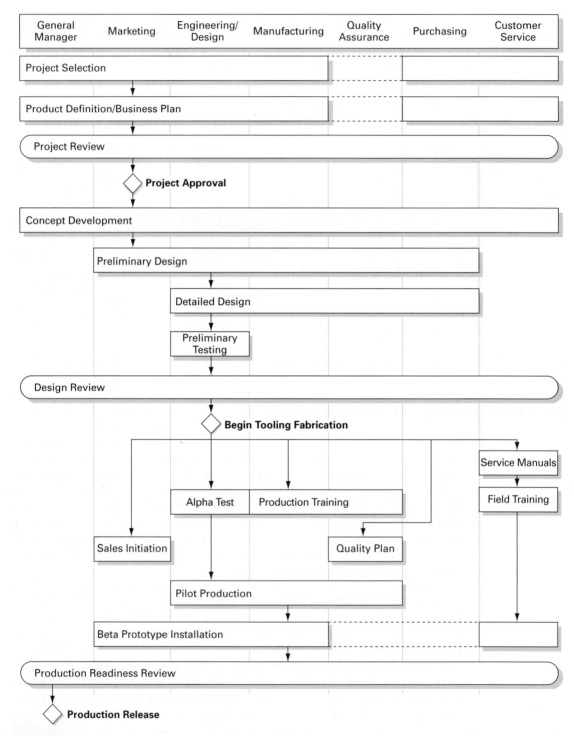

EXHIBIT 2-6 AMF Bowling's standard development process.

Any particular individual may be linked in several different ways to other individuals. For example, an engineer may be linked by a reporting relationship to another engineer in a different building, while being linked by physical layout to a marketing person sitting in the next office. The strongest organizational links are typically those involving performance evaluation, budgets, and other resource allocations.

Organizational Links May Be Aligned with Functions, Projects, or Both

Regardless of their organizational links, particular individuals can be classified in two different ways: according to their *function* and according to the *projects* they work on.

- A function (in organizational terms) is an area of responsibility usually involving specialized education, training, or experience. The classic functions in product development organizations are marketing, design, and manufacturing. Finer divisions than these are also possible and may include, for example, market research, market strategy, stress analysis, industrial design, human factors engineering, process development, and operations management.

- Regardless of their functions, individuals apply their expertise to specific projects. In product development, a project is the set of activities in the development process for a particular product and includes, for example, identifying customer needs and generating product concepts.

Note that these two classifications must overlap: individuals from several different functions will work on the same project. Also, while most individuals are associated with only one function, they may contribute to more than one project. Two classic organizational structures arise from aligning the organizational links according to function or according to projects. In *functional organizations,* the organizational links are primarily among those who perform similar functions. In *project organizations,* the organizational links are primarily among those who work on the same project.

For example, a strict functional organization might include a group of marketing professionals, all sharing similar training and expertise. These people would all report to the same manager, who would evaluate them and set their salaries. The group would have its own budget and the people would sit in the same part of a building. This marketing group would be involved in many different projects, but there would be no strong organizational links to the other members of each project team. There would be similarly arranged groups corresponding to design and to manufacturing.

A strict project organization would be made up of groups of people from several different functions, with each group focused on the development of a specific product (or product line). These groups would each report to an experienced project manager, who might be drawn from any of the functional areas. Performance evaluation would be handled by the project manager, and members of the team would typically be colocated as much as possible so that they all work in the same office or part of a building. New ventures, or "start-ups," are among the most extreme examples of project organizations: every individual, regardless of function, is linked together by a single project—the growth of the new company and the creation of its product(s). In these settings, the president or CEO can be viewed as the project manager. Established firms will sometimes form a "tiger team" with dedicated resources for a single project when special focus is required to complete an important development project.

The *matrix organization* was conceived as a hybrid of functional and project organizations. In the matrix organization, individuals are linked to others according to both the project they work on and their function. Typically each individual has two supervisors, one a project manager and one a functional manager. The practical reality is that either the project or the function tends to have stronger links. This is because, for example, both functional and project managers cannot have independent budget authority, they cannot independently evaluate and determine the salaries of their subordinates, and both functional and project organizations cannot easily be grouped together physically. As a result, either the functional or the project organization tends to dominate.

Two variants of the matrix organization are called the *heavyweight project organization* and *lightweight project organization* (Hayes et al., 1988). A heavyweight project organization contains strong project links. The heavyweight project manager has complete budget authority, is heavily involved in performance evaluation of the team members, and makes most of the major resource allocation decisions. Although each participant in a project also belongs to a functional organization, the functional managers have relatively little authority and control. A heavyweight project team in various industries may be called an *integrated product team* (IPT), a *design-build team* (DBT), or simply a *product development team* (PDT). Each of these terms emphasizes the cross-functional nature of these teams.

A lightweight project organization contains weaker project links and relatively stronger functional links. In this scheme, the project manager is more of a coordinator and administrator. The lightweight project manager updates schedules, arranges meetings, and facilitates coordination, but the manager has no real authority and control in the project organization. The functional managers are responsible for budgets, hiring and firing, and performance evaluation. Exhibit 2-7 illustrates the pure functional and project organizations, along with the heavyweight and lightweight variants of the matrix organization.

In this book we refer to the *project team* as the primary organizational unit. In this context, the team is the set of all people involved in the project, regardless of the organizational structure of the product development staff. In a functional organization, the team consists of individuals distributed throughout the functional groups without any organizational linkages other than their common involvement in a project. In the other organizations, the team corresponds to a formal organizational entity, the project group, and has a formally appointed manager. For this reason the notion of a team has much more meaning in matrix and project organizations than it does in functional organizations.

Choosing an Organizational Structure

The most appropriate choice of organizational structure depends on which organizational performance factors are most critical to success. Functional organizations tend to breed specialization and deep expertise in the functional areas. Project organizations tend to enable rapid and effective coordination among diverse functions. Matrix organizations, being hybrids, have the potential to exhibit some of each of these characteristics. The following questions help guide the choice of organizational structure:

- *How important is cross-functional integration?* Functional organizations may exhibit difficulty in coordinating project decisions which span the functional areas. Project or-

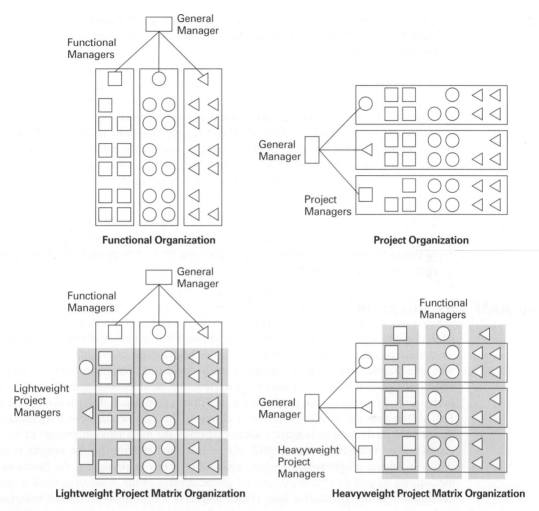

EXHIBIT 2-7 Various product development organizations. For simplicity, three functions and three projects are shown.

Adapted from Hayes et al., 1988

ganizations tend to enable strong cross-functional integration because of the organizational links of the team members across the functions.

- *How critical is cutting-edge functional expertise to business success?* When disciplinary expertise must be developed and retained over several product generations, then some functional links are necessary. For example, in some aerospace companies, computational fluid dynamics is so critical that the fluid dynamicists are organized functionally to ensure the firm will have the best possible capability in this area.
- *Can individuals from each function be fully utilized for most of the duration of a project?* For example, a project may require only a portion of an industrial designer's time for a fraction of the duration of a project. In order to use industrial design

resources efficiently, the firm may choose to organize the industrial designers functionally, so that several projects can draw on the industrial design resource in exactly the amount needed for a particular project.

* *How important is product development speed?* Project organizations tend to allow for conflicts to be resolved quickly and for individuals from different functions to coordinate their activities efficiently. Relatively little time is spent transferring information, assigning responsibilities, and coordinating tasks. For this reason, project organizations are usually faster than functional organizations in developing innovative products. For example, portable computer manufacturers almost always organize their product development teams by project. This allows the teams to develop new products within the extremely short periods required by the fast-paced computer market.

Dozens of other issues confound the choice between functional and project organizations. Exhibit 2-8 summarizes some of the strengths and weaknesses of each organizational type, examples of the types of firms pursuing each strategy, and the major issues associated with each approach.

The AMF Organization

AMF chose to organize its product development staff in a matrix structure. The functions involved in product development at AMF include engineering, manufacturing, marketing, sales, purchasing, and quality assurance. Each of these functions has a manager who reports to the general manager of the division. However, product development projects are led by project managers, and project teams are drawn from each of the functional areas. The AMF matrix organization is probably closest to the lightweight project organization. This is because the project managers are not typically the most senior managers in the division and do not have direct control of resources and staffing for the project teams. While in general a lightweight project organization tends to strengthen the functions at the expense of project efficiency, several characteristics of the AMF organization make the lightweight organization a wise choice and have led to good product development performance.

The most significant factor leading to the choice of a lightweight project organization is that AMF carries out many small product development projects along with one or two large projects. The result of this mix of projects is that many of the team members on smaller projects contribute on a part-time basis. By having relatively strong functional links between individuals, the assignment of staff to smaller projects and the balancing of workload within a function are more easily accomplished.

Another factor allowing AMF to use a lightweight project organization and still achieve high performance in product development is that AMF is an extraordinarily lean company. The Capital Equipment Division has fewer than 100 salaried employees generating and supporting sales of over $100 million per year. Everyone in the division works in the same building, and most of the key employees earn substantial financial rewards when the division is highly profitable. As a result, members of project teams are motivated to look beyond their own functions and work together to develop successful products.

	Functional Organization	Matrix Organization		Project Organization
		Lightweight Product Organization	Heavyweight Project Organization	
Strengths	Fosters development of deep specialization and expertise.	Coordination and administration of projects is explicitly assigned to a single project manager. Maintains development of specialization and expertise.	Provides integration and speed benefits of the project organization. Some of the specialization of a functional organization is retained.	Resources can be optimally allocated within the project team. Technical and market trade-offs can be evaluated quickly.
Weaknesses	Coordination among different functional groups can be slow and bureaucratic.	Requires more managers and administrators than a nonmatrix organization.	Requires more managers and administrators than a nonmatrix organization.	Individuals may have difficulty maintaining cutting-edge functional capabilities.
Typical Examples	Customization development—firms in which development involves slight variations to a standard design (e.g., custom motors, bearings, packaging).	Traditional automobile, electronics, and aerospace companies.	Many recently successful projects in automobile, electronics, and aerospace companies.	Start-up companies. "Tiger teams" and "skunk works" intended to achieve breakthroughs. Firms competing in extremely dynamic markets.
Major Issues	How to integrate different functions (e.g., marketing and design) to achieve a common goal.	How to balance functions and projects. How to simultaneously evaluate project and functional performance.		How to maintain functional expertise over time. How to share technical learning from one project to another.

EXHIBIT 2-8   Characteristics of different organizational structures.

A slight deviation from the standard lightweight project organization also facilitates project completion. The engineering manager is held personally responsible for all aspects of successful completion of projects and not for engineering excellence alone. Although he is responsible for the engineering function, he is primarily responsible for developing successful products. He therefore works daily to ensure that the appropriate coordination occurs, for example, between marketing and engineering.

Finally, the emphasis that the senior management places on product development encourages effective teamwork. The general manager takes a personal interest in every product development project and devotes several days each month to monitoring the progress of these projects. The message communicated to the project teams is that successful products are more important than strong functions.

Summary

An enterprise must make two important decisions about the way it carries out product development. It must define both a product development process and a product development organization.

- A product development process is the sequence of steps an enterprise employs to conceive, design, and commercialize a product.
- A well-defined development process helps to ensure product quality, facilitate coordination among team members, plan the development project, and continuously improve the process.
- The generic process presented in this chapter includes six phases: planning, concept development, system-level design, detail design, testing and refinement, and production ramp-up.
- The concept development phase requires tremendous integration across the different functions on the development team. This front-end process includes identifying customer needs, analyzing competitive products, establishing target specifications, generating product concepts, selecting one or more final concepts, setting final specifications, testing the concept(s), performing an economic analysis, and planning the remaining project activities. The results of the concept development phase are documented in a contract book.
- The development process employed by a particular firm may differ somewhat from the generic process described here. The generic process is most appropriate for market-pull products. Other types of products, which may require variants of the generic process, include technology-push products, platform products, process-intensive products, customized products, high-risk products, quick-build products, and complex systems.
- Regardless of the development process, tasks are completed by individuals residing in organizations. Organizations are defined by linking individuals through reporting relationships, financial relationships, and/or physical layout.
- Functional organizations are those in which the organizational links correspond to the development functions. Project organizations are those in which the organizational links correspond to the development projects. Two types of hybrid, or matrix, organization are the heavyweight project organization and the lightweight project organization.
- The classic trade-off between functional organizations and project organizations is between deep functional expertise and coordination efficiency.

References and Bibliography

Many current resources are available on the Internet via
www.ulrich-eppinger.net

Stage-gate product development processes have been dominant in manufacturing firms for the past 30 years. Cooper describes the modern stage-gate process and many of its enabling practices.

Cooper, Robert G., *Winning at New Products: Accelerating the Process from Idea to Launch,* Perseus Books, Cambridge, MA, 2001.

The spiral product development process has evolved primarily within the software industry: however, many aspects of spiral development can be applied in manufacturing and other industries. McConnell describes spiral software development, along with several other processes used to develop software products.

McConnell, Steve, *Rapid Development: Taming Wild Software Schedules,* Microsoft Press, Redmond, WA, 1996.

The concept of heavyweight and lightweight project organizations is articulated by Hayes, Wheelwright, and Clark. Wheelwright and Clark also discuss product strategy, planning, and technology development activities which generally precede the product development process.

Hayes, Robert H., Steven C. Wheelwright, and Kim B. Clark, *Dynamic Manufacturing: Creating the Learning Organization,* The Free Press, New York, 1988.

Wheelwright, Steven C., and Kim B. Clark, *Revolutionizing Product Development: Quantum Leaps in Speed, Efficiency, and Quality,* The Free Press, New York, 1992.

Andreasen and Hein provide some good ideas on how to integrate different functions in product development. They also show several conceptual models of product development organizations.

Andreasen, M. Myrup, and Lars Hein, *Integrated Product Development,* Springer-Verlag, New York, 1987.

Allen provides strong empirical evidence that physical layout can be used to create significant, although informal, organizational links. He also discusses the use of matrix organizations to mitigate the weaknesses of functional and project organizations.

Allen, Thomas J., *Managing the Flow of Technology: Technology Transfer and the Dissemination of Technological Information within the R&D Organization,* MIT Press, Cambridge, MA, 1977.

Galbraith's seminal book on organizational design contains much useful information which can be applied to product development. His 1994 book is an update of his earlier writing.

Galbraith, Jay R., *Designing Complex Organizations,* Addison-Wesley, Reading, MA, 1973.

Galbraith, Jay R., *Competing with Flexible Lateral Organizations,* second edition, Addison-Wesley, Reading, MA, 1994.

Exercises

1. Diagram a process for planning and cooking a family dinner. Does your process resemble the generic product development process? Is cooking dinner analogous to a market-pull, technology-push, platform, process-intensive, customization, high-risk, quick-build, or complex system process?

2. Define a process for finding a job. For what types of endeavor does a well-defined process enhance performance?

3. What type of development process would you expect to find in an established company successful at developing residential air-conditioning units? How about for a small company that is trying to break into the market for racing wheelchairs?

4. Sketch the organization (in some appropriate graphical representation) of a consulting firm that develops new products for clients on a project-by-project basis. Assume that the individuals in the firm represent all of the different functions required to develop a new product. Would this organization most likely be aligned with functions, be aligned by projects, or be a hybrid?

Thought Questions

1. What role does basic technological research play in the product development process? How would you modify Exhibit 2-3 to better represent the research and technology development activities in product development?

2. Is there an analogy between a university and a product development organization? Is a university a functional or project organization?

3. What is the product development organization for students engaged in projects as part of a product development class?

4. Is it possible for some members of a product development organization to be organized functionally, while others are organized by project? If so, which members of the team would be the most likely candidates for the functional organization?

CHAPTER THREE

Product Planning

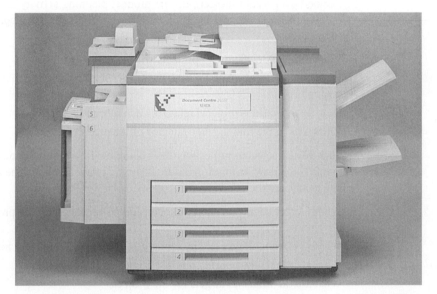

Courtesy of Xerox

EXHIBIT 3-1
The Lakes project developed a new copier platform, including this new product, the Xerox
Document Centre 265.

This chapter was developed in collaboration with Christopher Hession.

33

Xerox Corporation is a global enterprise offering a wide array of document-related products, services, and business solutions. Its mission is to be the leader in the global document market, providing document solutions that enhance business productivity. A key element of Xerox's competitive strategy is to exploit technological innovation in a rapidly changing market. Pursuing this strategy requires the ability to choose the right set of development projects and to define the scope of these projects in such a way that the projects are complementary. Exhibit 3-1 is a photo of the Xerox Document Centre 265, a product resulting from a Xerox project code-named Lakes.

The *product planning* process takes place before a product development project is formally approved, before substantial resources are applied, and before the larger development team is formed. Product planning is an activity that considers the portfolio of projects that an organization might pursue and determines what subset of these projects will be pursued over what time period. The product planning activity ensures that product development projects support the broader business strategy of the company and addresses these questions:

- What product development projects will be undertaken?
- What mix of fundamentally new products, platforms, and derivative products should be pursued?
- How do the various projects relate to each other as a portfolio?
- What will be the timing and sequence of the projects?

Each of the selected projects is then completed by a product development team. The team needs to know its mission before beginning development. The answers to these critical questions are included in a mission statement for the team:

- What market segments should be considered in designing the product and developing its features?
- What new technologies (if any) should be incorporated into the new product?
- What are the manufacturing and service goals and constraints?
- What are the financial targets for the project?
- What are the budget and time frame for the project?

This chapter explains how an organization can maximize the effectiveness of its product development efforts by first considering the set of potential projects it might pursue, deciding which projects are most desirable, and then launching each project with a focused mission. We present a five-step planning process beginning with the identification of opportunities and resulting in a mission statement for the project team.

The Product Planning Process

The *product plan* identifies the portfolio of products to be developed by the organization and the timing of their introduction to the market. The planning process considers product development opportunities identified by many sources, including suggestions from marketing, research, customers, current product development teams, and benchmarking of competitors. From among these opportunities, a portfolio of projects is chosen, timing of

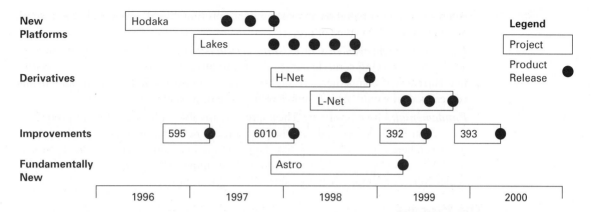

EXHIBIT 3-2   The product plan identifies the portfolio of projects to be pursued by the development organization. This plan divides projects into four categories: new platforms, derivatives of existing platforms, product improvements, and fundamentally new products.

projects is outlined, and resources are allocated. Exhibit 3-2 is an example of a product plan listing products to be developed and indicating the time frame for each.

The product plan is regularly updated to reflect changes in the competitive environment, changes in technology, and information on the success of existing products. Product plans are developed with the company's goals, capabilities, constraints, and competitive environment in mind. Product planning decisions generally involve the senior management of the organization and may take place only annually or a few times each year. Some organizations have a director of planning who manages this process.

Organizations that do not carefully plan the portfolio of development projects to pursue are often plagued with inefficiencies such as:

- Inadequate coverage of target markets with competitive products.
- Poor timing of market introductions of products.
- Mismatches between aggregate development capacity and the number of projects pursued.
- Poor distribution of resources, with some projects overstaffed and others understaffed.
- Initiation and subsequent cancellation of ill-conceived projects.
- Frequent changes in the directions of projects.

Four Types of Product Development Projects

Product development projects can be classified as four types:

- *New product platforms:* This type of project involves a major development effort to create a new family of products based on a new, common platform. The new product family would address familiar markets and product categories. The Xerox Lakes project, aimed at the development of a new, digital copier platform, is an example of this type of project.
- *Derivatives of existing product platforms:* These projects extend an existing product platform to better address familiar markets with one or more new products. To de-

velop a new copier based on an existing light-lens (not digital) product platform would be an example of this type of project.

- *Incremental improvements to existing products:* These projects may only involve adding or modifying some features of existing products in order to keep the product line current and competitive. A slight change to remedy minor flaws in an existing copier product would be an example of this type of project.
- *Fundamentally new products:* These projects involve radically different product or production technologies and may help to address new and unfamiliar markets. Such projects inherently involve more risk; however, the long-term success of the enterprise may depend on what is learned through these important projects. The first digital copier Xerox developed is an example of this type of project.

The Process

Exhibit 3-3 illustrates the steps in the product planning process. First, multiple opportunities are prioritized and a set of promising projects is selected. Resources are allocated to these projects and they are scheduled. These planning activities focus on a *portfolio* of opportunities and potential projects and are sometimes referred to as portfolio management, aggregate product planning, product line planning, or product management. Once projects have been selected and resources allocated, a mission statement is developed for each project. The formulation of a product plan and the development of a mission statement therefore precede the actual product development process.

Although we show the planning process as essentially linear, the activities of selecting promising projects and allocating resources are inherently iterative. The realities of schedules and budgets often force a reassessment of priorities and further refinement and culling of potential projects. The product plan is therefore reevaluated frequently and should be modified based on the latest information from development teams, research laboratories, production, marketing, and service organizations. People involved later in the process are often the first to realize that something about the overall plan or a project's mission is inconsistent, infeasible, or out of date. The ability to adjust the product plan over time is vital to the long-term success of the enterprise.

To develop a product plan and project mission statements, we suggest a five-step process:

1. Identify opportunities.
2. Evaluate and prioritize projects.

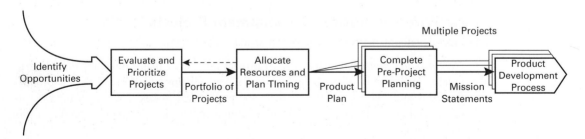

EXHIBIT 3-3   The product planning process. These activities address a portfolio of product development projects, resulting in a product plan and, for each selected project, a mission statement.

3. Allocate resources and plan timing.
4. Complete pre-project planning.
5. Reflect on the results and the process.

Step 1: Identify Opportunities

The planning process begins with the identification of product development opportunities. Such opportunities may involve any of the four types of projects defined above. This step can be thought of as the *opportunity funnel* because it brings together inputs from across the enterprise. Ideas for new products or features of products may come from several sources, including (among others):

- Marketing and sales personnel.
- Research and technology development organizations.
- Current product development teams.
- Manufacturing and operations organizations.
- Current or potential customers.
- Third parties such as suppliers, inventors, and business partners.

Opportunities may be collected passively, but we also recommend that the firm explicitly attempt to generate opportunities. The identification of product development opportunities is closely related to the activity of identifying customer needs. (See Chapter 4.) Some proactive approaches include:

- Document frustrations and complaints that current customers experience with existing products.
- Interview lead users, with attention devoted to innovations by these users and modifications these users may have made to existing products.
- Consider implications of trends in lifestyles, demographics, and technology for existing product categories and for opportunities for new product categories.
- Systematically gather suggestions of current customers, perhaps through the sales force or customer service system.
- Carefully study competitors' products on an ongoing basis (competitive benchmarking).
- Track the status of emerging technologies to facilitate transfer of the appropriate technologies from basic research and technology development into product development.

When employed actively, the opportunity funnel collects ideas continuously, and new product opportunities may arise at any time. As a way of tracking, sorting, and refining these opportunities, we recommend that each promising opportunity be described in a short, coherent statement and that this information be collected in a database. This database can be as simple as a list in a spreadsheet. Some of these opportunities may be expanded, refined, and explored. Often this exploration is done informally by someone who emerges as the "champion" of a particular idea.

At Xerox, many opportunities had been gathered and discussed. Some were simple enhancements to existing products, and others were proposals for products based on entirely new technologies. Following are some examples of opportunity statements similar to those proposed at Xerox:

- Create a document distribution system in which a networked printing device resides on each office worker's desk and automatically delivers mail and other documents.
- Create document delivery software that allows the digital delivery and storage of most intraorganizational documents via a worker's personal computer.

This opportunity statement eventually became the Lakes project:

- Develop a new black and white (B&W), digital, networkable, document center platform for the office market, including scanning, storage, fax, distribution, and printing capabilities.

Step 2: Evaluate and Prioritize Projects

If managed actively, the opportunity funnel can collect hundreds or even thousands of opportunities during a year. Some of these opportunities do not make sense in the context of the firm's other activities, and in most cases, there are simply too many opportunities for the firm to pursue at once. The second step in the product planning process is therefore to select the most promising projects to pursue. Four basic perspectives are useful in evaluating and prioritizing opportunities for new products in existing product categories: competitive strategy, market segmentation, technological trajectories, and product platforms. After discussing these four perspectives, we then discuss evaluating opportunities for fundamentally new products, and how to balance the portfolio of projects.

Competitive Strategy

An organization's *competitive strategy* defines a basic approach to markets and products with respect to competitors. The choice of which opportunities to pursue can be guided by this strategy. Most firms devote much discussion at senior management levels to their strategic competencies and the ways in which they aim to compete. Several strategies are possible, such as:

- *Technology leadership:* To implement this strategy, the firm places great emphasis on basic research and development of new technologies and on the deployment of these technologies through product development.
- *Cost leadership:* This strategy requires the firm to compete on production efficiency, either through economies of scale, use of superior manufacturing methods, low-cost labor, or better management of the production system. Design for manufacturing methods (see Chapter 11) are therefore emphasized in the product (and process) development activities under this strategy.
- *Customer focus:* To follow this strategy, the firm works closely with new and existing customers to assess their changing needs and preferences. Carefully designed product platforms facilitate the rapid development of derivative products with new features or functions of interest to customers. This strategy may result in a broad product line featuring high product variety in order to address the needs of heterogeneous customer segments.

- *Imitative:* This strategy involves closely following trends in the market, allowing competitors to explore which new products are successful for each segment. When viable opportunities have been identified, the firm quickly launches new products to imitate the successful competitors. A fast development process is essential to effectively implement this strategy.

At Xerox, strategic discussions centered around how the company would participate in the digital revolution of the office associated with growth of the Internet. Xerox believed that the Internet would enable a paradigm shift in business practices from one of "print and then distribute" to one of "distribute and then print." The Lakes project would need to support this corporate vision.

Market Segmentation

Customers can be usefully thought of as belonging to distinct market segments. Dividing a market into segments allows the firm to consider the actions of competitors and the strength of the firm's existing products with respect to each well-defined group of customers. By mapping competitors' products and the firm's own products onto segments, the firm can assess which product opportunities best address weaknesses in its own product line and which exploit weaknesses in the offerings of competitors. Exhibit 3-4 shows a *product segment map* of this type for some Xerox products in which markets are segmented according to the number of users sharing office equipment.

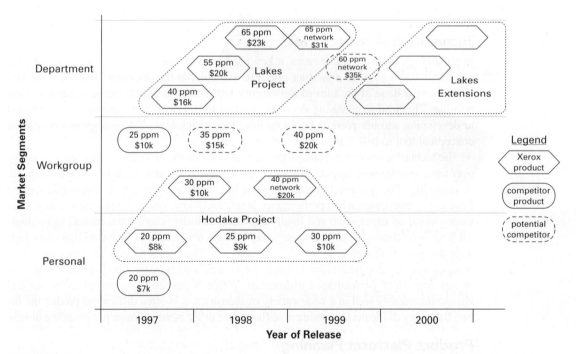

EXHIBIT 3-4   Product segment map showing Xerox B&W digital products and the competition in three market segments: personal, workgroup, and department machines. Key performance dimensions (pages per minute, networking capability) and price point are listed for each product in the map, along with the time of its market introduction.

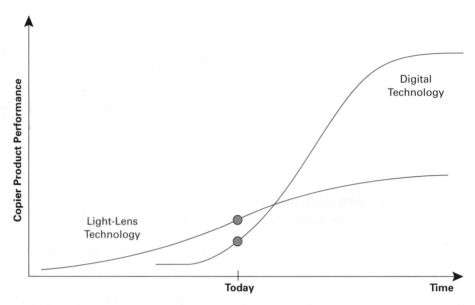

EXHIBIT 3-5   This technology S-curve illustrates that Xerox believed digital copier technologies were just emerging and will improve product performance in the coming years. Xerox believed that it could develop a full-featured digital copier in the near future with performance exceeding that of light-lens copiers.

Technological Trajectories

In technology-intensive businesses, a key product planning decision is when to adopt a new basic technology in a product line. For example, in the document business, the key technological issue at the turn of the century is the shift to digital image processing and printing. The product planning decision is when to develop digital products, as opposed to developing another product based on light-lens technology. *Technology S-curves* are a conceptual tool to help think about such decisions.

The technology S-curve displays the performance of the products in a product category over time, usually with respect to a single performance variable such as resolution, speed, or reliability. The S-curve illustrates a basic but important concept: Technologies evolve from initial emergence when performance is relatively low, through rapid growth in performance based on experience, and finally approach maturity where some natural technological limit is reached and the technology may become obsolete. The S-shaped trajectory captures this general dynamic, as shown in Exhibit 3-5. The horizontal axis may be cumulative research and development effort or time; the vertical axis may be a performance/cost ratio or any important performance dimension. While S-curves characterize technological change remarkably well in a wide variety of industries, it is often difficult to predict the future trajectory of the performance curve (how near or far is the ultimate performance limit).

Product Platform Planning

The product platform is the set of assets shared across a set of products. Components and subassemblies are often the most important of these assets. An effective platform can allow a variety of derivative products to be created more rapidly and easily, with each

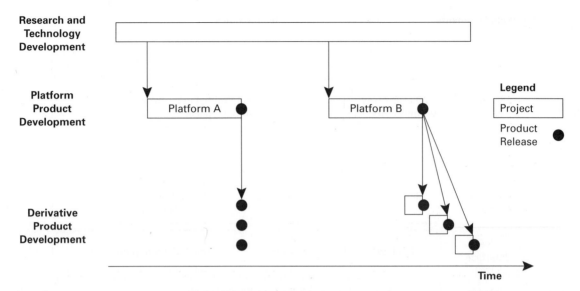

EXHIBIT 3-6   A platform development project creates the architecture of a family of products. Derivative products may be included in the initial platform development effort (Platform A) or derivative products may follow thereafter (Platform B).

product providing the features and functions desired by a particular market segment. See Chapter 9, Product Architecture, for more discussion of the underlying architecture enabling the product platform and for a platform planning method.

Since platform development projects can take from 2 to 10 times as much time and money as derivative product development projects, a firm cannot afford to make every project a new platform. Exhibit 3-6 illustrates the leverage of an effective product platform. The critical strategic decision at this stage is whether a project will develop a derivative product from an existing platform or develop an entirely new platform. Decisions about product platforms are very closely related to the technology development efforts of the firm and to decisions about which technologies to employ in new products.

One technique for coordinating technology development with product planning is the *technology roadmap.* A technology roadmap is a way to represent the expected availability and future use of various technologies relevant to the product being considered. This method has been used by Motorola, Philips, Xerox, and other leaders in fast-moving high-technology industries. The method is particularly useful for planning products in which the critical functional elements are well known in advance.

To create a technology roadmap, multiple generations of technologies are labeled and arranged along a time line, as shown in Exhibit 3-7. The technology roadmap can be augmented with the timing of projects and projects that would utilize these technological developments. (This is sometimes then called a *product-technology roadmap.*) The result is a diagram showing a product's key functional elements and the sequence of technologies expected to implement these elements over a given period of time. Technology roadmapping can serve as a planning tool to create a joint strategy between technology development and product development.

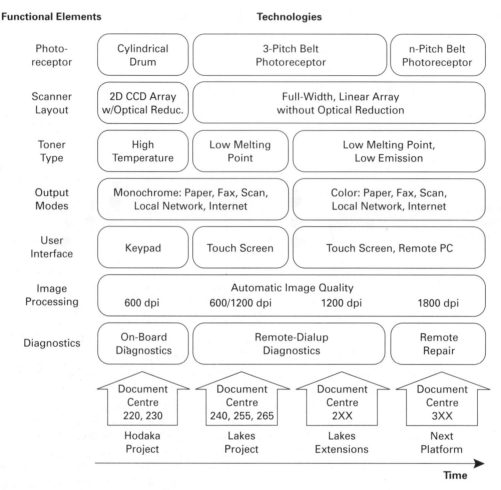

Functional Elements   **Technologies**

| Photo-receptor | Cylindrical Drum | 3-Pitch Belt Photoreceptor | n-Pitch Belt Photoreceptor |

| Scanner Layout | 2D CCD Array w/Optical Reduc. | Full-Width, Linear Array without Optical Reduction |

| Toner Type | High Temperature | Low Melting Point | Low Melting Point, Low Emission |

| Output Modes | Monochrome: Paper, Fax, Scan, Local Network, Internet | Color: Paper, Fax, Scan, Local Network, Internet |

| User Interface | Keypad | Touch Screen | Touch Screen, Remote PC |

| Image Processing | Automatic Image Quality: 600 dpi, 600/1200 dpi, 1200 dpi, 1800 dpi |

| Diagnostics | On-Board Diagnostics | Remote-Dialup Diagnostics | Remote Repair |

Document Centre 220, 230 — Hodaka Project

Document Centre 240, 255, 265 — Lakes Project

Document Centre 2XX — Lakes Extensions

Document Centre 3XX — Next Platform

Time

EXHIBIT 3-7 This technology roadmap shows the life cycles of several digital photocopying technologies and identifies which technologies would be used in each product. For the Lakes platform, Xerox selected technologies for critical functions which could be extended to the higher speeds and color capability required of its derivative products.

Evaluating Fundamentally New Product Opportunities

In addition to new versions of products in existing product categories, the firm faces many opportunities in either new markets or fundamentally new technologies. While investing scarce resources in the development of products using new technologies or for new markets is quite risky, some such investments are necessary to periodically rejuvenate the product portfolio (Christensen, 1997). Some criteria for evaluating fundamentally new product opportunities include:

- Market size (units/year × average price).
- Market growth rate (percent per year).
- Competitive intensity (number of competitors and their strengths).
- Depth of the firm's existing knowledge of the market.

- Depth of the firm's existing knowledge of the technology.
- Fit with the firm's other products.
- Fit with the firm's capabilities.
- Potential for patents, trade secrets, or other barriers to competition.
- Existence of a product champion within the firm.

While these criteria are particularly useful in evaluating fundamentally new product opportunities, they also apply generally to evaluating any product opportunity. These criteria can be used in a simple screening matrix to evaluate the overall attractiveness and types of risk for any given opportunity. Chapter 7, Concept Selection, describes screening matrices for selecting product concepts, but this method is directly applicable to selecting product opportunities as well.

Balancing the Portfolio

There are many methods to help managers balance an organization's portfolio of development projects. Several of these methods involve mapping the portfolio along useful dimensions so that managers may consider the strategic implications of their planning decisions. Cooper et al. (1998) describe numerous mapping approaches involving dimensions such as technical risk, financial return, market attractiveness, and the like. One particularly useful mapping, suggested by Wheelwright and Clark (1992), plots the portfolio of projects along two specific dimensions: the extent to which the project involves a change in the product line and the extent to which the project involves a change in production processes. Exhibit 3-8 illustrates this mapping, called a *product-process change matrix.* This perspective can be useful to illuminate imbalances in the portfolio of projects under consideration and in assessing the consistency between a portfolio of projects and the competitive strategy. For example, a firm may discover that it has identified essentially no breakthrough opportunities or that it has no projects aimed at incremental improvements to existing products.

Although there are no general procedures for deciding exactly what the portfolio should look like, in most cases a firm benefits from a diverse set of projects, just as an investment portfolio benefits from diversification. Furthermore, the firm's choice of competitive strategy should affect the shape of the product development portfolio. For example, a firm pursuing a low-cost strategy would expect the portfolio to contain more production process improvement projects. Firms following a strategy requiring high product variety would need to develop many derivative products based upon existing platforms. Firms implementing a strategy based on technological superiority may need to have a portfolio including more technology development and breakthrough projects in anticipation that not all of these risky projects will result in marketable new products. Note that planning research and technology development activities is closely coupled to, but generally outside the purview of, the product planning process.

Step 3: Allocate Resources and Plan Timing

It is likely that the firm cannot afford to invest in every product development opportunity in its desired balanced portfolio of projects. As timing and resource allocation are determined for the most promising projects, too many projects will invariably compete for too few resources. As a result, the attempt to assign resources and plan timing almost always

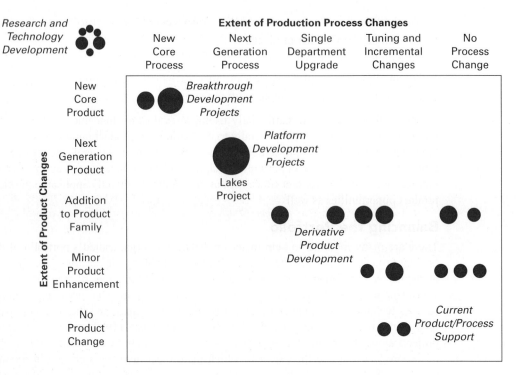

EXHIBIT 3-8 Product-process change matrix. The size of the circles indicates the relative cost of the development projects.

Adapted from Wheelright and Clark, 1992

results in a return to the prior evaluation and prioritization step to prune the set of projects to be pursued.

Resource Allocation

Many organizations take on too many projects without regard for the limited availability of development resources. As a result, skilled engineers and managers are assigned to more and more projects, productivity drops off dramatically, projects take longer to complete, products become late to the market, and profits are lower. *Aggregate planning* helps an organization make efficient use of its resources by pursuing only those projects that can reasonably be completed with the budgeted resources.

The Lakes project was only one of many projects proposed at Xerox. However, since Lakes involved the development of an entirely new platform, this project was substantially larger than the other projects being considered at the time. Any feasible portfolio of projects would be dominated by the resource demands of the Lakes platform development effort. In fact, for the managers at Xerox to find the resources necessary to execute Lakes, many other projects had to be eliminated or postponed until engineers were finished working on Lakes.

Estimating the resources required for each of the projects in the plan by month, quarter, or year forces the organization to face the realities of finite resources. In most cases, the primary resource to be managed is the effort of the development staff, usually expressed in person-hours or person-months. Other critical resources may also require careful planning, such as model shop facilities, rapid prototyping equipment, pilot production lines, testing facilities, and so on. Estimates of required resources in each period can be compared with available resources to compute an overall capacity utilization ratio (demand/capacity) as well as utilizations by resource types, as shown in Exhibit 3-9. Where utilization exceeds 100 percent, there are not sufficient resources to execute all of the projects in the plan on schedule. In fact, to allow for contingencies and to enable responsiveness, planned capacity utilization may be below 100 percent.

In the aggregate planning process, an organization may find that it is in danger of overcommitting resources (often by as much as 100 percent or more, according to Wheelwright and Clark, 1992). Therefore the organization must decide in the planning stage which projects are most important to the success of the firm, and pursue those with adequate resources. Other projects may need to be eliminated from the plan or shifted in time.

Project Timing

Determining the timing and sequence of projects, sometimes called *pipeline management,* must consider a number of factors, including:

- *Timing of product introductions:* Generally the sooner a product is brought to market the better. However, launching a product before it is of adequate quality can damage the reputation of the firm.
- *Technology readiness:* The robustness of the underlying technologies plays a critical role in the planning process. A proven, robust technology can be integrated into products much more quickly and reliably.
- *Market readiness:* The sequence of product introductions determines whether early adopters buy the low-end product and may trade up or whether they buy the high-end product offered at a high initial price. Releasing improvements too quickly can frustrate customers who want to keep up; on the other hand, releasing new products too slowly risks lagging behind competitors.
- *Competition:* The anticipated release of competing products may accelerate the timing of development projects.

The Product Plan

The set of projects approved by the planning process, sequenced in time, becomes the *product plan,* as shown earlier in Exhibit 3-2. The plan may include a mix of fundamentally new products, platform projects, and derivative projects of varying size. Product plans are updated on a periodic basis, perhaps quarterly or annually, as part of the firm's strategic planning activity.

Step 4: Complete Pre-Project Planning

Once the project has been approved, but before substantial resources are applied, a preproject planning activity takes place. This activity involves a small, cross-functional team of people, often known as the *core team.* The Lakes core team consisted of approximately 30 people representing a wide range of technical expertise, marketing, manufacturing, and service functions.

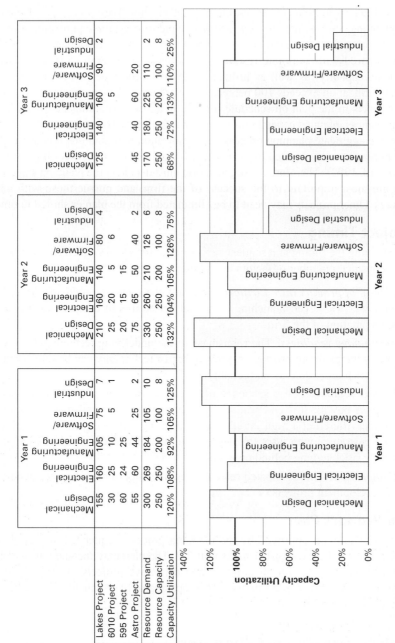

Year 1

	Mechanical Design	Electrical Engineering	Manufacturing Engineering	Software/Firmware	Industrial Design
Lakes Project	155	160	105	75	7
6010 Project	30	25	10	5	1
595 Project	60	24	25	25	2
Resource Demand	300	269	184	105	10
Resource Capacity	250	250	200	100	8
Capacity Utilization	120%	108%	92%	105%	125%

Year 2

	Mechanical Design	Electrical Engineering	Manufacturing Engineering	Software/Firmware	Industrial Design
Lakes Project	210	160	140	80	4
6010 Project	25	20	5	6	2
595 Project	20	15	15	40	
Resource Demand	330	260	210	126	6
Resource Capacity	250	250	200	100	8
Capacity Utilization	132%	104%	105%	126%	75%

Year 3

	Mechanical Design	Electrical Engineering	Manufacturing Engineering	Software/Firmware	Industrial Design
Lakes Project	125	140	160	90	2
6010 Project			5		
595 Project					
Resource Demand	170	180	225	110	2
Resource Capacity	250	250	200	100	8
Capacity Utilization	68%	72%	113%	110%	25%

Capacity Utilization (bar chart, 0%–140% scale with 100% reference line)

Year 1: Mechanical Design, Electrical Engineering, Manufacturing Engineering, Software/Firmware, Industrial Design
Year 2: Mechanical Design, Electrical Engineering, Manufacturing Engineering, Software/Firmware, Industrial Design
Year 3: Mechanical Design, Electrical Engineering, Manufacturing Engineering, Software/Firmware, Industrial Design

EXHIBIT 3-9   Aggregate resource planning can be achieved using a simple spreadsheet method based on estimates of resource demands over time. This example spreadsheet uses units of person-years, although smaller time units (quarters or months) are commonly used in practice. The associated chart highlights where capacity is insufficient to handle all of the projects.

EXHIBIT 3-10
Mission
statement for the
Lakes project.
This document
summarizes the
direction to be
followed by the
product
development
team. Many
more details are
appended to this
mission
statement,
including the
environmental
goals, service
objectives, and
specific
technologies
identified for
use in the Lakes
platform.

Mission Statement: Multifunctional Office Document Machine	
Product Description	• Networkable, digital machine with copy, print, fax, and scan functions
Key Business Goals	• Support Xerox strategy of leadership in digital office equipment • Serve as platform for all future B&W digital products and solutions
Primary Market	• Capture 50% of digital product sales in primary market • Environmentally friendly • First product introduction 4th Q 1997 • Office departments, mid-volume (40–65 ppm, above 42,000 avg. copies/mo.)
Secondary Markets	• Quick-print market • Small "satellite" operations
Assumptions and Constraints	• New product platform • Digital imaging technology • Compatible with CentreWare software • Input devices manufactured in Canada • Output devices manufactured in Brazil • Image processing engine manufactured in both the United States and Europe
Stakeholders	• Purchasers and users • Manufacturing operations • Service operations • Distributors and resellers

At this point, the earlier opportunity statement may be rewritten as a *product vision statement*. The Lakes concept team began with the following product vision statement:

Develop a networked, mid-range, digital platform for imaging, marking, and finishing.

The objective defined by a product vision statement may be very general. It may not say which specific new technologies should be used, nor does it necessarily specify the goals and constraints of functions such as production and service operations. In order to provide clear guidance for the product development organization, generally the team formulates a more detailed definition of the target market and of the assumptions under which the development team will operate. These decisions are captured in a *mission statement*, a summary of which is illustrated in Exhibit 3-10.

Mission Statements

The mission statement may include some or all of the following information:

- *Brief (one-sentence) description of the product:* This description typically includes the key customer benefit of the product but avoids implying a specific product concept. It may, in fact, be the product vision statement.
- *Key business goals:* In addition to the project goals which support the corporate strategy, these goals generally include goals for time, cost, and quality (e.g., timing of the product introduction, desired financial performance, market share targets).

- *Target market(s) for the product:* There may be several target markets for the product. This part of the mission statement identifies the primary market as well as any secondary markets that should be considered in the development effort.
- *Assumptions and constraints that guide the development effort:* Assumptions must be made carefully; although they restrict the range of possible product concepts, they help to maintain a manageable project scope. Information may be attached to the mission statement to document decisions about assumptions and constraints.
- *Stakeholders:* One way to ensure that many of the subtle development issues are addressed is to explicitly list all of the product's stakeholders, that is, all of the groups of people who are affected by the product's success or failure. The stakeholder list begins with the end user (the ultimate external customer) and the external customer who makes the buying decision about the product. Stakeholders also include the customers of the product who reside within the firm, such as the sales force, the service organization, and the production departments. The list of stakeholders serves as a reminder for the team to consider the needs of everyone who will be influenced by the product.

Assumptions and Constraints

In creating the mission statement, the team considers the strategies of several functional areas within the firm. Of the many possible functional strategies to consider, the manufacturing, service, and environmental strategies had the largest influence on the Lakes project. In fact, these strategies guided the core technical developments of the product.

One could reasonably ask why manufacturing, service, and environmental strategies (for example) should be part of the mission statement for a new product. An alternative view is that decisions about these issues should arise from the customer needs for the new product and should not be determined in advance. First, for extremely complex projects, like Lakes, the design of the manufacturing system is a project of similar magnitude to the design of the product itself. As a result, the manufacturing facilities involved in the product need to be identified very early in the process. Second, some product requirements may not be derived strictly from customer needs. For example, most customers will not directly express a need for low environmental impact. However, Xerox chose to adopt a corporate policy of environmentally responsible design. In such cases, the mission statement should reflect such corporate objectives and constraints.

Following are some of the issues that Xerox considered in establishing assumptions and constraints for the Lakes project.

- *Manufacturing:* Even at this very preliminary stage, it is important to consider the capabilities, capacities, and constraints of the manufacturing operations. A broad array of questions may be relevant, including: Which internal production facilities might be used to manufacture and assemble the product? What key suppliers should be involved in the development, and when? Are the existing production systems capable of producing the new technologies which have been identified for the product? For Lakes, Xerox assumed that input devices would be manufactured at production sites in Canada, output devices in Brazil, and the digital image processing engine in both the United States and Europe.
- *Service:* In a business where customer service and service revenue are critical to the success of the firm, it is necessary to also state strategic goals for levels of service

quality. Efforts to improve service include a strategic commitment to designing products that contain few parts, which can be serviced quickly. For Lakes, serviceability goals included reducing both the number of field-replaceable modules required to fully service the machine and the time to install them by an order of magnitude.

• *Environment:* Many corporations today are developing new products with environmental sustainability in mind. The Lakes concept team adopted Xerox's first "zero to landfill" policy, an aggressive goal even for a leader in environmental design practices such as Xerox. The stated goal was that no components from a Lakes product should ever go to a landfill. All components would be either remanufacturable or recyclable, or both. No parts should be disposed of by customers. The Lakes environmental design strategy also included an energy efficiency goal to be the "most efficient machine in its class."

Staffing and Other Pre-Project Planning Activities

The pre-project planning activity also generally addresses project staffing and leadership. This may involve getting key members of the development staff to "sign up" for a new project, that is, to agree to commit to leading the development of the product or of a critical element of the product. Budgets are also generally established during pre-project planning.

For fundamentally new products, budgets and staffing plans will be for the concept development phase of development only. This is because the details of the project are highly uncertain until the basic concept for the new product has been established. More detailed planning will occur when and if the concept is developed further.

Step 5: Reflect on the Results and the Process

In this final step of the planning and strategy process, the team should ask several questions to assess the quality of both the process and the results. Some suggested questions are:

• Is the opportunity funnel collecting an exciting and diverse set of product opportunities?
• Does the product plan support the competitive strategy of the firm?
• Does the product plan address the most important current opportunities facing the firm?
• Are the total resources allocated to product development sufficient to pursue the firm's competitive strategy?
• Have creative ways of leveraging finite resources been considered, such as the use of product platforms, joint ventures, and partnerships with suppliers?
• Does the core team accept the challenges of the resulting mission statement?
• Are the elements of the mission statement consistent?
• Are the assumptions listed in the mission statement really necessary or is the project overconstrained? Will the development team have the freedom to develop the best possible product?
• How can the product planning process be improved?

Because the mission statement is the handoff to the development team, a "reality check" must be performed before proceeding with the development process. This early stage is the time to remedy known flaws, lest they become more severe and expensive as the development process progresses.

This chapter explains the product planning method as a stepwise process, largely for simplicity of the presentation. However, reflection and criticism of consistency and fit should be an ongoing process. Steps in the process can and should be executed simultaneously to make sure that the many plans and decisions are consistent with one another and with the goals, capabilities, and constraints of the firm.

Summary

- Product planning is a periodic process that considers the portfolio of product development projects to be executed.
- Product planning involves a five-step process:

    1. Identify opportunities.
    2. Evaluate and prioritize projects.
    3. Allocate resources and plan timing.
    4. Complete pre-project planning.
    5. Reflect on the results and the process.

- The opportunity funnel collects possibilities for new product platforms, enhancements, and fundamentally new products from several sources within and outside the firm.
- Potential product development projects are evaluated based on the organization's competitive strategy, technological trajectories, and product platform plans.
- A balanced portfolio of development projects may include investments in breakthrough products, new platforms, derivatives, and current product support.
- Aggregate planning ensures that selected projects have adequate resources for successful completion.
- A mission statement for each product development project documents the product vision, business goals, target markets, critical assumptions, and the product's stakeholders.

References and Bibliography

Many current resources are available on the Internet via
www.ulrich-eppinger.net

There are many excellent books on competitive strategy. These selections include discussions related to product planning.

Porter, Michael E., *Competitive Advantage: Creating and Sustaining Superior Performance,* The Free Press, New York, 1985.

Day, George S., *Market Driven Strategy: Processes for Creating Value,* The Free Press, New York, 1990.

Moore, Geoffrey A., *Crossing the Chasm: Marketing and Selling Technology Products to Mainstream Customers,* Harper Business, New York, 1991.

Treacy, Michael, and Fred Wiersema, *The Discipline of Market Leaders,* Addison-Wesley, Reading, MA, 1995.

Wheelwright and Clark discuss several of the dimensions of product planning presented here, including aggregate planning and some mapping methods.

Wheelwright, Steven C., and Kim B. Clark, "Creating Plans to Focus Product Development," *Harvard Business Review,* March–April 1992, pp. 70–82.

Cooper, Edgett, and Kleinschmidt describe a wide range of product portfolio management methods, including financial analysis, scoring techniques, and visual mapping methods.

Cooper, Robert G., Scott J. Edgett, and Elko J. Kleinschmidt, *Portfolio Management for New Products,* Perseus Books, Reading, MA, 1998.

Fine relates product planning and competitive strategy to the decisions of supply-chain design and strategic partnerships with suppliers.

Fine, Charles. H., *Clockspeed: Winning Control in the Age of Temporary Advantage,* Perseus Books, Reading, MA, 1998.

McGrath emphasizes planning of product platforms and strategy for technology-based products.

McGrath, Michael E., *Product Strategy for High-Technology Companies,* McGraw-Hill, New York, 1995.

Reinertsen focuses particular attention on the issue of excess utilization of aggregate development capacity.

Reinertsen, Donald G., *Managing the Design Factory: The Product Developer's Toolkit,* The Free Press, New York, 1997.

Marketing texts include a more detailed treatment of market strategy, market analysis, and product planning.

Crawford, C. Merle, and Anthony Di Benedetto, *New Products Management,* seventh edition, McGraw-Hill, New York, 2003.

Urban, Glen L., and John R. Hauser, *Design and Marketing of New Products,* second edition, Prentice Hall, Englewood Cliffs, NJ, 1993.

Foster developed the S-curve concept and provides many interesting examples from a diverse set of industries.

Foster, Richard N., *Innovation: The Attacker's Advantage,* Summit Books, New York, 1986.

Burgelman and Maidique provide a very good discussion of technology S-curves and stress that these life-cycle curves are not predetermined; rather, they can be influenced by technical development efforts.

Burgelman, Robert A., and Modesto A. Maidique, *Strategic Management of Technology and Innovation,* third edition, Irwin Professional Publishing, Homewood, IL, 2001.

Several authors present a more thorough discussion of product platform planning in various industries.

Meyer, Marc H., and Alvin P. Lehnerd, *The Power of Product Platforms,* The Free Press, New York, 1997.

Sanderson, Susan W., and Mustafa Uzumeri, *Managing Product Families,* Irwin, Chicago, 1997.

Managers at Motorola and Philips describe their use of several technology roadmapping methods for integrating the planning of technology development and product development.

Willyard, Charles H., and Cheryl W. McClees, "Motorola's Technology Roadmap Process," *Research Management,* Vol. 30, No. 5, Sep./Oct. 1987, pp. 13–19.

Groenveld, Pieter, "Roadmapping Integrates Business and Technology," *Research-Technology Management,* Vol. 40, No. 5, Sep./Oct. 1997, pp. 48–55.

Christensen provides evidence that firms must invest in fundamentally new products, technologies, and markets in order to remain at the leading edge of their industries.

Christensen, Clayton M., *The Innovator's Dilemma: When New Technologies Cause Great Firms to Fail,* Harvard Business School Press, Boston, 1997.

Exercises

1. Conduct a search using the Internet or published corporate annual reports to identify the corporate strategy of a company in which you might be interested in investing. Learn about the firm's product lines and its newest products. How do these products support the corporate strategy? What types of projects would you expect to see in the product plan?

2. Create a product-technology roadmap illustrating the availability of technologies for a class of products you understand well, such as personal computers.

Thought Questions

1. How might a portfolio of development projects differ if the firm believes a particular product technology is currently at position A or B on the technology S-curve shown below?

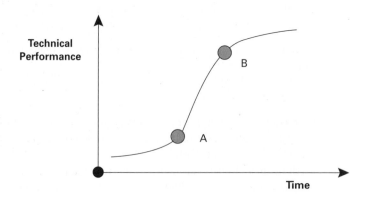

2. How might Xerox be able to address the shortage of mechanical design engineers identified by the aggregate project planning analysis shown in Exhibit 3-9? List five ways Xerox could increase the capacity and five ways to reduce the demand for mechanical design engineers.

CHAPTER FOUR

Identifying Customer Needs

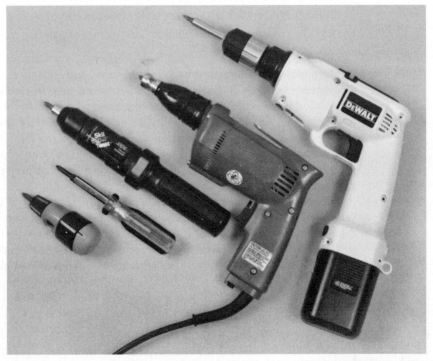

Photo by Stuart Cohen

EXHIBIT 4-1
Existing products used to drive screws: manual screwdrivers, cordless screwdriver, screw gun, cordless drill with driver bit.

This chapter was developed in collaboration with Jonathan Sterrett.

A successful hand tool manufacturer was exploring the growing market for hand-held power tools. After performing initial research, the firm decided to enter the market with a cordless screwdriver. Exhibit 4-1 shows several existing products used to drive screws. After some initial concept work, the manufacturer's development team fabricated and field-tested several prototypes. The results were discouraging. Although some of the products were liked better than others, each one had some feature that customers objected to in one way or another. The results were quite mystifying since the company had been successful in related consumer products for years. After much discussion, the team decided that its process for identifying customer needs was inadequate.

This chapter presents a method for comprehensively identifying a set of customer needs. The goals of the method are to:

- Ensure that the product is focused on customer needs.
- Identify latent or hidden needs as well as explicit needs.
- Provide a fact base for justifying the product specifications.
- Create an archival record of the needs activity of the development process.
- Ensure that no critical customer need is missed or forgotten.
- Develop a common understanding of customer needs among members of the development team.

The philosophy behind the method is to create a high-quality information channel that runs directly between customers in the target market and the developers of the product. This philosophy is built on the premise that those who directly control the details of the product, including the engineers and industrial designers, must interact with customers and experience the *use environment* of the product. Without this direct experience, technical trade-offs are not likely to be made correctly, innovative solutions to customer needs may never be discovered, and the development team may never develop a deep commitment to meeting customer needs.

The process of identifying customer needs is an integral part of the larger product development process and is most closely related to concept generation, concept selection, competitive benchmarking, and the establishment of product specifications. The customer-needs activity is shown in Exhibit 4-2 in relation to these other front-end product development activities, which collectively can be thought of as the *concept development* phase.

The concept development process illustrated in Exhibit 4-2 implies a distinction between customer needs and product specifications. This distinction is subtle but important.

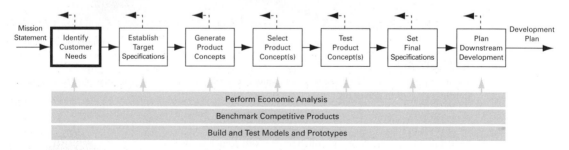

EXHIBIT 4-2   The customer-needs activity in relation to other concept development activities.

Needs are largely independent of any particular product we might develop; they are not specific to the concept we eventually choose to pursue. A team should be able to identify customer needs without knowing if or how it will eventually address those needs. On the other hand, *specifications* do depend on the concept we select. The specifications for the product we finally choose to develop will depend on what is technically and economically feasible and on what our competitors offer in the marketplace, as well as on customer needs. (See Chapter 5, Product Specifications, for a more detailed discussion of this distinction.) Also note that we choose to use the word *need* to label any attribute of a potential product that is desired by the customer; we do not distinguish here between a want and a need. Other terms used in industrial practice to refer to customer needs include *customer attributes* and *customer requirements.*

Identifying customer needs is itself a process, for which we present a five-step method. We believe that a little structure goes a long way in facilitating effective product development practices, and we hope and expect that this method will be viewed by those who employ it not as a rigid process but rather as a starting point for continuous improvement and refinement. The five steps are:

1. Gather raw data from customers.
2. Interpret the raw data in terms of customer needs.
3. Organize the needs into a hierarchy of primary, secondary, and (if necessary) tertiary needs.
4. Establish the relative importance of the needs.
5. Reflect on the results and the process.

We treat each of the five steps in turn and illustrate the key points with the cordless screwdriver example. We chose the screwdriver because it is simple enough that the method is not hidden by the complexity of the example. However, note that the same method, with minor adaptation, has been successfully applied to hundreds of products ranging from kitchen utensils costing less than $10 to machine tools costing hundreds of thousands of dollars.

Before beginning the development project, the firm typically specifies a particular market opportunity and lays out the broad constraints and objectives for the project. This information is frequently formalized as a *mission statement* (also sometimes called a *charter* or a *design brief*). The mission statement specifies which direction to go in but generally does not specify a precise destination or a particular way to proceed. The mission statement is the result of the product planning activities described in Chapter 3, Product Planning. The mission statement for the cordless screwdriver is shown in Exhibit 4-3.

The cordless screwdriver category of products is already relatively well developed. Such products are particularly well suited to a structured process for gathering customer needs. One could reasonably ask whether a structured method is effective for completely new categories of products with which customers have no experience. Satisfying needs is just as important in revolutionary products as in incremental products. A necessary condition for product success is that a product offer perceived benefits to the customer. Products offer benefits when they satisfy needs. This is true whether the product is an incremental variation on an existing product or whether it is a completely new product based on a revolutionary invention. Developing an entirely new category of product is a risky undertaking, and to some extent the only real indication of whether customer needs have

EXHIBIT 4-3
Mission
statement for
the cordless
screwdriver.

Mission Statement: Screwdriver Project	
Product Description	• A hand-held, power-assisted device for installing threaded fasteners
Key Business Goals	• Product introduced in fourth quarter of 2006 • 50% gross margin • 10% share of cordless screwdriver market by 2008
Primary Market	• Do-it-yourself consumer
Secondary Markets	• Casual consumer • Light-duty professional
Assumptions	• Hand-held • Power-assisted • Nickel-metal-hydride rechargeable battery technology
Stakeholders	• User • Retailer • Sales force • Service center • Production • Legal department

been identified correctly is whether customers like the team's first prototypes. Nevertheless, in our opinion, a structured method for gathering data from customers remains useful and can lower the inherent risk in developing a radically new product. Whether or not customers are able to fully articulate their latent needs, interaction with customers in the target market will help the development team build a personal understanding of the user's environment and point of view. This information is always useful, even if it does not result in the identification of every need the new product will address.

Step 1: Gather Raw Data from Customers

Consistent with our basic philosophy of creating a high-quality information channel directly from the customer, gathering data involves contact with customers and experience with the use environment of the product. Three methods are commonly used:

1. *Interviews:* One or more development team members discuss needs with a single customer. Interviews are usually conducted in the customer's environment and typically last one to two hours.

2. *Focus groups:* A moderator facilitates a two-hour discussion with a group of 8 to 12 customers. Focus groups are typically conducted in a special room equipped with a two-way mirror allowing several members of the development team to observe the group. In most cases, the moderator is a professional market researcher, but a member of the development team sometimes moderates. The proceedings are usually videotaped. Participants are usually paid a modest fee ($50 to $100 each) for their attendance. The total cost of a focus group, including rental of the room, participant fees, videotaping, and refreshments, is about $2,500. In most U.S. cities, firms that recruit participants, moderate focus groups, and/or rent facilities are listed in directories under "Market Research."

EXHIBIT 4-4
Comparison of the percentages of customer needs that are revealed for focus groups and interviews as a function of the number of sessions. Note that a focus group lasts two hours, while an interview lasts one hour.

Source: Griffin and Hauser, 1993

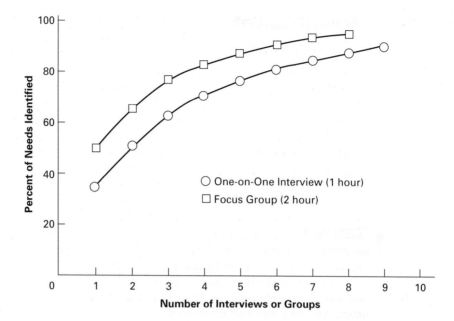

3. *Observing the product in use:* Watching customers use an existing product or perform a task for which a new product is intended can reveal important details about customer needs. For example, a customer painting a house may use a screwdriver to open paint cans in addition to driving screws. Observation may be completely passive, without any direct interaction with the customer, or may involve working side by side with a customer, allowing members of the development team to develop firsthand experience using the product. Ideally, team members observe the product in the actual use environment. For some products, such as do-it-yourself tools, actually using the products is simple and natural; for others, such as surgical instruments, the team may have to use the products on surrogate tasks (e.g., cutting fruit instead of human tissue when developing a new scalpel).

Some practitioners also rely on written surveys for gathering raw data. While a mail or web-based survey is quite useful later in the process, we cannot recommend this approach for initial efforts to identify customer needs; written surveys simply do not provide enough information about the use environment of the product, and they are generally ineffective in revealing unanticipated needs.

Research by Griffin and Hauser shows that one 2-hour focus group reveals about the same number of needs as two 1-hour interviews (Griffin and Hauser, 1993). (See Exhibit 4-4.) Because interviews are usually less costly (per hour) than focus groups and because an interview often allows the product development team to experience the use environment of the product, we recommend that interviews be the primary data collection method. Interviews may be supplemented with one or two focus groups as a way to allow top management to observe a group of customers or as a mechanism for sharing a common customer experience (via video) with the members of a larger team. Some practitioners believe that for certain products and customer groups, the interactions among the participants of focus groups can elicit more varied needs than are revealed through interviews, although this belief is not strongly supported by research findings.

Choosing Customers

Griffin and Hauser also addressed the question of how many customers to interview in order to reveal most of the customer needs. In one study, they estimated that 90 percent of the customer needs for picnic coolers were revealed after 30 interviews. In another study, they estimated that 98 percent of the customer needs for a piece of office equipment were revealed after 25 hours of data collection in both focus groups and interviews. As a practical guideline for most products, conducting fewer than 10 interviews is probably inadequate and 50 interviews are probably too many. However, interviews can be conducted sequentially and the process can be terminated when no new needs are revealed by additional interviews. These guidelines apply to cases in which the development team is addressing a single market segment. If the team wishes to gather customer needs from multiple distinct segments, then the team may need to conduct 10 or more interviews in each segment. Concept development teams consisting of more than 10 people usually collect data from plenty of customers simply by involving much of the team in the process. For example, if a 10-person team is divided into five pairs and each pair conducts 6 interviews, the team conducts 30 interviews in total.

Needs can be identified more efficiently by interviewing a class of customers called *lead users.* According to von Hippel, lead users are customers who experience needs months or years ahead of the majority of the market and stand to benefit substantially from product innovations (von Hippel, 1988). These customers are particularly useful sources of data for two reasons: (1) they are often able to articulate their emerging needs, because they have had to struggle with the inadequacies of existing products, and (2) they may have already invented solutions to meet their needs. By focusing a portion of the data collection efforts on lead users, the team may be able to identify needs which, although explicit for lead users, are still latent for the majority of the market. Developing products to meet these latent needs allows a firm to anticipate trends and to leapfrog competitive products.

The choice of which customers to interview is complicated when several different groups of people can be considered "the customer." For many products, one person (the buyer) makes the buying decision and another person (the user) actually uses the product. A good approach is to gather data from the end user of the product in all situations, and in cases where other types of customers and stakeholders are clearly important, to gather data from these people as well.

A customer selection matrix is useful for planning exploration of both market and customer variety. Burchill suggests that market segments be listed on the left side of the matrix while the different types of customers are listed across the top (Burchill et al., 1997), as shown in Exhibit 4-5. The number of intended customer contacts is entered in each cell to indicate the depth of coverage.

For industrial and commercial products, actually locating customers is usually a matter of making telephone calls. In developing such products within an existing firm, a field sales force can often provide names of customers, although the team must be careful about biasing the selection of customers toward those with allegiances to a particular manufacturer. The web or the telephone book can be used to identify names of some types of customers for some classes of products (e.g., building contractors or insurance agents). For products that are integral to a customer's job, getting someone to agree to an interview is usually simple; these customers are eager to discuss their needs. For con-

EXHIBIT 4-5
Customer
selection matrix
for the cordless
screwdriver
project.

	Lead Users	Users	Retailer or Sales Outlet	Service Centers
Homeowner (occasional use)	0	5	2	3
Handy person (frequent use)	3	10		
Professional (heavy-duty use)	3	2	2	

sumer products, customers can also be located by making telephone calls. However, arranging a set of interviews for consumer products generally requires more inquiries than for industrial or commercial products because the benefit of participating in an interview is less direct for these customers.

The Art of Eliciting Customer Needs Data

The techniques we present here are aimed primarily at interviewing end users, but these methods do apply to all of the three data-gathering modes and to all types of stakeholders. The basic approach is to be receptive to information provided by customers and to avoid confrontations or defensive posturing. Gathering needs data is very different from a sales call: the goal is to elicit an honest expression of needs, not to convince a customer of what he or she needs. In most cases customer interactions will be verbal; interviewers ask questions and the customer responds. A prepared interview guide is valuable for structuring this dialogue. Some helpful questions and prompts for use after the interviewers introduce themselves and explain the purpose of the interview are:

- When and why do you use this type of product?
- Walk us through a typical session using the product.
- What do you like about the existing products?
- What do you dislike about the existing products?
- What issues do you consider when purchasing the product?
- What improvements would you make to the product?

Here are some general hints for effective interaction with customers:

- *Go with the flow.* If the customer is providing useful information, do not worry about conforming to the interview guide. The goal is to gather important data on customer needs, not to complete the interview guide in the allotted time.
- *Use visual stimuli and props.* Bring a collection of existing and competitors' products, or even products that are tangentially related to the product under development. At the end of a session, the interviewers might even show some preliminary product concepts to get customers' early reactions to various approaches.
- *Suppress preconceived hypotheses about the product technology.* Frequently customers will make assumptions about the product concept they expect would meet their

needs. In these situations, the interviewers should avoid biasing the discussion with assumptions about how the product will eventually be designed or produced. When customers mention specific technologies or product features, the interviewer should probe for the underlying need the customer believes the suggested solution would satisfy.

- *Have the customer demonstrate the product and/or typical tasks related to the product.* If the interview is conducted in the use environment, a demonstration is usually convenient and invariably reveals new information.

- *Be alert for surprises and the expression of latent needs.* If a customer mentions something surprising, pursue the lead with follow-up questions. Frequently, an unexpected line of questioning will reveal *latent needs*—important dimensions of the customers' needs that are neither fulfilled nor commonly articulated and understood.

- *Watch for nonverbal information.* The process described in the chapter is aimed at developing better physical products. Unfortunately, words are not always the best way to communicate needs related to the physical world. This is particularly true of needs involving the human dimensions of the product, such as comfort, image, or style. The development team must be constantly aware of the nonverbal messages provided by customers. What are their facial expressions? How do they hold competitors' products?

Note that many of our suggested questions and guidelines assume that the customer has some familiarity with products similar to the new product under development. This is almost always true. For example, even before the first cordless screwdriver became available, people installed fasteners. Developing an understanding of customer needs as they relate to the general fastening task would still have been beneficial in developing the first cordless tool. Similarly, understanding the needs of customers using other types of cordless appliances, such as electric razors, would also have been useful. We can think of no product so revolutionary that there would be no analogous products or tasks from which the development team could learn. However, in gathering needs relating to truly revolutionary products with which customers have no experience, the interview questions should be focused on the task or situation in which the new product will be applied, rather than on the product itself.

Documenting Interactions with Customers

Four methods are commonly used for documenting interactions with customers:

   1. *Audio recording:* Making an audio recording of the interview is very easy. Unfortunately, transcribing the recording into text is very time consuming, and hiring someone to do it can be expensive. Also, audio recording has the disadvantage of being intimidating to some customers.

   2. *Notes:* Handwritten notes are the most common method of documenting an interview. Designating one person as the primary notetaker allows the other person to concentrate on effective questioning. The notetaker should strive to capture some of the wording of every customer statement verbatim. These notes, if transcribed immediately after the interview, can be used to create a description of the interview that is very close to an actual transcript. This debriefing immediately after the interview also facilitates sharing of insights between the interviewers.

   3. *Video recording:* Video recording is almost always used to document a focus group session. It is also very useful for documenting observations of the customer in the use environment and/or using existing products. The video recording is useful for bringing new team members "up to speed" and is also useful as raw material for presentations to upper management. Multiple viewings of video recordings of customers in action often facilitate the identification of latent customer needs. Video recording is also useful for capturing many aspects of the end user's environment.

4. *Still photography:* Taking photographs provides many of the benefits of video recording. The primary advantages of still photography are ease of display of the photos, excellent image quality, and readily available equipment. The primary disadvantage is the relative inability to record dynamic information.

The final result of the data-gathering phase of the process is a set of raw data, usually in the form of *customer statements* but frequently supplemented by video recordings or photographs. A data template implemented in a spreadsheet is useful for organizing these raw data. Exhibit 4-6 is an example of a portion of such a template. We recommend that the template be filled in as soon as possible after the interaction with the customer and edited by the other development team members present during the interaction. The first column in the main body of the template indicates the question or prompt that elicited the customer data. The second column is a list of verbatim statements the customer made or an observation of a customer action (from a video recording or from direct observation). The third column contains the customer needs implied by the raw data. Some emphasis should be placed on investigating clues that may identify potential latent needs. Such clues may be in the form of humorous remarks, less serious suggestions, frustrations, nonverbal information, or observations and descriptions of the use environment. The symbol (!) is used in Exhibit 4-6 to flag potential latent needs. Techniques for interpreting the raw data in terms of customer needs are given in the next section.

The final task in step 1 is to write thank-you notes to the customers involved in the process. Invariably, the team will need to solicit further customer information, so developing and maintaining a good rapport with a set of users is important.

Step 2: Interpret Raw Data in Terms of Customer Needs

Customer needs are expressed as written statements and are the result of interpreting the need underlying the raw data gathered from the customers. Each statement or observation (as listed in the second column of the data template) may be translated into any number of customer needs. Griffin and Hauser found that multiple analysts may translate the same interview notes into different needs, so it is useful to have more than one team member conducting the translation process. Below we provide five guidelines for writing need statements. The first two guidelines are fundamental and are critical to effective translation; the remaining three guidelines ensure consistency of phrasing and style across all team members. Exhibit 4-7 provides examples to illustrate each guideline.

- ***Express the need in terms of* what *the product has to do, not in terms of how it might do it.*** Customers often express their preferences by describing a solution concept or an implementation approach; however, the need statement should be expressed in terms independent of a particular technological solution.

Customer:	Bill Esposito	Interviewer(s):	Jonathan and Lisa
Address:	100 Memorial Drive	Date:	19 December 2002
	Cambridge, MA 02139		
Telephone:	617-864-1274	Currently uses:	Craftsman Model A3
Willing to do follow-up?	Yes	Type of user:	Building maintenance

Question/Prompt	Customer Statement	Interpreted Need
Typical uses	I need to drive screws fast, faster than by hand.	The SD drives screws faster than by hand.
	I sometimes do duct work; use sheet metal screws.	The SD drives sheet metal screws into metal duct work.
	A lot of electrical; switch covers, outlets, fans, kitchen appliances.	The SD can be used for screws on electrical devices.
Likes—current tool	I like the pistol grip; it feels the best.	The SD is comfortable to grip.
	I like the magnetized tip.	The SD tip retains the screw before it is driven.
Dislikes—current tool	I don't like it when the tip slips off the screw.	The SD tip remains aligned with the screw head without slipping.
	I would like to be able to lock it so I can use it with a dead battery.	The user can apply torque manually to the SD to drive a screw. (!)
	Can't drive screws into hard wood.	The SD can drive screws into hard wood.
	Sometimes I strip tough screws.	The SD does not strip screw heads.
Suggested improvements	An attachment to allow me to reach down skinny holes.	The SD can access screws at the end of deep, narrow holes.
	A point so I can scrape paint off of screws.	The SD allows the user to work with screws that have been painted over.
	Would be nice if it could punch a pilot hole.	The SD can be used to create a pilot hole. (!)

EXHIBIT 4-6   Customer data template filled in with sample customer statements and interpreted needs. SD is an abbreviation for screwdriver. (Note that this template represents a partial list from a single interview. A typical interview session may elicit more than 50 customer statements and interpreted needs.)

- *Express the need as specifically as the raw data.* Needs can be expressed at many different levels of detail. To avoid loss of information, express the need at the same level of detail as the raw data.
- *Use positive, not negative, phrasing.* Subsequent translation of a need into a product specification is easier if the need is expressed as a positive statement. This is not a rigid guideline, because sometimes positive phrasing is difficult and awkward. For example, one of the need statements in Exhibit 4-6 is "the screwdriver does not strip screw heads." This need is more naturally expressed in a negative form.
- *Express the need as an attribute of the product.* Wording needs as statements about the product ensures consistency and facilitates subsequent translation into product specifications. Not all needs can be cleanly expressed as attributes of the product, however, and in most of these cases the needs can be expressed as attributes of the

Guideline	Customer Statement	Need Statement— Right	Need Statement— Wrong
"What" not "how"	"Why don't you put protective shields around the battery contacts?"	The screwdriver battery is protected from accidental shorting.	The screwdriver battery contacts are covered by a plastic sliding door.
Specificity	"I drop my screwdriver all the time."	The screwdriver operates normally after repeated dropping.	The screwdriver is rugged.
Positive not negative	"It doesn't matter if it's raining; I still need to work outside on Saturdays."	The screwdriver operates normally in the rain.	The screwdriver is not disabled by the rain.
An attribute of the product	"I'd like to charge my battery from my cigarette lighter."	The screwdriver battery can be charged from an automobile cigarette lighter.	An automobile cigarette lighter adapter can charge the screwdriver battery.
Avoid "must" and "should"	"I hate it when I don't know how much juice is left in the batteries of my cordless tools."	The screwdriver provides an indication of the energy level of the battery.	The screwdriver should provide an indication of the energy level of the battery.

EXHIBIT 4-7   Examples illustrating the guidelines for writing need statements.

user of the product (e.g., "the user can apply torque manually to the screwdriver to drive a screw").

- *Avoid the words* **must** *and* **should.** The words *must* and *should* imply a level of importance for the need. Rather than casually assigning a binary importance rating (*must* versus *should*) to the needs at this point, we recommend deferring the assessment of the importance of each need until step 4.

The list of customer needs is the superset of all the needs elicited from all the interviewed customers in the target market. Some needs may not be technologically realizable. The constraints of technical and economic feasibility are incorporated into the process of establishing product specifications in subsequent development steps. (See Chapter 5, Product Specifications.) In some cases customers will have expressed conflicting needs. At this point in the process the team does not attempt to resolve such conflicts, but simply documents both needs. Deciding how to address conflicting needs is one of the challenges of the subsequent concept development activities.

Step 3: Organize the Needs into a Hierarchy

The result of steps 1 and 2 should be a list of 50 to 300 *need statements.* Such a large number of detailed needs is awkward to work with and difficult to summarize for use in subsequent development activities. The goal of step 3 is to organize these needs into a hierarchical list. The list will typically consist of a set of *primary needs,* each one of which will be further characterized by a set of *secondary needs.* In cases of very complex products, the secondary needs may be broken down into tertiary needs as well. The primary needs are the most general needs, while the secondary and tertiary needs express needs in more detail. Exhibit 4-8 shows the resulting hierarchical list of needs for the screwdriver

The SD provides plenty of power to drive screws.
* The SD maintains power for several hours of heavy use.
** The SD can drive screws into hardwood.
The SD drives sheet metal screws into metal ductwork.
*** The SD drives screws faster than by hand.

The SD makes it easy to start a screw.
* The SD retains the screw before it is driven.
*! The SD can be used to create a pilot hole.

The SD works with a variety of screws.
** The SD can turn Phillips, Torx, socket, and hex head screws.
** The SD can turn many sizes of screws.

The SD can access most screws.
The SD can be maneuvered in tight areas.
** The SD can access screws at the end of deep, narrow holes.

The SD turns screws that are in poor condition.
The SD can be used to remove grease and dirt from screws.
The SD allows the user to work with painted screws.

The SD feels good in the user's hand.
*** The SD is comfortable when the user pushes on it.
*** The SD is comfortable when the user resists twisting.
* The SD is balanced in the user's hand.
! The SD is equally easy to use in right or left hands.
The SD weight is just right.
The SD is warm to touch in cold weather.
The SD remains comfortable when left in the sun.

The SD is easy to control while turning screws.
*** The user can easily push on the SD.
*** The user can easily resist the SD twisting.
The SD can be locked "on."
**! The SD speed can be controlled by the user while turning a screw.
* The SD remains aligned with the screw head without slipping.
** The user can easily see where the screw is.
* The SD does not strip screw heads.
* The SD is easily reversible.

The SD is easy to set up and use.
* The SD is easy to turn on.
* The SD prevents inadvertent switching off.
* The user can set the maximum torque of the SD.
*! The SD provides ready access to bits or accessories.
* The SD can be attached to the user for temporary storage.

The SD power is convenient.
* The SD is easy to recharge.
The SD can be used while recharging.
*** The SD recharges quickly.
The SD batteries are ready to use when new.
**! The user can apply torque manually to the SD to drive a screw.

The SD lasts a long time.
** The SD tip survives heavy use.
The SD can be hammered.
* The SD can be dropped from a ladder without damage.

The SD is easy to store.
* The SD fits in a toolbox easily.
** The SD can be charged while in storage.
The SD resists corrosion when left outside or in damp places.
*! The SD maintains its charge after long periods of storage.
The SD maintains its charge when wet.

The SD prevents damage to the work.
* The SD prevents damage to the screw head.
The SD prevents scratching of finished surfaces.

The SD has a pleasant sound when in use.

The SD looks like a professional quality tool.

The SD is safe.
The SD can be used on electrical devices.
*** The SD does not cut the user's hands.

EXHIBIT 4-8 Hierarchical list of primary and secondary customer needs for the cordless screwdriver. Importance ratings for the secondary needs are indicated by the number of *'s, with *** denoting critically important needs. Latent needs are denoted by !.

example. For the screwdriver, there are 15 primary needs and 49 secondary needs. Note that two of the primary needs have no associated secondary needs.

The procedure for organizing the needs into a hierarchical list is intuitive, and many teams can successfully complete the task without detailed instructions. For completeness, we provide a step-by-step procedure here. This activity is best performed on a wall or a large table by a small group of team members.

1. *Print or write each need statement on a separate card or self-stick note.* A print macro can be easily written to print the need statements directly from the data template. A nice feature of this approach is that the need can be printed in a large font in the center of the card and then the original customer statement and other relevant information can be printed in a small font at the bottom of the card for easy reference. Four cards can be cut from a standard printed sheet.

2. *Eliminate redundant statements.* Those cards expressing redundant need statements can be stapled together and treated as a single card. Be careful to consolidate only those statements that are identical in meaning.

3. *Group the cards according to the similarity of the needs they express.* At this point, the team should attempt to create groups of roughly three to seven cards that express similar needs. The logic by which groups are created deserves special attention. Novice development teams often create groups according to a technological perspective, clustering needs relating to, for example, materials, packaging, or power. Or they create groups according to assumed physical components such as enclosure, bits, switch, and battery. Both of these approaches are dangerous. Recall that the goal of the process is to create a description of the needs of the customer. For this reason, the groupings should be consistent with the way customers think about their needs and not with the way the development team thinks about the product. The groups should correspond to needs customers would view as similar. In fact, some practitioners use a process in which customers actually organize the need statements.

4. *For each group, choose a label.* The label is itself a statement of need that generalizes all of the needs in the group. It can be selected from one of the needs in the group, or the team can write a new need statement.

5. *Consider creating supergroups consisting of two to five groups.* If there are fewer than 20 groups, then a two-level hierarchy is probably sufficient to organize the data. In this case, the group labels are primary needs and the group members are secondary needs. However, if there are more than 20 groups, the team may consider creating supergroups, and therefore a third level in the hierarchy. The process of creating supergroups is identical to the process of creating groups. As with the previous step, cluster groups according to similarity of the need they express and then create or select a supergroup label. These supergroup labels become the primary needs, the group labels become the secondary needs, and the members of the groups become tertiary needs.

6. *Review and edit the organized needs statements.* The arrangement of needs in a hierarchy is not unique in terms of being correct. At this point, the team may wish to consider alternative groupings or labels and may engage another group to suggest alternative arrangements.

The process is more complicated when the team attempts to reflect the needs of two or more distinct market segments. There are at least two approaches that can be taken to address this challenge. First, the team can label each need with the segment (and possibly the name) of the customer from whom the need was elicited. This way, differences in needs across segments can be observed directly. One practical visual technique for this labeling is to use different colors of paper for the cards on which the needs statements are written, with each color corresponding to a different market segment. The other approach to multiple market segments is to perform the clustering process separately for each market segment. Using this approach, the team can observe differences both in the needs themselves and in the ways in which these needs are best organized. We recommend that the team adopt this parallel, independent approach when the segments are very different in their needs and when there is some doubt about the ability of the team to address the different segments with the same product.

Step 4: Establish the Relative Importance of the Needs

The hierarchical list alone does not provide any information on the relative importance that customers place on different needs. Yet the development team will have to make trade-offs and allocate resources in designing the product. A sense of the relative importance of the various needs is essential to making these trade-offs correctly. Step 4 in the needs process establishes the relative importance of the customer needs identified in steps 1 through 3. The outcome of this step is a numerical importance weighting for a subset of the needs. There are two basic approaches to the task: (1) relying on the consensus of the team members based on their experience with customers, or (2) basing the importance assessment on further customer surveys. The obvious trade-off between the two approaches is cost and speed versus accuracy: the team can make an educated assessment of the relative importance of the needs in one meeting, while a customer survey generally takes a minimum of two weeks. In most cases we believe the customer survey is important and worth the time required to complete it. Other development tasks, such as concept generation and analysis of competitive products, can begin before the relative importance surveys are complete.

The team should at this point have developed a rapport with a group of customers. These same customers can be surveyed to rate the relative importance of the needs that have been identified. The survey can be done in person, by telephone, via the Internet, or by mail. Few customers will respond to a survey asking them to evaluate the importance of 100 needs, so typically the team will work with only a subset of the needs. A practical limit on how many needs can be addressed in a customer survey is about 50. This limit is not too severe, however, because many of the needs are either obviously important (e.g., the screwdriver fits in a toolbox easily) or are easy to implement (e.g., the screwdriver prevents inadvertent switching off). The team can therefore limit the scope of the survey by querying customers only about needs that are likely to give rise to difficult technical trade-offs or costly features in the product design. Such needs would include the need to vary speed, the need to drive screws into hardwood, and the need to have the screwdriver emit a pleasant sound. Alternatively the team could develop a set of surveys to ask a variety of customers each about different subsets of the needs list. There are many survey designs for establishing the relative importance of customer needs. One good design is illustrated by the portion of the cordless screwdriver survey shown in Exhibit 4-9. In addition to asking for importance ratings, this survey asks the respondent to explicitly identify the needs that are unique or unexpected. This information can be used to help the team identify latent needs.

The survey responses for each need statement can be characterized in a variety of ways: by the mean, by the standard deviation, or by the number of responses in each cate-

---

Cordless Screwdriver Survey

For each of the following cordless screwdriver features, please indicate on a scale of 1 to 5 how important the feature is to you. Please use the following scale:

1. Feature is undesirable. I would not consider a product with this feature.
2. Feature is not important, but I would not mind having it.
3. Feature would be nice to have, but is not necessary.
4. Feature is highly desirable, but I would consider a product without it.
5. Feature is critical. I would not consider a product without this feature.

Also indicate by checking the box to the right if you feel that the feature is unique, exciting, and/or unexpected.

Importance of feature on scale of 1 to 5		Check box if feature is unique, exciting, and/or unexpected.
_____	The screwdriver maintains power for several hours of heavy use.	❑
_____	The screwdriver can drive screws into hardwood.	❑
_____	The screwdriver speed can be controlled by the user while turning a screw.	❑
_____	The screwdriver has a pleasant sound when in use.	❑

And so forth.

EXHIBIT 4-9   Example importance survey (partial).

gory. The responses can then be used to assign an importance weighting to the need statements. The same scale of 1 to 5 can be used to summarize the importance data. The needs in Exhibit 4-8 are rated according to the survey data, with the importance ratings denoted by the number of *'s next to each need statement and the latent needs denoted by !. Note that no critical needs are also latent needs. This is because if a need were critical, customers would not be surprised or excited by it; they would expect it to be met.

Step 5: Reflect on the Results and the Process

The final step in the method is to reflect on the results and the process. While the process of identifying customer needs can be usefully structured, it is not an exact science. The team must challenge its results to verify that they are consistent with the knowledge and intuition the team has developed through many hours of interaction with customers. Some questions to ask include:

- Have we interacted with all of the important types of customers in our target market?
- Are we able to see beyond needs related only to existing products in order to capture the latent needs of our target customers?
- Are there areas of inquiry we should pursue in follow-up interviews or surveys?
- Which of the customers we spoke to would be good participants in our ongoing development efforts?
- What do we know now that we didn't know when we started? Are we surprised by any of the needs?
- Did we involve everyone within our own organization who needs to deeply understand customer needs?
- How might we improve the process in future efforts?

Summary

Identifying customer needs is an integral part of the concept development phase of the product development process. The resulting customer needs are used to guide the team in establishing product specifications, generating product concepts, and selecting a product concept for further development.

- The process of identifying customer needs includes five steps:

    1. Gather raw data from customers.
    2. Interpret the raw data in terms of customer needs.
    3. Organize the needs into a hierarchy.
    4. Establish the relative importance of the needs.
    5. Reflect on the results and the process.

- Creating a high-quality information channel from customers to the product developers ensures that those who directly control the details of the product, including the product designers, fully understand the needs of the customer.

- Lead users are a good source of customer needs because they experience new needs months or years ahead of most customers and because they stand to benefit substantially from new product innovations. Furthermore, they are frequently able to articulate their needs more clearly than typical customers.

- Latent needs may be even more important than explicit needs in determining customer satisfaction. Latent needs are those that many customers recognize as important in a final product but do not or cannot articulate in advance.

- Customer needs should be expressed in terms of what the product has to do, not in terms of how the product might be implemented. Adherence to this principle leaves the development team with maximum flexibility to generate and select product concepts.

- The key benefits of the method are: ensuring that the product is focused on customer needs and that no critical customer need is forgotten; developing a clear understanding among members of the development team of the needs of the customers in the target market; developing a fact base to be used in generating concepts, selecting a product concept, and establishing product specifications; and creating an archival record of the needs phase of the development process.

References and Bibliography

Many current resources are available on the Internet via
www.ulrich-eppinger.net

Concept engineering is a method developed by Burchill at MIT in collaboration with the Center for Quality of Management. This chapter benefits from our observations of the development and application of concept engineering. For a complete and detailed description of concept engineering, see:

Burchill, Gary, et al., *Concept Engineering,* Center for Quality of Management, Cambridge, MA, Document No. ML0080, 1997.

The research by Griffin and Hauser is one of the only rigorous efforts to validate different methods for extracting needs from interview data. Their study of the fraction of needs identified as a function of the number of customers interviewed is particularly interesting.

Griffin, Abbie, and John R. Hauser, "The Voice of the Customer," *Marketing Science,* Vol. 12, No. 1, Winter 1993, pp. 1–27.

Kinnear and Taylor thoroughly discuss data collection methods and survey design.

Kinnear, Thomas C., and James R. Taylor, *Marketing Research: An Applied Approach,* fifth edition, McGraw-Hill, New York, 1995.

Norman has written extensively on user needs, especially as related to the cognitive challenges of using products.

Norman, Donald A., *The Design of Everyday Things,* Doubleday, New York, 1990.

Payne's book is a detailed and interesting discussion of how to pose questions in surveys.

Payne, Stanley L., *The Art of Asking Questions,* Princeton University Press, Princeton, NJ, 1980.

Total quality management (TQM) provides a valuable perspective on how identifying customer needs fits into an overall effort to improve the quality of goods and services.

Shiba, Shoji, Alan Graham, and David Walden, *A New American TQM: Four Practical Revolutions in Management,* Productivity Press, Cambridge, MA, and The Center for Quality of Management, Cambridge, MA, 1993.

Urban and Hauser provide a thorough discussion of how to create hierarchies of needs (along with many other topics).

Urban, Glen L., and John R. Hauser, *Design and Marketing of New Products,* second edition, Prentice Hall, Englewood Cliffs, NJ, 1993.

Von Hippel describes many years of research on the role of lead users in innovation. He provides useful guidelines for identifying lead users.

von Hippel, Eric, *The Sources of Innovation,* Oxford University Press, New York, 1988.

Exercises

1. Translate the following customer statements about a student book bag into proper needs statements:

   a. "See how the leather on the bottom of the bag is all scratched; it's ugly."
   b. "When I'm standing in line at the cashier trying to find my checkbook while balancing my bag on my knee, I feel like a stork."
   c. "This bag is my life; if I lose it I'm in big trouble."
   d. "There's nothing worse than a banana that's been squished by the edge of a textbook."
   e. "I never use both straps on my knapsack; I just sling it over one shoulder."

2. Observe someone performing an everyday task. (Ideally, you should choose a task for which you can observe different users performing the task repeatedly.) Identify frustrations and difficulties encountered by these people. Identify the latent customer needs.

3. Choose a product that continually annoys you. Identify the needs the developers of this product missed. Why do you think these needs were not met? Do you think the developers deliberately ignored these needs?

Thought Questions

1. One of the reasons the method is effective is that it involves the entire development team. Unfortunately, the method can become unwieldy with a team of more than 10 people. How might you modify the method to maximize involvement yet maintain a focused and decisive effort given a large development team?

2. Can the process of identifying customer needs lead to the creation of innovative product concepts? In what ways? Could a structured process of identifying customer needs lead to a fundamentally new product concept like the Post-it Note?

CHAPTER FIVE

Product Specifications

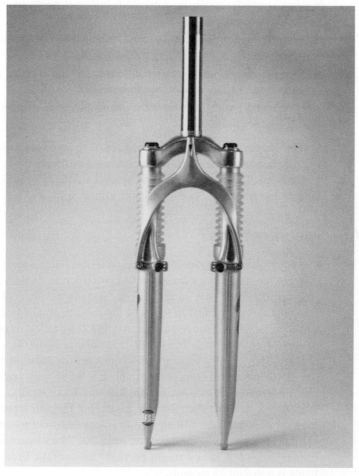

Photo by Stuart Cohen

EXHIBIT 5-1
One of Specialized's existing suspension forks.

Specialized Bicycle Components was interested in developing a front suspension fork for the mountain bike market. Although the firm was already selling a suspension fork (Exhibit 5-1), it was successful primarily in the high-performance segment of the market—racing cyclists with less concern for cost or long-term durability. The firm wished to broaden the sales of suspension forks and therefore was interested in developing a product that would provide high value for the recreational cyclist.

The development team had spent a great deal of time identifying customer needs. In addition to logging many hours of riding on suspended bikes themselves, the members of the team had interviewed lead users at mountain bike races and recreational cyclists on local trails, and they also had spent time working with dealers in their stores. As a result of this process they had assembled a list of customer needs. They now faced several challenges:

- How could the relatively subjective customer needs be translated into precise targets for the remaining development effort?
- How could the team and its senior management agree on what would constitute success or failure of the resulting product design?
- How could the team develop confidence that its intended product would garner a substantial share of the suspension fork market?
- How could the team resolve the inevitable trade-offs among product characteristics like cost and weight?

This chapter presents a method for establishing product specifications. We assume that the customer needs are already documented as described in Chapter 4, Identifying Customer Needs. The method employs several simple information systems, all of which can be constructed using conventional spreadsheet software.

What Are Specifications?

Customer needs are generally expressed in the "language of the customer." The primary customer needs for the suspension fork are listed in Exhibit 5-2. Customer needs such as "the suspension is easy to install" or "the suspension enables high-speed descents on bumpy trails" are typical in terms of the subjective quality of the expressions. However, while such expressions are helpful in developing a clear sense of the issues of interest to customers, they provide little specific guidance about how to design and engineer the product. They simply leave too much margin for subjective interpretation. For this reason, development teams usually establish a set of specifications, which spell out in precise, measurable detail *what* the product has to do. Product specifications do not tell the team *how* to address the customer needs, but they do represent an unambiguous agreement on what the team will attempt to achieve in order to satisfy the customer needs. For example, in contrast to the customer need that "the suspension is easy to install," the corresponding specification might be that "the average time to assemble the fork to the frame is less than 75 seconds."

We intend the term *product specifications* to mean the precise description of what the product has to do. Some firms use the terms "product requirements" or "engineering characteristics" in this way. Other firms use "specifications" or "technical specifications" to

EXHIBIT 5-2
Customer needs for the suspension fork and their relative importance (shown in a convenient spreadsheet format).

No.		Need	Imp.
1	The suspension	reduces vibration to the hands.	3
2	The suspension	allows easy traversal of slow, difficult terrain.	2
3	The suspension	enables high-speed descents on bumpy trails.	5
4	The suspension	allows sensitivity adjustment.	3
5	The suspension	preserves the steering characteristics of the bike.	4
6	The suspension	remains rigid during hard cornering.	4
7	The suspension	is lightweight.	4
8	The suspension	provides stiff mounting points for the brakes.	2
9	The suspension	fits a wide variety of bikes, wheels, and tires.	5
10	The suspension	is easy to install.	1
11	The suspension	works with fenders.	1
12	The suspension	instills pride.	5
13	The suspension	is affordable for an amateur enthusiast.	5
14	The suspension	is not contaminated by water.	5
15	The suspension	is not contaminated by grunge.	5
16	The suspension	can be easily accessed for maintenance.	3
17	The suspension	allows easy replacement of worn parts.	1
18	The suspension	can be maintained with readily available tools.	3
19	The suspension	lasts a long time.	5
20	The suspension	is safe in a crash.	5

refer to key design variables of the product such as the oil viscosity or spring constant of the suspension system. These are just differences in terminology. For clarity, let us be precise about a few definitions. A *specification* (singular) consists of a *metric* and a *value*. For example, "average time to assemble" is a metric, while "less than 75 seconds" is the value of this metric. Note that the value may take on several forms, including a particular number, a range, or an inequality. Values are always labeled with the appropriate units (e.g., seconds, kilograms, joules). Together, the metric and value form a specification. The *product specifications* (plural) are simply the set of the individual specifications.

When Are Specifications Established?

In an ideal world, the team would establish the product specifications once early in the development process and then proceed to design and engineer the product to exactly meet those specifications. For some products, such as soap or soup, this approach works quite well; the technologists on the team can reliably concoct a formulation that satisfies almost any reasonable specifications. However, for technology-intensive products this is rarely possible. For such products, specifications are established at least twice. Immediately after identifying the customer needs, the team sets *target specifications*. These specifications

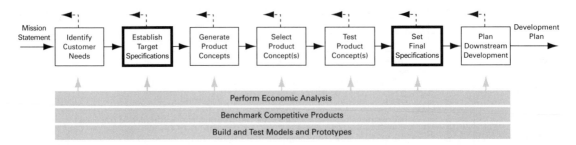

EXHIBIT 5-3   The concept development process. The target specifications are set early in the process, but setting the final specifications must wait until after the product concept has been selected.

represent the hopes and aspirations of the team, but they are established before the team knows what constraints the product technology will place on what can be achieved. The team's efforts may fail to meet some of these specifications and may exceed others, depending on the product concept the team eventually selects. For this reason, the target specifications must be refined after a product concept has been selected. The team revisits the specifications while assessing the actual technological constraints and the expected production costs. To set the *final specifications,* the team must frequently make hard trade-offs among different desirable characteristics of the product. For simplicity, we present a two-stage process for establishing specifications, but we note that in some organizations specifications are revisited many times throughout the development process.

The two stages in which specifications are established are shown as part of the concept development process in Exhibit 5-3. Note that the final specifications are one of the key elements of the development plan, which is usually documented in the project's *contract book.* The contract book (described in Chapter 16, Managing Projects) specifies what the team agrees to achieve, the project schedule, the required resources, and the economic implications for the business. The list of product specifications is also one of the key information systems used by the team throughout the development process.

This chapter presents two methods: the first is for establishing the target specifications and the second is for setting the final specifications after the product concept has been selected.

Establishing Target Specifications

As Exhibit 5-3 illustrates, the target specifications are established after the customer needs have been identified but before product concepts have been generated and the most promising one(s) selected. An arbitrary setting of the specifications may not be technically feasible. For example, in designing a suspension fork, the team cannot assume in advance that it will be able to achieve simultaneously a mass of 1 kilogram, a manufacturing cost of $30, and the best descent time on the test track, as these are three quite aggressive specifications. Actually meeting the specifications established at this point is contingent upon the details of the product concept the team eventually selects. For this reason, such preliminary specifications are labeled "target specifications." They are the goals of the development team, describing a product that the team believes would succeed in the marketplace. Later these specifications will be refined based on the limitations of the product concept actually selected.

The process of establishing the target specifications contains four steps:

1. Prepare the list of metrics.
2. Collect competitive benchmarking information.
3. Set ideal and marginally acceptable target values.
4. Reflect on the results and the process.

Step 1: Prepare the List of Metrics

The most useful metrics are those that reflect as directly as possible the degree to which the product satisfies the customer needs. The relationship between needs and metrics is central to the entire concept of specifications. The working assumption is that a translation from customer needs to a set of precise, measurable specifications is possible and that meeting specifications will therefore lead to satisfaction of the associated customer needs.

A list of metrics is shown in Exhibit 5-4. A good way to generate the list of metrics is to contemplate each need in turn and to consider what precise, measurable characteristic of the product will reflect the degree to which the product satisfies that need. In the ideal case, there is one and only one metric for each need. In practice, this is frequently not possible.

For example, consider the need that the suspension be "easy to install." The team may conclude that this need is largely captured by measuring the time required for assembly of the fork to the frame. However, note the possible subtleties in this translation. Is assembly time really identical to ease of installation? The installation could be extremely fast but require an awkward and painful set of finger actions, which ultimately may lead to worker injury or dealer frustration. Because of the imprecise nature of the translation process, those establishing the specifications should have been directly involved in identifying the customer needs. In this way the team can rely on its understanding of the meaning of each need statement derived from firsthand interactions with customers.

The need for the fork to reduce vibration to the user's hands may be even more difficult to translate into a single metric, because there are many different conditions under which vibration can be transmitted, including small bumps on level roads and big bumps on rough trails. The team may conclude that several metrics are required to capture this need, including, for example, the metrics "attenuation from dropout to handlebar at 10 Hz" and "maximum value from the Monster." (The "Monster" is a suspension test developed by *Mountain Bike* magazine.)

A simple needs-metrics matrix represents the relationship between needs and metrics. An example needs-metrics matrix is shown in Exhibit 5-5. The rows of the matrix correspond to the customer needs, and the columns of the matrix correspond to the metrics. A mark in a cell of the matrix means that the need and the metric associated with the cell are related; performance relative to the metric will influence the degree to which the product satisfies the customer need. This matrix is a key element of the *House of Quality*, a graphical technique used in *Quality Function Deployment*, or *QFD* (Hauser and Clausing, 1988). In many cases, we find the information in the needs-metrics matrix is just as

Metric No.	Need Nos.	Metric	Imp.	Units
1	1, 3	Attenuation from dropout to handlebar at 10 Hz	3	dB
2	2, 6	Spring preload	3	N
3	1, 3	Maximum value from the Monster	5	g
4	1, 3	Minimum descent time on test track	5	s
5	4	Damping coefficient adjustment range	3	N-s/m
6	5	Maximum travel (26-in. wheel)	3	mm
7	5	Rake offset	3	mm
8	6	Lateral stiffness at the tip	3	kN/m
9	7	Total mass	4	kg
10	8	Lateral stiffness at brake pivots	2	kN/m
11	9	Headset sizes	5	in.
12	9	Steertube length	5	mm
13	9	Wheel sizes	5	List
14	9	Maximum tire width	5	in.
15	10	Time to assemble to frame	1	s
16	11	Fender compatibility	1	List
17	12	Instills pride	5	Subj.
18	13	Unit manufacturing cost	5	US$
19	14	Time in spray chamber without water entry	5	s
20	15	Cycles in mud chamber without contamination	5	k-cycles
21	16, 17	Time to disassemble/assemble for maintenance	3	s
22	17, 18	Special tools required for maintenance	3	List
23	19	UV test duration to degrade rubber parts	5	hr
24	19	Monster cycles to failure	5	Cycles
25	20	Japan Industrial Standards test	5	Binary
26	20	Bending strength (frontal loading)	5	kN

EXHIBIT 5-4   List of metrics for the suspension. The relative importance of each metric and the units for the metric are also shown. "Subj." is an abbreviation indicating that a metric is subjective.

easily communicated by listing the numbers of the needs related to each metric alongside the list of metrics (the second column in Exhibit 5-4). There are some cases, however, in which the mapping from needs to metrics is complex, and the matrix can be quite useful for representing this mapping.

A few guidelines should be considered when constructing the list of metrics:

- *Metrics should be complete.* Ideally each customer need would correspond to a single metric, and the value of that metric would correlate perfectly with satisfaction of that need. In practice, several metrics may be necessary to completely reflect a single customer need.

Need	1 Attenuation from dropout to handlebar at 10 Hz	2 Spring preload	3 Maximum value from the Monster	4 Minimum descent time on test track	5 Damping coefficient adjustment range	6 Maximum travel (26 in. wheel)	7 Rake offset	8 Lateral stiffness at the tip	9 Total mass	10 Lateral stiffness at brake pivots	11 Headset sizes	12 Steertube length	13 Wheel sizes	14 Maximum tire width	15 Time to assemble to frame	16 Fender compatibility	17 Instills pride	18 Unit manufacturing cost	19 Time in spray chamber without water entry	20 Cycles in mud chamber without contamination	21 Time to disassemble/assemble for maintenance	22 Special tools required for maintenance	23 UV test duration to degrade rubber parts	24 Monster cycles to failure	25 Japan Industrial Standards test	26 Bending strength (frontal loading)
1 Reduces vibration to the hands	●		●	●																						
2 Allows easy traversal of slow, difficult terrain		●																								
3 Enables high-speed descents on bumpy trails	●		●	●																						
4 Allows sensitivity adjustment					●																					
5 Preserves the steering characteristics of the bike						●	●																			
6 Remains rigid during hard cornering	●							●																		
7 Is lightweight									●																	
8 Provides stiff mounting points for the brakes										●																
9 Fits a wide variety of bikes, wheels, and tires											●	●	●	●												
10 Is easy to install															●											
11 Works with fenders																●										
12 Instills pride																	●									
13 Is affordable for an amateur enthusiast																		●								
14 Is not contaminated by water																			●							
15 Is not contaminated by grunge																				●						
16 Can be easily accessed for maintenance																					●					
17 Allows easy replacement of worn parts																					●	●				
18 Can be maintained with readily available tools																						●				
19 Lasts a long time																							●	●		
20 Is safe in a crash																									●	●

EXHIBIT 5-5   The needs-metrics matrix.

- ***Metrics should be dependent, not independent, variables.*** This guideline is a variant of the *what-not-how* principle introduced in Chapter 4. As do customer needs, specifications also indicate *what* the product must do, but not *how* the specifications will be achieved. Designers use many types of variables in product development; some are *dependent,* such as the mass of the fork, and some are *independent,* such as the material used for the fork. In other words, designers cannot control mass directly because it arises from other independent decisions the designers will make, such as dimensions and materials choices. Metrics specify the overall performance of a product and should therefore be the dependent variables (i.e., the performance measures or output variables) in the design problem. By using dependent variables for the specifications, designers are left with the freedom to achieve the specifications using the best approach possible.

- *Metrics should be practical.* It does not serve the team to devise a metric for a bicycle suspension that can only be measured by a scientific laboratory at a cost of $100,000. Ideally, metrics will be directly observable or analyzable properties of the product that can be easily evaluated by the team.

- *Some needs cannot easily be translated into quantifiable metrics.* The need that the suspension instills pride may be quite critical to success in the fashion-conscious mountain bike market, but how can pride be quantified? In these cases, the team simply repeats the need statement as a specification and notes that the metric is subjective and would be evaluated by a panel of customers. (We indicate this by entering "Subj." in the units column.)

- *The metrics should include the popular criteria for comparison in the marketplace.* Many customers in various markets buy products based on independently published evaluations. Such evaluations are found, for example, in *Popular Science, Consumer Reports,* on various Internet sites, or, in our case, in *Bicycling* and *Mountain Bike* magazines. If the team knows that its product will be evaluated by the trade media and knows what the evaluation criteria will be, then it should include metrics corresponding to these criteria. *Mountain Bike* magazine uses a test machine called the Monster, which measures the vertical acceleration (in g's) of the handlebars as a bicycle equipped with the fork runs over a block 50 millimeters tall. For this reason, the team included "maximum value from the Monster" as a metric. If the team cannot find a relationship between the criteria used by the media and the customer needs it has identified, then it should ensure that a need has not been overlooked and/or should work with the media to revise the criteria. In a few cases, the team may conclude that high performance in the media evaluations is in itself a customer need and choose to include a metric used by the media that has little intrinsic technical merit.

In addition to denoting the needs related to each metric, Exhibit 5-4 contains the units of measurement and an importance rating for each metric. The units of measurement are most commonly conventional engineering units such as kilograms and seconds. However, some metrics will not lend themselves to numerical values. The need that the suspension "works with fenders" is best translated into a specification listing the models of fenders with which the fork is compatible. In this case, the value of the metric is actually a list of fenders rather than a number. For the metric involving the standard safety test, the value is pass/fail. (We indicate these two cases by entering "List" and "Binary" in the units column.)

The importance rating of a metric is derived from the importance ratings of the needs it reflects. For cases in which a metric maps directly to a single need, the importance rating of the need becomes the importance rating of the metric. For cases in which a metric is related to more than one need, the importance of the metric is determined by considering the importances of the needs to which it relates and the nature of these relationships. We believe that there are enough subtleties in this process that importance weightings can best be determined through discussion among the team members, rather than through a formal algorithm. When there are relatively few specifications and establishing the relative importance of these specifications is critically important, *conjoint analysis* may be useful. Conjoint analysis is described briefly later in this chapter and publications explaining the technique are referenced at the end of the chapter.

Step 2: Collect Competitive Benchmarking Information

Unless the team expects to enjoy a total monopoly, the relationship of the new product to competitive products is paramount in determining commercial success. While the team will have entered the product development process with some idea of how it wishes to compete in the marketplace, the target specifications are the language the team uses to discuss and agree on the detailed positioning of its product relative to existing products, both its own and competitors'. Information on competing products must be gathered to support these positioning decisions.

An example of a competitive benchmarking chart is shown in Exhibit 5-6. The columns of the chart correspond to the competitive products and the rows are the metrics established in step 1. Note that the competitive benchmarking chart can be constructed as a simple appendage to the spreadsheet containing the list of metrics. (This information is one of the "rooms" in the *House of Quality,* described by Hauser and Clausing.)

The benchmarking chart is conceptually very simple. For each competitive product, the values of the metrics are simply entered down a column. Gathering these data can be very time consuming, involving (at the least) purchasing, testing, disassembling, and estimating the production costs of the most important competitive products. However, this investment of time is essential, as no product development team can expect to succeed without having this type of information. A word of warning: Sometimes the data contained in competitors' catalogs and supporting literature are not accurate. Where possible, values of the key metrics should be verified by independent testing or observation.

An alternative competitive benchmarking chart can be constructed with rows corresponding to the customer needs and columns corresponding to the competitive products (see Exhibit 5-7). This chart is used to compare customers' perceptions of the relative degree to which the products satisfy their needs. Constructing this chart requires collecting customer perception data, which can also be very expensive and time consuming. Some techniques for measuring customers' perceptions of satisfaction of needs are contained in a book by Urban and Hauser (1993). Both charts can be useful and any discrepancies between the two are instructive. At a minimum, a chart showing the competitive values of the metrics (Exhibit 5-6) should be created.

Step 3: Set Ideal and Marginally Acceptable Target Values

In this step, the team synthesizes the available information in order to actually set the *target values* for the metrics. Two types of target value are useful: an *ideal value* and a *marginally acceptable value.* The ideal value is the best result the team could hope for. The marginally acceptable value is the value of the metric that would just barely make the product commercially viable. Both of these targets are useful in guiding the subsequent stages of concept generation and concept selection, and for refining the specifications after the product concept has been selected.

There are five ways to express the values of the metrics:

- *At least X:* These specifications establish targets for the lower bound on a metric, but higher is still better. For example, the value of the brake mounting stiffness is specified to be at least 325 kilonewtons/meter.

Metric No.	Need Nos.	Metric	Imp.	Units	ST Tritrack	Maniray 2	Rox Tahx Quadra	Rox Tahx Ti 21	Tonka Pro	Gunhill Head Shox
1	1, 3	Attenuation from dropout to handlebar at 10 Hz	3	dB	8	15	10	15	9	13
2	2, 6	Spring preload	3	N	550	760	500	710	480	680
3	1, 3	Maximum value from the Monster	5	g	3.6	3.2	3.7	3.3	3.7	3.4
4	1, 3	Minimum descent time on test track	5	s	13	11.3	12.6	11.2	13.2	11
5	4	Damping coefficient adjustment range	3	N-s/m	0	0	0	200	0	0
6	5	Maximum travel (26-in. wheel)	3	mm	28	48	43	46	33	38
7	5	Rake offset	3	mm	41.5	39	38	38	43.2	39
8	6	Lateral stiffness at the tip	3	kN/m	59	110	85	85	65	130
9	7	Total mass	4	kg	1.409	1.385	1.409	1.364	1.222	1.100
10	8	Lateral stiffness at brake pivots	2	kN/m	295	550	425	425	325	650
11	9	Headset sizes	5	in.	1.000 1.125	1.000 1.125 1.250	1.000 1.125	1.000 1.125 1.250	1.000 1.125	NA
12	9	Steertube length	5	mm	150 180 210 230 255	140 165 190 215	150 170 190 210	150 170 190 210 230	150 190 210 220	NA
13	9	Wheel sizes	5	List	26 in.	26 in.	26 in.	26 in. 700C	26 in.	26 in.

EXHIBIT 5-6   Competitive benchmarking chart based on metrics.

- *At most X:* These specifications establish targets for the upper bound on a metric, with smaller values being better. For example, the value for the mass of the suspension fork is set to be at most 1.4 kilograms.
- *Between X and Y:* These specifications establish both upper and lower bounds for the value of a metric. For example, the value for the spring preload is set to be between 480 and 800 newtons. Any more and the suspension is harsh; any less and the suspension is too bouncy.
- *Exactly X:* These specifications establish a target of a particular value of a metric, with any deviation degrading performance. For example, the ideal value for the rake offset metric is set to 38 millimeters. This type of specification is to be avoided if possible because such specifications substantially constrain the design. Often, upon recon-

Metric No.	Need Nos.	Metric	Imp.	Units	ST Tritrack	Maniray 2	Rox Tahx Quadra	Rox Tahx Ti 21	Tonka Pro	Gunhill Head Shox
14	9	Maximum tire width	5	in.	1.5	1.75	1.5	1.75	1.5	1.5
15	10	Time to assemble to frame	1	s	35	35	45	45	35	85
16	11	Fender compatibility	1	List	Zefal	None	None	None	None	All
17	12	Instills pride	5	Subj.	1	4	3	5	3	5
18	13	Unit manufacturing cost	5	US$	65	105	85	115	80	100
19	14	Time in spray chamber without water entry	5	s	1300	2900	>3600	>3600	2300	>3600
20	15	Cycles in mud chamber without contamination	5	k-cycles	15	19	15	25	18	35
21	16, 17	Time to disassemble/ assemble for maintenance	3	s	160	245	215	245	200	425
22	17, 18	Special tools required for maintenance	3	List	Hex	Hex	Hex	Hex	Long hex	Hex, pin wrench
23	19	UV test duration to degrade rubber parts	5	hr	400+	250	400+	400+	400+	250
24	19	Monster cycles to failure	5	Cycles	500k+	500k+	500k+	480k	500k+	330k
25	20	Japan Industrial Standards test	5	Binary	Pass	Pass	Pass	Pass	Pass	Pass
26	20	Bending strength (frontal loading)	5	kN	5.5	8.9	7.5	7.5	6.2	10.2

EXHIBIT 5-6   *Continued*

sideration, the team realizes that what initially appears as an "exactly X" specification can be expressed as a "between X and Y" specification.

- *A set of discrete values:* Some metrics will have values corresponding to several discrete choices. For example, the headset diameters are 1.000, 1.125, or 1.250 inches. (Industry practice is to use English units for these and several other critical bicycle dimensions.)

The desirable range of values for one metric may depend on another. In other words, we may wish to express a target as, for example, "the fork tip lateral stiffness is no more than 20 percent of the lateral stiffness at the brake pivots." In applications where the team feels this level of complexity is warranted, such targets can easily be included, although we recommend that this level of complexity not be introduced until the final phase of the specifications process.

No.	Need	Imp.	ST Tritrack	Maniray 2	Rox Tahx Quadra	Rox Tahx Ti 21	Tonka Pro	Gunhill Head Shox
1	Reduces vibration to the hands	3	•	••••	••	•••••	••	•••
2	Allows easy traversal of slow, difficult terrain	2	••	••••	•••	•••••	•••	••••••
3	Enables high-speed descents on bumpy trails	5	•	•••••	••	•••••	••	•••
4	Allows sensitivity adjustment	3	•	••••	••	•••••	••	•••
5	Preserves the steering characteristics of the bike	4	••••	••	•	••	•••••	••••••
6	Remains rigid during hard cornering	4	•	•••	•	•••••	•	••••••
7	Is lightweight	4	•	•••	•	•••	••••	•••••
8	Provides stiff mounting points for the brakes	2	•	••••	•••	•••	•••••	••
9	Fits a wide variety of bikes, wheels, and tires	5	••••	•••••	•••	•••••	•••	•
10	Is easy to install	1	••••	•••••	••••	••••	•••••	•
11	Works with fenders	1	•••	•	•	•	•	•••••
12	Instills pride	5	•	••••	•••	•••••	•••	•••••
13	Is affordable for an amateur enthusiast	5	•••••	•	•••	•	•••	••
14	Is not contaminated by water	5	•	•••	••••	••••	••	•••••
15	Is not contaminated by grunge	5	•	•••	•	••••	••	•••••
16	Can be easily accessed for maintenance	3	••••	•••••	••••	••••	•••••	•
17	Allows easy replacement of worn parts	1	••••	•••••	••••	••••	•••••	•
18	Can be maintained with readily available tools	3	•••••	•••••	•••••	•••••	••	•
19	Lasts a long time	5	•••••	•••••	•••••	•••	•••••	•
20	Is safe in a crash	5	•••••	•••••	•••••	•••••	•••••	•••••

EXHIBIT 5-7   Competitive benchmarking chart based on perceived satisfaction of needs. (Scoring more "dots" corresponds to greater perceived satisfaction of the need.)

Using these five different types of expressions for values of the metrics, the team sets the target specifications. The team simply proceeds down the list of metrics and assigns both the marginally acceptable and ideal target values for each metric. These decisions are facilitated by the metric-based competitive benchmarking chart shown in Exhibit 5-6. To set the target values, the team has many considerations, including the capability of competing products available at the time, competitors' future product capabilities (if these are predictable), and the product's mission statement and target market segment. Exhibit 5-8 shows the targets assigned for the suspension fork.

Because most of the values are expressed in terms of bounds (upper or lower or both), the team is establishing the boundaries of the competitively viable product space. The team hopes that the product will meet some of the ideal targets but is confident that a product can be commercially viable even if it exhibits one or more marginally acceptable characteristics. Note that these specifications are preliminary because until a product concept is chosen and some of the design details are worked out, many of the exact trade-offs are uncertain.

Step 4: Reflect on the Results and the Process

The team may require some iteration to agree on the targets. Reflection after each iteration helps to ensure that the results are consistent with the goals of the project. Questions to consider include:

- Are members of the team "gaming"? For example, is the key marketing representative insisting that an aggressive value is required for a particular metric in the hopes that by setting a high goal, the team will actually achieve more than if his or her true, and more lenient, beliefs were expressed?

- Should the team consider offering multiple products or at least multiple options for the product in order to best match the particular needs of more than one market segment, or will one "average" product suffice?

- Are any specifications missing? Do the specifications reflect the characteristics that will dictate commercial success?

Once the targets have been set, the team can proceed to generate solution concepts. The target specifications then can be used to help the team select a concept and will help the team know when a concept is commercially viable. (See Chapter 6, Concept Generation, and Chapter 7, Concept Selection.)

Setting the Final Specifications

As the team finalizes the choice of a concept and prepares for subsequent design and development, the specifications are revisited. Specifications which originally were only targets expressed as broad ranges of values are now refined and made more precise.

Finalizing the specifications is difficult because of trade-offs—inverse relationships between two specifications that are inherent in the selected product concept. Trade-offs frequently occur between different technical performance metrics and almost always occur between technical performance metrics and cost. For example, one trade-off is between brake mounting stiffness and mass of the fork. Because of the basic mechanics of

Metric No.	Need Nos.	Metric	Imp.	Units	Marginal Value	Ideal Value
1	1, 3	Attenuation from dropout to handlebar at 10 Hz	3	dB	>10	>15
2	2, 6	Spring preload	3	N	480–800	650–700
3	1, 3	Maximum value from the Monster	5	g	<3.5	<3.2
4	1, 3	Minimum descent time on test track	5	s	<13.0	<11.0
5	4	Damping coefficient adjustment range	3	N-s/m	0	>200
6	5	Maximum travel (26-in. wheel)	3	mm	33–50	45
7	5	Rake offset	3	mm	37–45	38
8	6	Lateral stiffness at the tip	3	kN/m	>65	>130
9	7	Total mass	4	kg	<1.4	<1.1
10	8	Lateral stiffness at brake pivots	2	kN/m	>325	>650
11	9	Headset sizes	5	in.	1.000 1.125	1.000 1.125 1.250
12	9	Steertube length	5	mm	150 170 190 210	150 170 190 210 230
13	9	Wheel sizes	5	List	26 in.	26 in. 700C
14	9	Maximum tire width	5	in.	>1.5	>1.75
15	10	Time to assemble to frame	1	s	<60	<35
16	11	Fender compatibility	1	List	None	All
17	12	Instills pride	5	Subj.	>3	>5
18	13	Unit manufacturing cost	5	US$	<85	<65
19	14	Time in spray chamber without water entry	5	s	>2300	>3600
20	15	Cycles in mud chamber without contamination	5	k-cycles	>15	>35
21	16, 17	Time to disassemble/assemble for maintenance	3	s	<300	<160
22	17, 18	Special tools required for maintenance	3	List	Hex	Hex
23	19	UV test duration to degrade rubber parts	5	hr	>250	>450
24	19	Monster cycles to failure	5	Cycles	>300k	>500k
25	20	Japan Industrial Standards test	5	Binary	Pass	Pass
26	20	Bending strength (frontal loading)	5	kN	>7.0	>10.0

EXHIBIT 5-8   The target specifications. Like the other information systems, this one is easily encoded with a spreadsheet as a simple extension to the list of specifications.

the fork structure, these specifications are inversely related, assuming other factors are held constant. Another trade-off is between cost and mass. For a given concept, the team may be able to reduce the mass of the fork by making some parts out of titanium instead of steel. Unfortunately, decreasing the mass in this way will most likely increase the manufacturing cost of the product. The difficult part of refining the specifications is choosing how such trade-offs will be resolved.

Here, we propose a five-step process:

1. Develop technical models of the product.
2. Develop a cost model of the product.
3. Refine the specifications, making trade-offs where necessary.
4. Flow down the specifications as appropriate.
5. Reflect on the results and the process.

Step 1: Develop Technical Models of the Product

A *technical model* of the product is a tool for predicting the values of the metrics for a particular set of design decisions. We intend the term *models* to refer to both analytical and physical approximations of the product. (See Chapter 12, Prototyping, for further discussion of such models.)

At this point, the team had chosen an oil-damped coil spring concept for the suspension fork. The design decisions facing the team included details such as the materials for the structural components, the orifice diameter and oil viscosity for the damper, and the spring constant. Three models linking such design decisions to the performance metrics are shown in conceptual form in Exhibit 5-9. Such models can be used to predict the product's performance along a number of dimensions. The inputs to these models are the independent design variables associated with the product concept, such as oil viscosity, orifice diameter, spring constant, and geometry. The outputs of the model are the values of the metrics, such as attenuation, stiffness, and fatigue life.

Ideally, the team will be able to accurately model the product analytically, perhaps by implementing the model equations in a spreadsheet or computer simulation. Such a model allows the team to predict rapidly what type of performance can be expected from a particular choice of design variables, without costly physical experimentation. In most cases, such analytical models will be available for only a small subset of the metrics. For example, the team was able to model attenuation analytically, based on the engineers' knowledge of dynamic systems.

Several independent models, each corresponding to a subset of the metrics, may be more manageable than one large integrated model. For example, the team developed a separate analytical model for the brake mounting stiffness that was completely independent of the dynamic model used to predict vibration attenuation. In some cases, no analytical models will be available at all. For example, the team was not able to model analytically the fatigue performance of the suspension, so physical models were built and tested. It is generally necessary to actually build a variety of different physical mock-ups or prototypes in order to explore the implications of several combinations of design variables. To reduce the number of models which must be constructed, it is useful to employ design-of-experiments (DOE) techniques, which can minimize the number

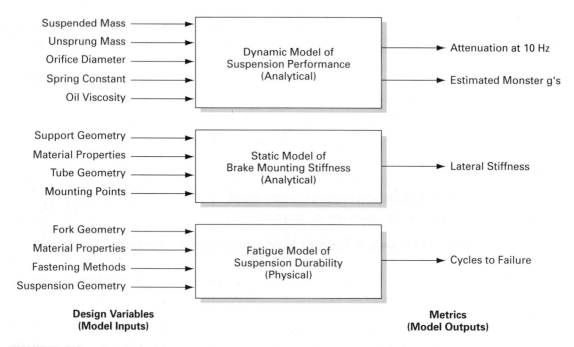

Suspended Mass ⟶
Unsprung Mass ⟶
Orifice Diameter ⟶
Spring Constant ⟶
Oil Viscosity ⟶

Dynamic Model of
Suspension Performance
(Analytical)

⟶ Attenuation at 10 Hz
⟶ Estimated Monster g's

Support Geometry ⟶
Material Properties ⟶
Tube Geometry ⟶
Mounting Points ⟶

Static Model of
Brake Mounting Stiffness
(Analytical)

⟶ Lateral Stiffness

Fork Geometry ⟶
Material Properties ⟶
Fastening Methods ⟶
Suspension Geometry ⟶

Fatigue Model of
Suspension Durability
(Physical)

⟶ Cycles to Failure

Design Variables
(Model Inputs)

Metrics
(Model Outputs)

EXHIBIT 5-9   Models used to assess technical feasibility. Technical models may be analytical or physical approximations of the product concept.

of experiments required to explore the design space. (See Chapter 13, Robust Design, for a summary of DOE methods.)

Armed with these technical models, the team can predict whether any particular set of specifications (such as the ideal target values) is technically feasible by exploring different combinations of design variables. This type of modeling and analysis prevents the team from setting a combination of specifications that cannot be achieved using the available latitude in the product concept.

Note that a technical model is almost always unique to a particular product concept. One of the models illustrated in Exhibit 5-9 is for an oil-damped suspension system; the model would be substantially different if the team had selected a concept employing a rubber suspension element. Thus, the modeling step can only be performed after the concept has been chosen.

Step 2: Develop a Cost Model of the Product

The goal of this step of the process is to make sure that the product can be produced at the *target cost.* The target cost is the manufacturing cost at which the company and its distribution partners can make adequate profits while still offering the product to the end customer at a competitive price. The appendix to this chapter provides an explanation of target costing. It is at this point that the team attempts to discover, for example, how much it will have to sacrifice in manufacturing cost to save 50 grams of mass.

For most products, the first estimates of manufacturing costs are completed by drafting a *bill of materials* (a list of all the parts) and estimating a purchase price or fabrication

Component	Qty/ Fork	High ($ ea.)	Low ($ ea.)	High Total ($/fork)	Low Total ($/fork)
Steertube	1	2.50	2.00	2.50	2.00
Crown	1	4.00	3.00	4.00	3.00
Boot	2	1.00	0.75	2.00	1.50
Lower tube	2	3.00	2.00	6.00	4.00
Lower tube top cover	2	2.00	1.50	4.00	3.00
Main lip seal	2	1.50	1.40	3.00	2.80
Slide bushing	4	0.20	0.18	0.80	0.72
Slide bushing spacer	2	0.50	0.40	1.00	0.80
Lower tube plug	2	0.50	0.35	1.00	0.70
Upper tube	2	5.50	4.00	11.00	8.00
Upper tube top cap	2	3.00	2.50	6.00	5.00
Upper tube adjustment knob	2	2.00	1.75	4.00	3.50
Adjustment shaft	2	4.00	3.00	8.00	6.00
Spring	2	3.00	2.50	6.00	5.00
Upper tube orifice cap	1	3.00	2.25	3.00	2.25
Orifice springs	4	0.50	0.40	2.00	1.60
Brake studs	2	0.40	0.35	0.80	0.70
Brake brace bolt	2	0.25	0.20	0.50	0.40
Brake brace	1	5.00	3.50	5.00	3.50
Oil (liters)	0.1	2.50	2.00	0.25	0.20
Misc. snap rings, o-rings	10	0.15	0.10	1.50	1.00
Decals	4	0.25	0.15	1.00	0.60
Assembly at $20/hr		30 min	20 min	10.00	6.67
Overhead at 25% of direct cost				20.84	15.74
Total				$104.19	$78.68

EXHIBIT 5-10   A bill of materials with cost estimates. This simple cost model allows early cost estimates to facilitate realistic trade-offs in the product specifications.

cost for each part. At this point in the development process the team does not generally know all of the components that will be in the product, but the team nevertheless makes an attempt to list the components it expects will be required. While early estimates generally focus on the cost of components, the team will usually make a rough estimate of assembly and other manufacturing costs (e.g., overhead) at this point as well. Efforts to develop these early cost estimates involve soliciting cost estimates from vendors and estimating the production costs of the components the firm will make itself. This process is often facilitated by a purchasing expert and a production engineer. A bill-of-materials cost model is shown in Exhibit 5-10 for the suspension fork. (See Chapter 11, Design for Manufacturing, for more details on estimating manufacturing cost.)

A useful way to record cost information is to list figures for the high and low estimates of each item. This helps the team to understand the range of uncertainty in the estimates. The bill of materials is typically used iteratively: the team performs a "what if" cost analysis for a set of design decisions and then revises these decisions based on what it learns. The bill of materials is itself a kind of performance model, but instead of predicting the value of a technical performance metric, it predicts cost performance. The bill of materials remains useful throughout the development process and is updated regularly (as frequently as once each week) to reflect the current status of the estimated manufacturing cost.

At this point in the development process, teams developing complex products containing hundreds or thousands of parts will not generally be able to include every part in the bill of materials. Instead, the team will list the major components and subsystems and place bounds on their costs based on past experience or on the judgment of suppliers.

Step 3: Refine the Specifications, Making Trade-Offs Where Necessary

Once the team has constructed technical performance models where possible and constructed a preliminary cost model, these tools can be used to develop final specifications. Finalizing specifications can be accomplished in a group session in which feasible combinations of values are determined through the use of the technical models and then the cost implications are explored. In an iterative fashion, the team converges on the specifications which will most favorably position the product relative to the competition, will best satisfy the customer needs, and will ensure adequate profits.

One important tool for supporting this decision-making process is the *competitive map*. An example competitive map is shown in Exhibit 5-11. This map is simply a scatter plot of the competitive products along two dimensions selected from the set of metrics and is sometimes called a trade-off map. The map displayed in Exhibit 5-11 shows esti-

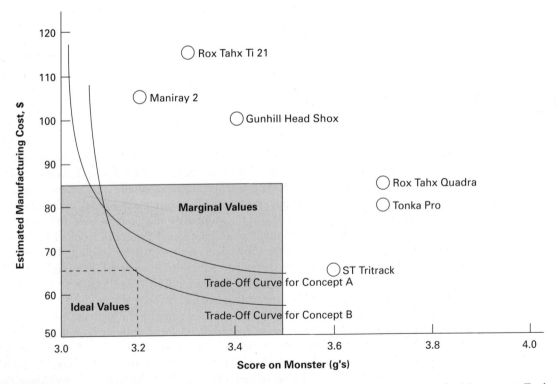

EXHIBIT 5-11 A competitive map showing estimated manufacturing cost versus score on the Monster test. Trade-off curves for two suspension concepts are also drawn on this map.

mated manufacturing cost versus score on the Monster test. The regions defined by the marginal and ideal values of the specifications are shown on the map. This map is particularly useful in showing that all of the high-performance suspensions (low Monster scores) have high estimated manufacturing costs. Armed with technical performance models and a cost model, the team can assess whether or not it will be able to "beat the trade-off" exhibited in the competitive map.

These maps can be constructed directly from the data contained in the competitive benchmarking chart using the plotting feature of the spreadsheet software. Generally the team will prepare three or four such maps corresponding to a handful of critical metrics. Additional maps may be created as needed to support subsequent decision making.

The competitive map is used to position the new product relative to the competition. Trade-off curves, showing performance of the product concept for a range of design variables, can be drawn directly on the competitive map, as shown in Exhibit 5-11. Using the technical and cost models of the product and the competitive maps, the team can refine the specifications in order to both satisfy the inherent constraints of the product concept and make the trade-offs in a way that will provide a performance advantage relative to the competitive products. The final specifications for the suspension fork are shown in Exhibit 5-12.

For relatively mature product categories in which competition is based on performance relative to a handful of well-understood performance metrics, *conjoint analysis* may be useful in refining product specifications. Conjoint analysis uses customer survey data to construct a model of customer preference. Essentially each respondent in a sample of potential customers is repeatedly asked to evaluate hypothetical products characterized by a set of attributes. These attributes must generally be metrics that are easily understood by customers (e.g., fuel economy and price for automobiles). Subjective attributes (e.g., styling) can be represented graphically. The hypothetical products are constructed using the statistical techniques of experimental design. Using customer responses, conjoint analysis infers the relative importance of each attribute to the customer. These data can then be used to predict which product a customer would choose when offered a hypothetical set of alternatives. By using these predictions for all of the customers in a sample, the market share of each product in the set of alternatives can be forecast. Using this approach, the specification values that maximize market share can be estimated. The details of conjoint analysis are fairly straightforward, but beyond the scope of this chapter. Relevant references are listed at the end of the chapter.

Step 4: Flow Down the Specifications as Appropriate

This chapter focuses on the specifications for a relatively simple component designed by a single, relatively small development team. Establishing specifications takes on additional importance and is substantially more challenging when developing a highly complex product consisting of multiple subsystems designed by multiple development teams. In such a context, specifications are used to define the development objectives of each of the subsystems as well as for the product as a whole. The challenge in this case is to *flow down* the overall specifications to specifications for each subsystem. For example, the overall specifications for an automobile contain metrics like fuel economy, 0–100 kilometer/hour acceleration time, and turning radius. However, specifications must also be created for the several dozen major subsystems that make up the automobile, including the body, engine, transmission, braking system, and suspension. The specifications for the engine include metrics like peak power, peak torque, and fuel consumption at peak

EXHIBIT 5-12

The final specifications.

No.	Metric	Unit	Value
1	Attenuation from dropout to handlebar at 10 Hz	dB>	12
2	Spring preload	N	600–650
3	Maximum value from the Monster	g	<3.4
4	Minimum descent time on test track	s	<11.5
5	Damping coefficient adjustment range	N-s/m	>100
6	Maximum travel (26-in. wheel)	mm	43
7	Rake offset	mm	38
8	Lateral stiffness at the tip	kN/m	>75
9	Total mass	kg	<1.4
10	Lateral stiffness at brake pivots	kN/m	>425
11	Headset sizes	in.	1.000 1.125
12	Steertube length	mm	150 170 190 210 230
13	Wheel sizes	List	26 in.
14	Maximum tire width	in.	>1.75
15	Time to assemble to frame	s	<45
16	Fender compatibility	List	Zefal
17	Instills pride	Subj.	>4
18	Unit manufacturing cost	US$	<80
19	Time in spray chamber without water entry	s	>3600
20	Cycles in mud chamber without contamination	k-cycles	>25
21	Time to disassemble/assemble for maintenance	s	<200
22	Special tools required for maintenance	List	Hex
23	UV test duration to degrade rubber parts	hr	>450
24	Monster cycles to failure	Cycles	>500k
25	Japan Industrial Standards test	Binary	Pass
26	Bending strength (frontal loading)	kN	>10.0

efficiency. One challenge in the flow-down process is to ensure that the subsystem specifications in fact reflect the overall product specifications—that if specifications for the subsystems are achieved, the overall product specifications will be achieved. A second challenge is to ensure that certain specifications for different subsystems are equally difficult to meet. That is, for example, that the mass specification for the engine is not inordinately more difficult to meet than is the mass specification for the body. Otherwise, the cost of the product will likely be higher than necessary.

Some overall component specifications can be established through *budget allocations.* For example, specifications for manufacturing cost, mass, and power consumption can be allocated to subsystems with the confidence that the overall cost, mass, and power consumption of the product will simply be the sum of these quantities for each subsystem. To some extent, geometric volume can be allocated this way as well. Other component specifications must be established through a more complex understanding of how subsystem performance relates to overall product performance. For example, fuel efficiency is a relatively complex function of vehicle mass, rolling resistance, aerodynamic drag coefficient, frontal area, and engine efficiency. Establishing specifications for the body, tires, and engine requires a model of how these variables relate to overall fuel efficiency.

A comprehensive treatment of flowing down specifications for complex products is beyond the scope of this chapter, and in fact is a major focus of the field of *systems engineering.* We refer the reader to several useful books on this subject in the reference list.

Step 5: Reflect on the Results and the Process

As always, the final step in the method is to reflect on the outcome and the process. Some questions the team may want to consider are:

- Is the product a winner? The product concept should allow the team to actually set the specifications so that the product will meet the customer needs and excel competitively. If not, then the team should return to the concept generation and selection phase or abandon the project.

- How much uncertainty is there in the technical and cost models? If competitive success is dictated by metrics around which much uncertainty remains, the team may wish to refine the technical or cost models in order to increase confidence in meeting the specifications.

- Is the concept chosen by the team best suited to the target market, or could it be best applied in another market (say, the low end or high end instead of the middle)? The selected concept may actually be too good. If the team has generated a concept that is dramatically superior to the competitive products, it may wish to consider employing the concept in a more demanding, and potentially more profitable, market segment.

- Should the firm initiate a formal effort to develop better technical models of some aspect of the product's performance for future use? Sometimes the team will discover that it does not really understand the underlying product technology well enough to create useful performance models. In such circumstances, an engineering effort to develop better understanding and models may be useful in subsequent development projects.

Summary

Customer needs are generally expressed in the "language of the customer." In order to provide specific guidance about how to design and engineer a product, development teams establish a set of specifications, which spell out in precise, measurable detail what the product has to do to be commercially successful. The specifications must reflect the customer needs, differentiate the product from the competitive products, and be technically and economically realizable.

- Specifications are typically established at least twice. Immediately after identifying the customer needs, the team sets *target specifications*. After concept selection and testing, the team develops *final specifications*.

- Target specifications represent the hopes and aspirations of the team, but they are established before the team knows the constraints the product technology will place on what can be achieved. The team's efforts may fail to meet some of these specifications and may exceed others, depending on the details of the product concept the team eventually selects.

- The process of establishing the target specifications entails four steps:

  1. Prepare the list of metrics.
  2. Collect competitive benchmarking information.
  3. Set *ideal* and *marginally acceptable* target values.
  4. Reflect on the results and the process.

- Final specifications are developed by assessing the actual technological constraints and the expected production costs using analytical and physical models. During this refinement phase the team must make difficult trade-offs among various desirable characteristics of the product.

- The five-step process for refining the specifications is:

  1. Develop technical models of the product.
  2. Develop a cost model of the product.
  3. Refine the specifications, making trade-offs where necessary.
  4. Flow down the specifications as appropriate.
  5. Reflect on the results and the process.

- The specifications process is facilitated by several simple information systems which can easily be created using conventional spreadsheet software. Tools such as the list of metrics, the needs-metrics matrix, the competitive benchmarking charts, and the competitive maps all support the team's decision making by providing the team with a way to represent and discuss the specifications.

- Because of the need to utilize the best possible knowledge of the market, the customers, the core product technology, and the cost implications of design alternatives, the specifications process requires active participation from team members representing the marketing, design, and manufacturing functions of the enterprise.

References and Bibliography

Many current resources are available on the Internet via
www.ulrich-eppinger.net

The process of translating customer needs into a set of specifications is also accomplished by the Quality Function Deployment (QFD) method. The key ideas behind QFD and the House of Quality are clearly presented by Hauser and Clausing in a popular article.

Hauser, John, and Don Clausing, "The House of Quality," *Harvard Business Review,* Vol. 66, No. 3, May–June 1988, pp. 63–73.

Urban and Hauser present several techniques for selecting combinations of product attributes in order to maximize customer satisfaction. Some of these techniques can serve as powerful analytical support for the general method described in this chapter.

Urban, Glen, and John Hauser, *Design and Marketing of New Products,* second edition, Prentice Hall, Englewood Cliffs, NJ, 1993.

Ramaswamy and Ulrich treat the use of engineering models in setting specifications in detail. They also identify some of the weaknesses in the conventional House of Quality method.

Ramaswamy, Rajan, and Karl Ulrich, "Augmenting the House of Quality with Engineering Models," *Research in Engineering Design,* Vol. 5, 1994, pp. 70–79.

Most marketing research textbooks discuss conjoint analysis. Here are two references.

Conjoint Analysis: A Guide for Designing and Interpreting Conjoint Studies, American Marketing Association, June 1992.

Aaker, David A., V. Kumar, and George S. Day, *Marketing Research,* sixth edition, John Wiley & Sons, New York, 1997.

Systems engineering and the flow down of specifications are treated comprehensively in the following books.

Hatley, Derek J., and Imtiaz A. Pirbhai, *Strategies for Real-Time System Specification,* Dorset House, New York, 1998.

Rechtin, Eberhardt, and Mark W. Maier, *The Art of Systems Architecting,* second edition, CRC Press, Boca Raton, FL, 2000.

More detail on the use of target costing is available in this article by Cooper and Slagmulder.

Cooper, Robin, and Regine Slagmulder, "Develop Profitable New Products with Target Costing," *Sloan Management Review,* Vol. 40, No. 4, Summer 1999, pp. 23–33.

Exercises

1. List a set of metrics corresponding to the need that a pen write smoothly.
2. Devise a metric and a corresponding test for the need that a roofing material last many years.
3. Some of the same metrics seem to be involved in trade-offs for many different products. Which metrics are these?

Thought Questions

1. How might you establish precise and measurable specifications for intangible needs such as "the front suspension looks great"?
2. Why are some customer needs difficult to map to a single metric?
3. How might you explain a situation in which customers' perceptions of the competitive products (as in Exhibit 5-7) are not consistent with the values of the metrics for those same products (as in Exhibit 5-6)?

4. Can poor performance relative to one specification always be compensated for by high performance on other specifications? If so, how can there ever really be a "marginally acceptable" value for a metric?

5. Why should independent design variables not be used as metrics?

Appendix

Target Costing

Target costing is a simple idea: set the value of the manufacturing cost specification based on the price the company hopes the end user will pay for the product and on the profit margins that are required for each stage in the distribution channel. For example, assume Specialized wishes to sell its suspension fork to its customers through bicycle shops. If the price it expected the customer to pay was $250 and if bicycle shops normally expect a gross profit margin of 45 percent on components, then Specialized would have to sell its fork to bicycle shops for $(1 - 0.45) \cdot 250 = \$137.50$. If Specialized wishes to earn a gross margin of at least 40 percent on its components, then its unit manufacturing cost must be less than $(1 - 0.40) \cdot 137.50 = \$82.50$.

Target costing is the reverse of the *cost-plus* approach to pricing. The cost-plus approach begins with what the firm expects its manufacturing costs to be and then sets its prices by adding its expected profit margin to the cost. This approach ignores the realities of competitive markets, in which prices are driven by market and customer factors. Target costing is a mechanism for ensuring that specifications are set in a way that allows the product to be competitively priced in the marketplace.

Some products are sold directly by a manufacturer to end users of the product. Frequently, products are distributed through one or more intermediate stages, such as distributors and retailers. Exhibit 5-13 provides some approximate values of gross profit margins for different product categories.

Let M be the gross profit *margin* of a stage in the distribution channel.

$$M = \frac{(P - C)}{P}$$

where P is the price this stage charges its customers and C is the cost this stage pays for the product it sells. (Note that *mark-up* is similar to margin, but is defined slightly differently as $P/C - 1$, so that a margin of 50 percent is equivalent to a mark-up of 100 percent.)

Target cost, C, is given by the following expression:

$$C = P \prod_{i=1}^{n}(1 - M_i)$$

where P is the price paid by the end user, n is the number of stages in the distribution channel, and M_i is the margin of the ith stage.

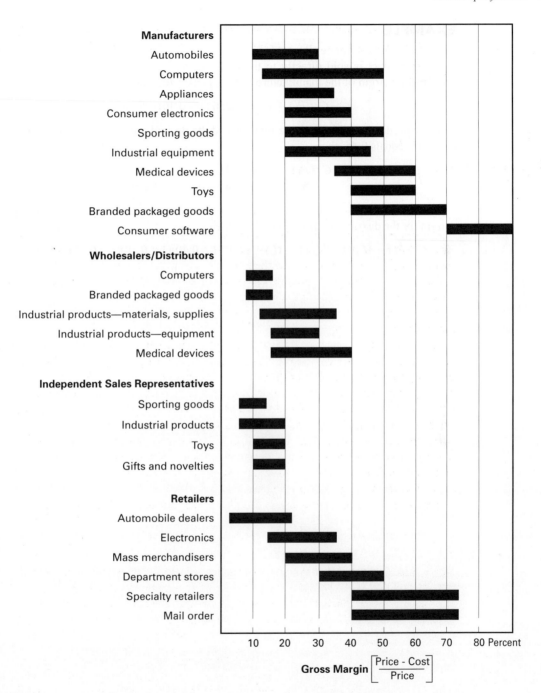

EXHIBIT 5-13 Approximate margins for manufacturers, wholesalers, distributors, sales representatives, and retailers. Note that these values are quite approximate. Actual margins depend on many idiosyncratic factors, including competitive intensity, the volume of units sold, and the level of customer support required.

EXAMPLE

Assume the end user price, P, equals $250.

If the product is sold directly to the end user by the manufacturer, and the desired gross profit margin of the manufacturer, M_m, equals 0.40, then the target cost is

$$C = P(1 - M_m) = \$250\,(1 - 0.40) = \$150$$

If the product is sold through a retailer, and the desired gross profit margin for the retailer, M_r, equals 0.45, then

$$C = P(1 - M_m)(1 - M_r)$$
$$= \$250\,(1 - 0.40)(1 - 0.45) = \$82.50$$

If the product is sold through a distributor and a retailer, and the desired gross profit margin for the distributor, M_d, equals 0.20, then

$$C = P(1 - M_m)(1 - M_d)(1 - M_r) = \$250\,(1 - 0.40)(1 - 0.20)(1 - 0.45) = \$66.00$$

CHAPTER SIX

Concept Generation

Courtesy of The Stanley Works

EXHIBIT 6-1
A cordless electric roofing nailer.

This chapter was developed in collaboration with Gavin Zau.

The president of Stanley-Bostitch commissioned a team to develop a new hand-held nailer for the roofing market. The product that eventually resulted from the effort is shown in Exhibit 6-1. The mission of the team was to consider broadly alternative product concepts, assuming only that the tool would employ conventional nails as the basic fastening technology. After identifying a set of customer needs and establishing target product specifications, the team faced the following questions:

- What existing solution concepts, if any, could be successfully adapted for this application?
- What new concepts might satisfy the established needs and specifications?
- What methods can be used to facilitate the concept generation process?

The Activity of Concept Generation

A product concept is an approximate description of the technology, working principles, and form of the product. It is a concise description of how the product will satisfy the customer needs. A concept is usually expressed as a sketch or as a rough three-dimensional model and is often accompanied by a brief textual description. The degree to which a product satisfies customers and can be successfully commercialized depends to a large measure on the quality of the underlying concept. A good concept is sometimes poorly implemented in subsequent development phases, but a poor concept can rarely be manipulated to achieve commercial success. Fortunately, concept generation is relatively inexpensive and can be done relatively quickly in comparison to the rest of the development process. For example, concept generation had typically consumed less than 5 percent of the budget and 15 percent of the development time in previous nailer development efforts. Because the concept generation activity is not costly, there is no excuse for a lack of diligence and care in executing a sound concept generation method.

The concept generation process begins with a set of customer needs and target specifications and results in a set of product concepts from which the team will make a final selection. The relation of concept generation to the other concept development activities is shown in Exhibit 6-2. In most cases, an effective development team will generate hundreds of concepts, of which 5 to 20 will merit serious consideration during the concept selection activity.

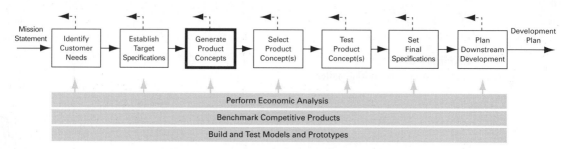

EXHIBIT 6-2   Concept generation is an integral part of the concept development phase.

Good concept generation leaves the team with confidence that the full space of alternatives has been explored. Thorough exploration of alternatives early in the development process greatly reduces the likelihood that the team will stumble upon a superior concept late in the development process or that a competitor will introduce a product with dramatically better performance than the product under development.

Structured Approaches Reduce the Likelihood of Costly Problems

Common dysfunctions exhibited by development teams during concept generation include:

- Consideration of only one or two alternatives, often proposed by the most assertive members of the team.
- Failure to consider carefully the usefulness of concepts employed by other firms in related and unrelated products.
- Involvement of only one or two people in the process, resulting in lack of confidence and commitment by the rest of the team.
- Ineffective integration of promising partial solutions.
- Failure to consider entire categories of solutions.

A structured approach to concept generation reduces the incidence of these problems by encouraging the gathering of information from many disparate information sources, by guiding the team in the thorough exploration of alternatives, and by providing a mechanism for integrating partial solutions. A structured method also provides a step-by-step procedure for those members of the team who may be less experienced in design-intensive activities, allowing them to participate actively in the process.

A Five-Step Method

This chapter presents a five-step concept generation method. The method, outlined in Exhibit 6-3, breaks a complex problem into simpler subproblems. Solution concepts are then identified for the subproblems by external and internal search procedures. Classification trees and concept combination tables are then used to systematically explore the space of solution concepts and to integrate the subproblem solutions into a total solution. Finally, the team takes a step back to reflect on the validity and applicability of the results, as well as on the process used.

This chapter will follow the recommended method and will describe each of the five steps in detail. Although we present the method in a linear sequence, concept generation is almost always iterative. Like our other development methods, these steps are intended to be a baseline from which product development teams can develop and refine their own unique problem-solving style.

Our presentation of the method is focused primarily on the overall concept for a new product; however, the method can and should be used at several different points in the development process. The process is useful not only for overall product concepts but also for concepts for subsystems and specific components as well. Also note that while the example in this chapter involves a relatively technical product, the same basic approach can be applied to nearly any product.

EXHIBIT 6-3
The five-step
concept
generation
method.

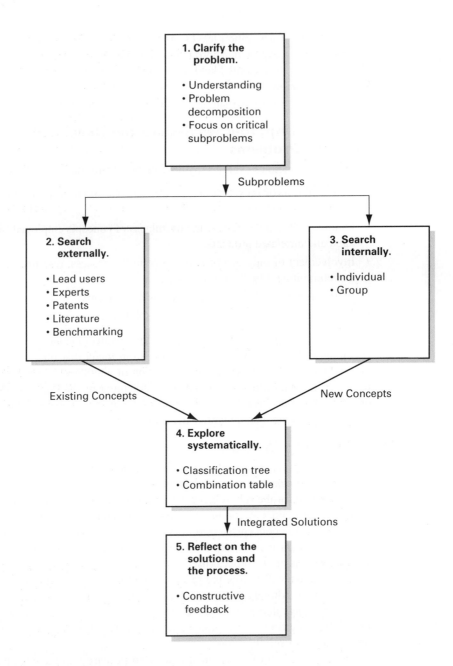

1. Clarify the problem.

• Understanding
• Problem decomposition
• Focus on critical subproblems

Subproblems

2. Search externally.

• Lead users
• Experts
• Patents
• Literature
• Benchmarking

3. Search internally.

• Individual
• Group

Existing Concepts

New Concepts

4. Explore systematically.

• Classification tree
• Combination table

Integrated Solutions

5. Reflect on the solutions and the process.

• Constructive feedback

Step 1: Clarify the Problem

Clarifying the problem consists of developing a general understanding and then breaking the problem down into subproblems if necessary.

The mission statement for the project, the customer needs list, and the preliminary product specification are the ideal inputs to the concept generation process, although often these pieces of information are still being refined as the concept generation phase begins. Ideally the team has been involved both in the identification of the customer needs and in the setting of the target product specifications. Those members of the team who were not involved in these preceding steps should become familiar with the processes used and their results before concept generation activities begin. (See Chapter 4, Identifying Customer Needs, and Chapter 5, Product Specifications.)

As stated before, the challenge was to "design a better hand-held roofing nailer." The scope of the design problem could have been defined more generally (e.g., "fasten roofing materials") or more specifically (e.g., "improve the speed of the existing pneumatic tool concept"). Some of the assumptions in the team's mission statement were:

- The nailer will use nails (as opposed to adhesives, screws, etc.).
- The nailer will be compatible with nail magazines on existing tools.
- The nailer will nail through roofing shingles into wood.
- The nailer will be hand held.

Based on the assumptions, the team had identified the customer needs for a hand-held nailer. These included:

- The nailer inserts nails in rapid succession.
- The nailer is lightweight.
- The nailer has no noticeable nailing delay after tripping the tool.

The team gathered supplemental information to clarify and quantify the needs, such as the approximate energy and speed of the nailing. These basic needs were subsequently translated into target product specifications. The target specifications included the following:

- Nail lengths from 25 millimeters to 38 millimeters.
- Maximum nailing energy of 40 joules per nail.
- Nailing forces of up to 2,000 newtons.
- Peak nailing rate of one nail per second.
- Average nailing rate of 12 nails per minute.
- Tool mass less than 4 kilograms.
- Maximum trigger delay of 0.25 second.

Decompose a Complex Problem into Simpler Subproblems

Many design challenges are too complex to solve as a single problem and can be usefully divided into several simpler subproblems. For example, the design of a complex product like a document copier can be thought of as a collection of more focused design problems, including, for example, the design of a document handler, the design of a paper feeder, the design of a printing device, and the design of an image capture device. In some cases, however, the design problem cannot readily be divided into subproblems. For example, the problem of designing a paper clip may be hard to divide into subproblems. As a general rule, we feel that teams should attempt to decompose design problems, but should be aware that such a decomposition may not be very useful for products with extremely simple functions.

Dividing a problem into simpler subproblems is called *problem decomposition*. There are many schemes by which a problem can be decomposed. Here we demonstrate a *functional* decomposition and also list several other approaches that are frequently useful.

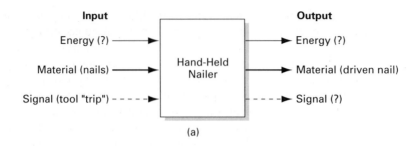

(a)

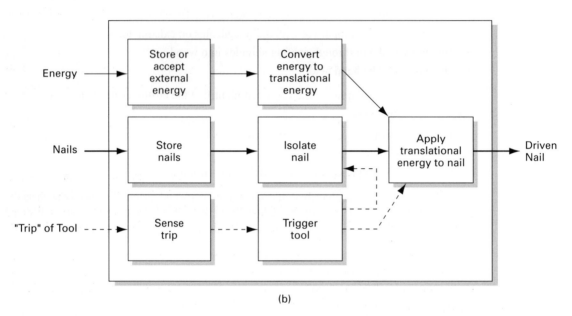

(b)

EXHIBIT 6-4 Function diagram of a hand-held nailer arising from a functional decomposition: (a) overall "black box"; (b) refinement showing subfunctions.

The first step in decomposing a problem functionally is to represent it as a single *black box* operating on material, energy, and signal flows, as shown in Exhibit 6-4(a). Thin solid lines denote the transfer and conversion of energy, thick solid lines signify the movement of material within the system, and dashed lines represent the flows of control and feedback signals within the system. This black box represents the overall function of the product.

The next step in functional decomposition is to divide the single black box into subfunctions to create a more specific description of what the elements of the product might do in order to implement the overall function of the product. Each subfunction can generally be further divided into even simpler subfunctions. The division process is repeated until the team members agree that each subfunction is simple enough to work with. A good rule of thumb is to create between 3 and 10 subfunctions in the diagram. The end result, shown in Exhibit 6-4(b), is a function diagram containing subfunctions connected by energy, material, and signal flows.

Note that at this stage the goal is to describe the functional elements of the product without implying a specific technological working principle for the product concept. For example, Exhibit 6-4(b) includes the subfunction "isolate nail." This subfunction is expressed in such a way that it does not imply any particular physical solution concept, such as indexing the coil of nails into a slot or breaking a nail sideways off of the stick. The team should consider each subfunction in turn and ask whether it is expressed in a way that does not imply a particular physical solution principle.

There is no single correct way of creating a function diagram and no single correct functional decomposition of a product. A helpful way to create the diagram is to quickly create several drafts and then work to refine them into a single diagram that the team is comfortable with. Some useful techniques for getting started are:

- Create a function diagram of an existing product.
- Create a function diagram based on an arbitrary product concept already generated by the team or based on a known subfunction technology. Be sure to generalize the diagram to the appropriate level of abstraction.
- Follow one of the flows (e.g., material) and determine what operations are required. The details of the other flows can be derived by thinking about their connections to the initial flow.

Note that the function diagram is typically not unique. In particular, subfunctions can often be ordered in different ways to produce different function diagrams. Also note that in some applications the material, energy, and signal flows are difficult to identify. In these cases, a simple list of the subfunctions of the product, without connections between them, is often sufficient.

Functional decomposition is most applicable to technical products, but it can also be applied to simple and apparently nontechnical products. For example, an ice cream scoop has material flow of ice cream being separated, formed, transported, and deposited. These subfunctions could form the basis of a problem decomposition.

Functional decomposition is only one of several possible ways to divide a problem into simpler subproblems. Two other approaches are:

- ***Decomposition by sequence of user actions:*** For example, the nailer problem might be broken down into three user actions: moving the tool to the gross nailing position, positioning the tool precisely, triggering the tool. This approach is often useful for products with very simple technical functions involving a lot of user interaction.
- ***Decomposition by key customer needs:*** For the nailer, this decomposition might include the following subproblems: fires nails in rapid succession, is lightweight, and has a large nail capacity. This approach is often useful for products in which form, and not working principles or technology, is the primary problem. Examples of such products include toothbrushes (assuming the basic brush concept is retained) and storage containers.

Focus Initial Efforts on the Critical Subproblems

The goal of all of these decomposition techniques is to divide a complex problem into simpler problems such that these simpler problems can be tackled in a focused way. Once problem decomposition is complete, the team chooses the subproblems that are most critical to the success of the product and that are most likely to benefit from novel or creative solutions. This approach involves a conscious decision to defer the solution of some of the subproblems. For example, the nailer team chose to focus on the subproblems of storing/accepting energy, converting the energy to translational energy, and applying the translational energy to the nail. The team felt confident that the nail handling and triggering issues could be solved after the energy storage and conversion issues were addressed. The team also deferred most of the user interaction issues of the tool. The team believed that the choice of a basic working principle for the tool would so constrain the eventual form of the tool that they had to begin with the core technology and then proceed to consider how to embody that technology in an attractive and user-friendly form. Teams can usually agree after a few minutes of discussion on which subproblems should be addressed first and which should be deferred for later consideration.

Step 2: Search Externally

External search is aimed at finding existing solutions to both the overall problem and the subproblems identified during the problem clarification step. While external search is listed as the second step in the concept generation method, this sequential labeling is deceptive; external search occurs continually throughout the development process. Implementing an existing solution is usually quicker and cheaper than developing a new solution. Liberal use of existing solutions allows the team to focus its creative energy on the critical subproblems for which there are no satisfactory prior solutions. Furthermore, a conventional solution to one subproblem can frequently be combined with a novel solution to another subproblem to yield a superior overall design. For this reason external search includes detailed evaluation not only of directly competitive products but also of technologies used in products with related subfunctions.

The external search for solutions is essentially an information-gathering process. Available time and resources can be optimized by using an expand-and-focus strategy: first *expand* the scope of the search by broadly gathering information that might be related to the problem and then *focus* the scope of the search by exploring the promising directions in more detail. Too much of either approach will make the external search inefficient.

There are at least five good ways to gather information from external sources: lead user interviews, expert consultation, patent searches, literature searches, and competitive benchmarking.

Interview Lead Users

While identifying customer needs, the team may have sought out or encountered lead users. *Lead users* are those users of a product who experience needs months or years before the majority of the market and stand to benefit substantially from a product innovation (von Hippel, 1988). Frequently these lead users will have already invented solutions to meet their needs. This is particularly true among highly technical user communities, such as those in the medical or scientific fields. Lead users may be sought out in the market for which the team is developing the new product, or they may be found in markets for products implementing some of the subfunctions of the product.

In the hand-held nailer case, the nailer team consulted with the building contractors from the PBS television series *This Old House* in order to solicit new concepts. These lead users, who are exposed to tools from many manufacturers, made many interesting observations about the weaknesses in existing tools, but in this case did not provide many new product concepts.

Consult Experts

Experts with knowledge of one or more of the subproblems not only can provide solution concepts directly but also can redirect the search in a more fruitful area. Experts may include professionals at firms manufacturing related products, professional consultants, university faculty, and technical representatives of suppliers. These people can be found by calling universities, by calling companies, and by looking up authors of articles. While finding experts can be hard work, it is almost always less time consuming than re-creating existing knowledge.

Most experts are willing to talk on the telephone or meet in person for an hour or so without charge. In general, consultants will expect to be paid for time they spend on a problem beyond an initial meeting or telephone conversation. Suppliers are usually willing to provide several days of effort without direct compensation if they anticipate that someone will use their product as a component in a design. Of course, experts at directly competing firms are in most cases unwilling to provide proprietary information about their product designs. A good habit to develop is to always ask people consulted to suggest others who should be contacted. The best information often comes from pursuing these "second generation" leads.

The nailer design team consulted dozens of experts, including a rocket fuel specialist, electric motor researchers at MIT, and engineers from a vendor of gas springs. Most of this consultation was done on the telephone, although the engineers from the spring vendor made two trips to visit the team, at their company's expense.

Search Patents

Patents are a rich and readily available source of technical information containing detailed drawings and explanations of how many products work. The main disadvantage of patent searches is that concepts found in recent patents are protected (generally for 20 years from the date of the patent application), so there may be a royalty involved in using them. However, patents are also useful to see what concepts are already protected and must be avoided or licensed. Concepts contained in foreign patents without global coverage and in expired patents can be used without payment of royalties.

The formal indexing scheme for patents is difficult for novices to navigate. Fortunately, several databases contain the actual text of all patents. These text databases can be searched electronically by key words. Key word searches can be conducted efficiently with only modest practice and are remarkably effective in finding patents relevant to a particular product. Copies of U.S. patents including illustrations can be obtained for a nominal fee from the U.S. Patent and Trademark Office and from several suppliers. (See the web site www.ulrich-eppinger.net for a current list of online patent databases and suppliers of patent documents.)

A U.S. patent search in the area of nailers revealed several interesting concepts. One of the patents described a motor-driven double-flywheel nailer. One of the illustrations from this patent is shown in Exhibit 6-5. The design in this patent uses the accumulation of rotational kinetic energy in a flywheel, which is then suddenly converted into translational energy by a friction clutch. The energy is then delivered to the nail with a single impact of a drive pin.

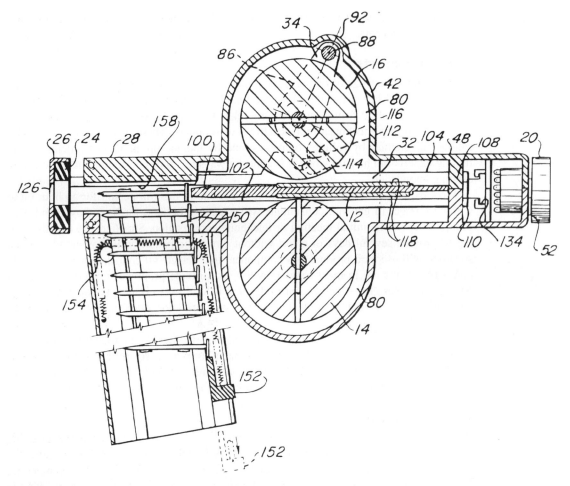

EXHIBIT 6-5 Concept from motor-driven double-flywheel nailer patent (U.S. Patent 4,042,036). The accompanying text describing the patent is nine pages long.

Search Published Literature

Published literature includes journals; conference proceedings; trade magazines; government reports; market, consumer, and product information; and new product announcements. Literature searches are therefore very fertile sources of existing solutions.

Electronic searches are frequently the most efficient way to gather information from published literature. Searching the Internet is often a good first step, although the quality of the results can be hard to assess. More structured databases are available from online sources or on mass storage devices (e.g., CD-ROM). Many databases store only abstracts of articles and not the full text and diagrams. A follow-up search for an actual article is often needed for complete information. The two main difficulties in conducting good database searches are determining the key words and limiting the scope of the search. There is a trade-off between the need to use more key words for complete coverage and the need to restrict the number of matches to a manageable number.

Handbooks cataloging technical information can also be very useful references for external search. Examples of such engineering references are *Marks' Standard Handbook of Mechanical Engineering, Perry's Chemical Engineers' Handbook,* and *Mechanisms and Mechanical Devices Sourcebook.*

The nailer team found several useful articles related to the subproblems, including articles on energy storage describing flywheel and battery technologies. In a handbook they found an impact tool mechanism that provided a useful energy conversion concept.

Benchmark Related Products

In the context of concept generation, *benchmarking* is the study of existing products with functionality similar to that of the product under development or to the subproblems on which the team is focused. Benchmarking can reveal existing concepts that have been implemented to solve a particular problem, as well as information on the strengths and weaknesses of the competition.

At this point the team will likely already be familiar with the competitive and closely related products. Products in other markets, but with related functionality, are more difficult to find. One of the most useful sources of this information is the *Thomas Register of American Manufacturers,* a directory of manufacturers of industrial products organized by product type. Often the hardest part of using the *Thomas Register* is finding out what related products are actually called and how they are cataloged. The *Thomas Register* can be accessed via the Internet.

For the nailer, the closely related products included a single-shot gunpowder-actuated tool for nailing into concrete, an electrical solenoid-actuated tacker, a pneumatic nailer for factory use, and a palm-held multiblow pneumatic nailer. The products with related functionality (in this case, energy storage and conversion) included air bags and the sodium azide propellant used as an energy source, chemical hand warmers for skiing, air rifles powered by carbon dioxide cartridges, and portable computers and their battery packs. The team obtained and disassembled most of these related products in order to discover the general concepts on which they were based, as well as other, more detailed information, including, for example, the names of the suppliers of specific components.

External search is an important method of gathering solution concepts. Skill in conducting external searches is therefore a valuable personal and organizational asset. This ability can be developed through careful observation of the world in order to develop a mental database of technologies and through the development of a network of professional contacts. Even with the aid of personal knowledge and contacts, external search remains "detective work" and is completed most effectively by those who are persistent and resourceful in pursuing leads and opportunities.

Step 3: Search Internally

Internal search is the use of personal and team knowledge and creativity to generate solution concepts. The search is *internal* in that all of the ideas to emerge from this step are created from knowledge already in the possession of the team. This activity may be the most open-ended and creative of any in new-product development. We find it useful to think of internal search as a process of retrieving a potentially useful piece of information from one's memory and then adapting that information to the problem at hand. This process can be carried out by individuals working in isolation or by a group of people working together.

Four guidelines are useful for improving both individual and group internal search:

1. *Suspend judgment.* In most aspects of daily life, success depends on an ability to quickly evaluate a set of alternatives and take action. For example, none of us would be very productive if deciding what to wear in the morning or what to eat for breakfast involved an extensive period of generating alternatives before making a judgment. Because most decisions in our day-to-day lives have implications of only a few minutes or hours, we are accustomed to making decisions quickly and moving on. Concept generation for product development is fundamentally different. We have to live with the consequences of product concept decisions for years. As a result, suspending evaluation for the days or weeks required to generate a large set of alternatives is critical to success. The imperative to suspend judgment is frequently translated into the rule that during group concept generation sessions no criticism of concepts is allowed. A better approach is for individuals perceiving weaknesses in concepts to channel any judgmental tendencies into suggestions for improvements or alternative concepts.

2. *Generate a lot of ideas.* Most experts believe that the more ideas a team generates, the more likely the team is to explore fully the solution space. Striving for quantity lowers the expectations of quality for any particular idea and therefore may encourage people to share ideas they may otherwise view as not worth mentioning. Further, each idea acts as a stimulus for other ideas, so a large number of ideas has the potential to stimulate even more ideas.

3. *Welcome ideas that may seem infeasible.* Ideas which initially appear infeasible can often be improved, "debugged," or "repaired" by other members of the team. The more infeasible an idea, the more it stretches the boundaries of the solution space and encourages the team to think of the limits of possibility. Therefore, infeasible ideas are quite valuable and their expression should be encouraged.

4. *Use graphical and physical media.* Reasoning about physical and geometric information with words is difficult. Text and verbal language are inherently inefficient vehicles for describing physical entities. Whether working as a group or as an individual, abundant sketching surfaces should be available. Foam, clay, cardboard, and other three-dimensional media may also be appropriate aids for problems requiring a deep understanding of form and spatial relationships.

Both Individual and Group Sessions Can Be Useful

Formal studies of group and individual problem solving suggest that a set of people working alone for a period of time will generate more and better concepts than the same people working together for the same time period (McGrath, 1984). This finding is contrary to the actual practices of the many firms that perform most of their concept generation activities in group sessions. Our observations confirm the formal studies, and we believe that team members should spend at least some of their concept generation time working alone. We also believe that group sessions are critical for building consensus, communicating information, and refining concepts. In an ideal setting, each individual on the team would spend several hours working alone and then the group would get together to discuss and improve the concepts generated by individuals.

However, we also know that there is a practical reason for holding group concept generation sessions: it is one way to guarantee that the individuals in the group will devote a certain amount of time to the task. Especially in very intense and demanding work environments, without scheduling a meeting, few people will allocate several hours for concentrated individual effort on generating new concepts. The phone rings, people interrupt, urgent problems demand attention. In certain environments, scheduled group sessions may be the only way to guarantee that enough attention is paid to the concept generation activity.

The nailer team used both individual effort and group sessions for internal search. For example, during one particular week each member was assigned one or two subproblems and was expected to develop at least 10 solution concepts. This divided the concept generation work among all members. The group then met to discuss and expand on the individually generated concepts. The more promising concepts were investigated further.

Hints for Generating Solution Concepts

Experienced individuals and teams can usually just sit down and begin generating good concepts for a subproblem. Often these people have developed a set of techniques they use to stimulate their thinking, and these techniques have become a natural part of their problem-solving process. Novice product development professionals may be aided by a set of hints that stimulate new ideas or encourage relationships among ideas. VanGundy (1988), von Oech (1998), and McKim (1980) give dozens of helpful suggestions. Here are some hints we have found to be helpful:

- *Make analogies.* Experienced designers always ask themselves what other devices solve a related problem. Frequently they will ask themselves if there is a natural or biological analogy to the problem. They will think about whether their problem exists at a much larger or smaller dimensional scale than that which they are considering. They will ask what devices do something similar in an unrelated area of application. The nailer team, when posing these questions, realized that construction pile drivers are similar to nailers in some respects. In following up on this idea, they developed the concept of a multiblow tool.

- *Wish and wonder.* Beginning a thought or comment with "I wish we could . . ." or "I wonder what would happen if . . ." helps to stimulate oneself or the group to consider new possibilities. These questions cause reflection on the boundaries of the problem. For example, a member of the nailer team, when confronted with the required length of a rail gun (an electromagnetic device for accelerating a projectile) for driving a nail, said, "I wish the tool could be 1 meter long." Discussion of this comment led to the idea that perhaps a long tool could be used like a cane for nailing decking, allowing users to remain on their feet.

- *Use related stimuli.* Most individuals can think of a new idea when presented with a new stimulus. Related stimuli are those stimuli generated in the context of the problem at hand. For example, one way to use related stimuli is for each individual in a group session to generate a list of ideas (working alone) and then pass the list to his or her neighbor. Upon reflection on someone else's ideas, most people are able to generate new ideas. Other related stimuli include customer needs statements and photographs of the use environment of the product.

- *Use unrelated stimuli.* Occasionally, random or unrelated stimuli can be effective in encouraging new ideas. An example of such a technique is to choose, at random, one of a collection of photographs of objects, and then to think of some way that the randomly generated object might relate to the problem at hand. In a variant of this idea, individuals can be sent out on the streets with a digital camera to capture random images for subsequent use in stimulating new ideas. (This may also serve as a good change of pace for a tired group.)

- *Set quantitative goals.* Generating new ideas can be exhausting. Near the end of a session, individuals and groups may find quantitative goals useful as a motivating force. The nailer team frequently issued individual concept generation assignments with quantitative targets of 10 to 20 concepts.

- *Use the gallery method.* The *gallery method* is a way to display a large number of concepts simultaneously for discussion. Sketches, usually one concept to a sheet, are taped or pinned to the walls of the meeting room. Team members circulate and look at each concept. The creator of the concept may offer explanation, and the group subsequently makes suggestions for improving the concept or spontaneously generates related concepts. This method is a good way to merge individual and group efforts.

In the 1990s, a Russian problem-solving methodology called TRIZ (a Russian acronym for *theory of inventive problem solving*) began to be disseminated in Europe and in the United States. The methodology is primarily useful in identifying physical working principles to solve technical problems. The key idea underlying TRIZ is to identify a contradiction that is implicit in a problem. For example, a contradiction in the nailer problem might be that increasing power (a desirable characteristic) would also tend to increase weight (an undesirable characteristic). One of the TRIZ tools is a matrix of 39 by 39 characteristics with each cell corresponding to a particular conflict between two characteristics. In each cell of the matrix, up to four physical principles are suggested as ways of resolving the corresponding conflict. There are 40 basic principles, including, for example, the *periodic action* principle (i.e., replace a continuous action with a periodic action, like an impulse). Using TRIZ, the nailer team might have arrived at the concept of using repeated smaller impacts to drive the nail. The idea of identifying a conflict in the design problem and then thinking about ways to resolve the conflict appears to be a very useful problem-solving heuristic. This approach can be useful in generating concepts even without adopting the entire TRIZ methodology.

Exhibit 6-6 shows some of the solutions the nailer team generated for the subproblems of (1) storing or accepting energy and (2) delivering translational energy to a nail.

Step 4: Explore Systematically

As a result of the external and internal search activities, the team will have collected tens or hundreds of concept *fragments*—solutions to the subproblems. Systematic exploration is aimed at navigating the space of possibilities by organizing and synthesizing these solution fragments. The nailer team focused on the energy storage, conversion, and delivery subproblems and had generated dozens of concept fragments for each subproblem. One approach to organizing and synthesizing these fragments would be to consider all of the

EXHIBIT 6-6

Some of the solutions to the subproblems of (1) storing or accepting energy and (2) delivering translational energy to a nail.

Solutions to Subproblem of Storing or Accepting Energy

- Self-regulating chemical reaction emitting high-pressure gas
- Carbide (as for lanterns)
- Combusting sawdust from job site
- Gun powder
- Sodium azide (air bag explosive)
- Fuel-air combustion (butane, propane, acetylene, etc.)
- Compressed air (in tank or from compressor)
- Carbon dioxide in tank
- Electric wall outlet and cord
- High-pressure oil line (hydraulics)
- Flywheel with charging (spin-up)
- Battery pack or tool, belt, or floor
- Fuel cell
- Human power: arms or legs
- Methane from decomposing organic materials
- "Burning" like that of chemical hand warmers
- Nuclear reactions
- Cold fusion
- Solar electric cells
- Solar-steam conversion
- Steam supply line
- Wind
- Geothermal

Solutions to Subproblem of Applying Translational Energy to Nail

Single impact

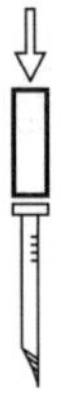

Multiple impacts (tens or hundreds)

Multiple impacts (hundreds or thousands)

Push

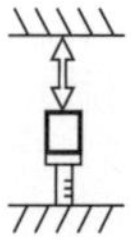

Twist-push

possible combinations of the fragments associated with each subproblem; however, a little arithmetic reveals the impossibility of this approach. Given the three subproblems on which the team focused and an average of 15 fragments for each subproblem, the team would have to consider 3,375 combinations of fragments ($15 \times 15 \times 15$). This would be a daunting task for even the most enthusiastic team. Furthermore, the team would quickly discover that many of the combinations do not even make sense. Fortunately, there are two specific tools for managing this complexity and organizing the thinking of the team: the *concept classification tree* and the *concept combination table*. The classification tree helps the team divide the possible solutions into independent categories. The combination table guides the team in selectively considering combinations of fragments.

EXHIBIT 6-7
A classification
tree for the
nailer energy
source concept
fragments.

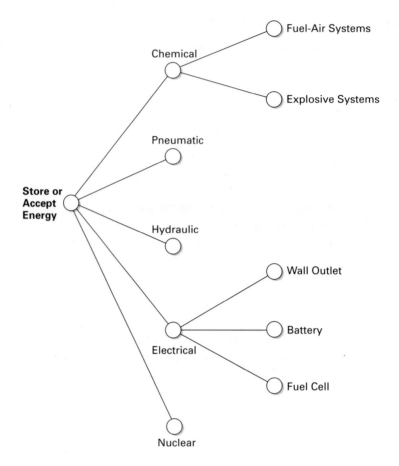

Concept Classification Tree

The concept classification tree is used to divide the entire space of possible solutions into several distinct classes which will facilitate comparison and pruning. An example of a tree for the nailer example is shown in Exhibit 6-7. The branches of this tree correspond to different energy sources.

The classification tree provides at least four important benefits:

1. *Pruning of less promising branches:* If by studying the classification tree the team is able to identify a solution approach that does not appear to have much merit, then this approach can be pruned and the team can focus its attention on the more promising branches of the tree. Pruning a branch of the tree requires some evaluation and judgment and should therefore be done carefully, but the reality of product development is that there are limited resources and that focusing the available resources on the most promising directions is an important success factor. For the nailer team, the nuclear energy source was pruned from consideration. Although the team had identified some very intriguing nuclear devices for use in powering artificial hearts, they felt that these devices would not be economically practical for at least a decade and would probably be hampered by regulatory requirements indefinitely.

EXHIBIT 6-8

A new problem decomposition assuming an electrical energy source and the accumulation of energy in the mechanical domain.

2. *Identification of independent approaches to the problem:* Each branch of the tree can be considered a different approach to solving the overall problem. Some of these approaches may be almost completely independent of each other. In these cases, the team can cleanly divide its efforts among two or more individuals or task forces. When two approaches both look promising, this division of effort can reduce the complexity of the concept generation activities. It also may engender some healthy competition among the approaches under consideration. The nailer team found that both the chemical/explosive branch and the electrical branch appeared quite promising. They assigned these two approaches to two different subteams and pursued them independently for several weeks.

3. *Exposure of inappropriate emphasis on certain branches:* Once the tree is constructed, the team is able to reflect quickly on whether the effort applied to each branch has been appropriately allocated. The nailer team recognized that they had applied very little effort to thinking about hydraulic energy sources and conversion technologies. This recognition guided them to focus on this branch of the tree for a few days.

4. *Refinement of the problem decomposition for a particular branch:* Sometimes a problem decomposition can be usefully tailored to a particular approach to the problem. Consider the branch of the tree corresponding to the electrical energy source. Based on additional investigation of the nailing process, the team determined that the instantaneous power delivered during the nailing process was about 10,000 watts for a few milliseconds and so exceeds the power which is available from a wall outlet, a battery, or a fuel cell (of reasonable size, cost, and mass). They concluded, therefore, that energy must be accumulated over a substantial period of the nailing cycle (say 100 milliseconds) and then suddenly released to supply the required instantaneous power to drive the nail. This quick analysis led the team to add a subfunction ("accumulate translational energy") to their function diagram (see Exhibit 6-8). They chose to add the subfunction after the conversion of electrical energy to mechanical energy, but briefly considered the possibility of accumulating the energy in the electrical domain with a capacitor. This kind of refinement of the function diagram is quite common as the team makes more assumptions about the approach and as more information is gathered.

The classification tree in Exhibit 6-7 shows the alternative solutions to the energy source subproblem. However, there are other possible trees. The team might have chosen to use a tree classifying the alternative solutions to the energy delivery subproblem, showing branches for single impact, multiple impact, or pushing. Trees can be constructed with branches corresponding to the solution fragments of any of the subproblems, but certain classifications are more useful. In general, a subproblem whose solution highly constrains the possible solutions to the remaining subproblems is a good candidate for a classification tree. For example, the choice of energy source (electrical, nuclear, pneumatic, etc.) constrains whether a motor or a piston-cylinder can be used to convert the energy to translational energy. In contrast, the choice of energy delivery mechanism

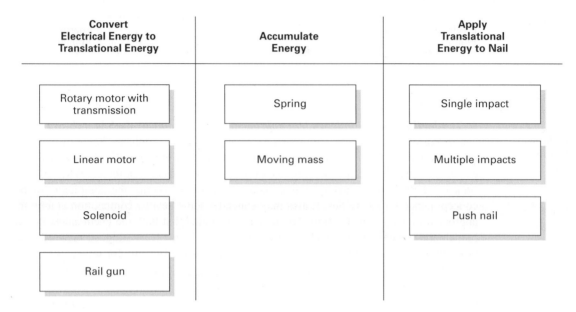

Convert Electrical Energy to Translational Energy	Accumulate Energy	Apply Translational Energy to Nail
Rotary motor with transmission	Spring	Single impact
Linear motor	Moving mass	Multiple impacts
Solenoid		Push nail
Rail gun		

EXHIBIT 6-9   Concept combination table for the hand-held nailer.

(single impact, multiple impact, etc.) does not greatly constrain the solutions to the other subproblems. Reflection on which subproblem is likely to most highly constrain the solutions to the remaining subproblems will usually lead to one or two clear ways to construct the classification tree.

Concept Combination Table

The concept combination table provides a way to consider combinations of solution fragments systematically. Exhibit 6-9 shows an example of a combination table that the nailer team used to consider the combinations of fragments for the electrical branch of the classification tree. The columns in the table correspond to the subproblems identified in Exhibit 6-8. The entries in each column correspond to the solution fragments for each of these subproblems derived from external and internal search. For example, the subproblem of converting electrical energy to translational energy is the heading for the first column. The entries in this column are a rotary motor with a transmission, a linear motor, a solenoid, and a rail gun.

Potential solutions to the overall problem are formed by combining one fragment from each column. For the nailer example, there are 24 possible combinations ($4 \times 2 \times 3$). Choosing a combination of fragments does not lead spontaneously to a solution to the overall problem. The combination of fragments must usually be developed and refined before an integrated solution emerges. This development may not even be possible or may lead to more than one solution, but at a minimum it involves additional creative thought. In some ways, the combination table is simply a way to make forced associations among fragments in order to stimulate further creative thinking; in no way does the mere act of selecting a combination yield a complete solution.

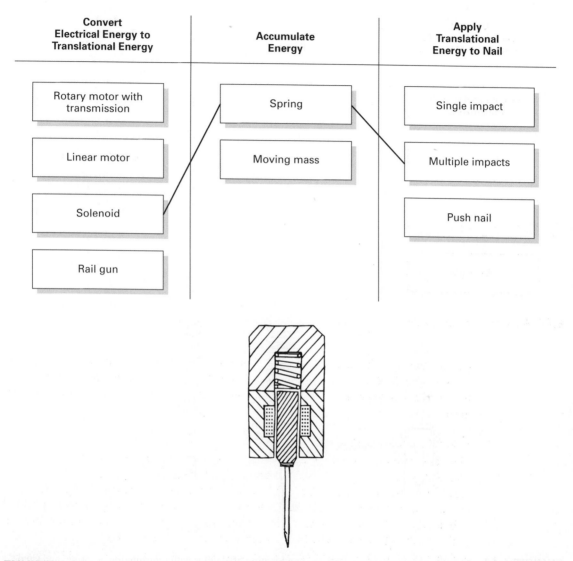

Convert Electrical Energy to Translational Energy	Accumulate Energy	Apply Translational Energy to Nail
Rotary motor with transmission	Spring	Single impact
Linear motor	Moving mass	Multiple impacts
Solenoid		Push nail
Rail gun		

EXHIBIT 6-10 In this solution concept, a solenoid compresses a spring and then releases it repeatedly in order to drive the nail with multiple impacts.

Exhibit 6-10 shows a sketch of a concept arising from the combination of the fragments "solenoid," "spring," and "multiple impacts." Exhibit 6-11 shows some sketches of concepts arising from the combination of the fragments "rotary motor with transmission," "spring," and "single impact." Exhibit 6-12 shows a sketch of a concept arising from the combination of "rotary motor with transmission," "spring," and "multiple impacts." Exhibit 6-13 shows some sketches of concepts arising from the combination of "linear motor," "moving mass," "and single impact."

Convert Electrical Energy to Translational Energy	Accumulate Energy	Apply Translational Energy to Nail
Rotary motor with transmission	Spring	Single impact
Linear motor	Moving mass	Multiple impacts
Solenoid		Push nail
Rail gun		

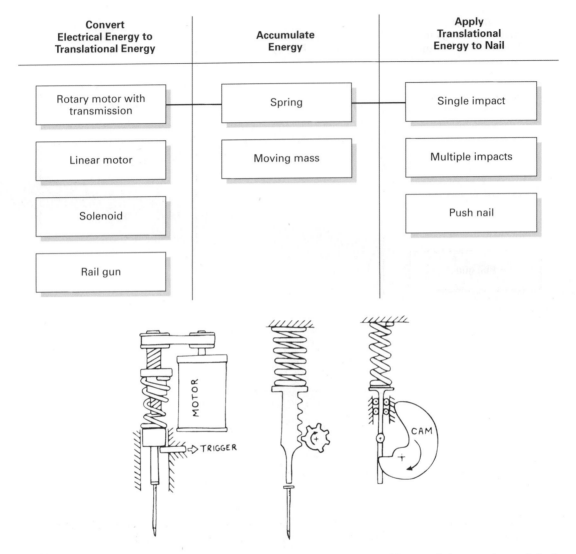

EXHIBIT 6-11 Multiple solutions arising from the combination of a motor with transmission, a spring, and single impact. The motor winds a spring, accumulating potential energy which is then delivered to the nail in a single blow.

Two guidelines make the concept combination process easier. First, if a fragment can be eliminated as being infeasible before combining it with other fragments, then the number of combinations the team needs to consider is dramatically reduced. For example, if the team could determine that the rail gun would not be feasible under any condition, they could reduce the number of combinations from 24 to 18. Second, the concept combination table should be concentrated on the subproblems that are coupled. Coupled subproblems are those whose solutions can be evaluated only in combination with the solutions to other subproblems. For example, the choice of the specific electrical energy source to be

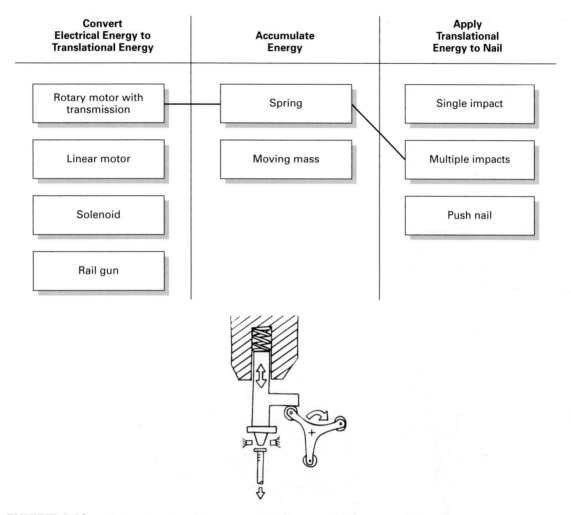

Convert Electrical Energy to Translational Energy	Accumulate Energy	Apply Translational Energy to Nail
Rotary motor with transmission	Spring	Single impact
Linear motor	Moving mass	Multiple impacts
Solenoid		Push nail
Rail gun		

EXHIBIT 6-12   Solution from the combination of a motor with transmission, a spring, and multiple impacts. The motor repeatedly winds and releases the spring, storing and delivering energy over several blows.

used (e.g., battery versus wall outlet), although extremely critical, is somewhat independent of the choice of energy conversion (e.g., motor versus solenoid). Therefore, the concept combination table does not need to contain a column for the different types of electrical energy sources. This reduces the number of combinations the team must consider. As a practical matter, concept combination tables lose their usefulness when the number of columns exceeds three or four.

Managing the Exploration Process

The classification tree and combination tables are tools that a team can use somewhat flexibly. They are simple ways to organize thinking and guide the creative energies of the team. Rarely do teams generate only one classification tree and one concept combination table. More typically the team will create several alternative classification trees and several

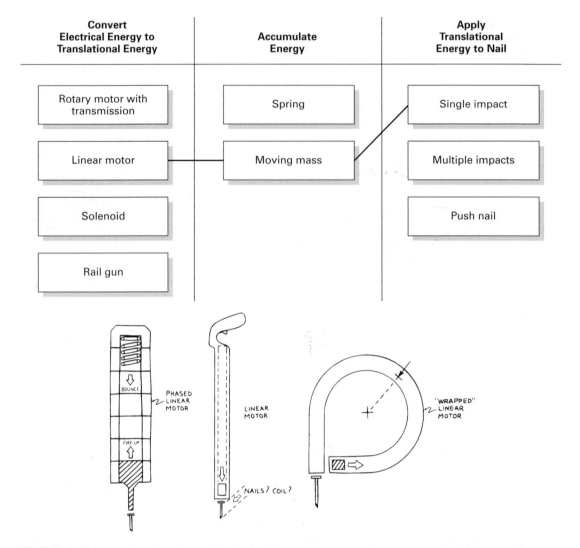

Convert Electrical Energy to Translational Energy	Accumulate Energy	Apply Translational Energy to Nail
Rotary motor with transmission	Spring	Single impact
Linear motor	Moving mass	Multiple impacts
Solenoid		Push nail
Rail gun		

EXHIBIT 6-13   Solutions from the combination of a linear motor, a moving mass, and single impact. A linear motor accelerates a massive hammer, accumulating kinetic energy which is delivered to the nail in a single blow.

concept combination tables. Interspersed with this exploratory activity may be a refining of the original problem decomposition or the pursuit of additional internal or external search. The exploration step of concept generation usually acts more as a guide for further creative thinking than as the final step in the process.

Recall that at the beginning of the process the team chooses a few subproblems on which to focus attention. Eventually the team must return to address all of the subproblems. This usually occurs after the team has narrowed the range of alternatives for the critical subproblems. The nailer team narrowed its alternatives to a few chemical and a few electric concepts and then refined them by working out the user interface, industrial design, and configuration issues. One of the resulting concept descriptions is shown in Exhibit 6-14.

EXHIBIT 6-14
One of several
refined solution
concepts.

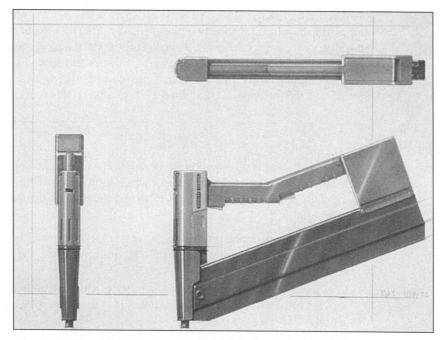

Courtesy of Product Genesis, Inc.

Step 5: Reflect on the Results and the Process

Although the reflection step is placed here at the end for convenience in presentation, reflection should in fact be performed throughout the whole process. Questions to ask include:

- Is the team developing confidence that the solution space has been fully explored?
- Are there alternative function diagrams?
- Are there alternative ways to decompose the problem?
- Have external sources been thoroughly pursued?
- Have ideas from everyone been accepted and integrated in the process?

The nailer team members discussed whether they had focused too much attention on the energy storage and conversion issues in the tool while ignoring the user interface and overall configuration. They decided that the energy issues remained at the core of the problem and that their decision to focus on these issues first was justified. They also wondered if they had pursued too many branches of the classification tree. Initially they had pursued electrical, chemical, and pneumatic concepts before ultimately settling on an electric concept. In hindsight, the chemical approach had some obvious safety and customer perception shortcomings (they were exploring the use of explosives as an energy source). They decided that although they liked some aspects of the chemical solution, they should have eliminated it from consideration earlier in the process, allowing more time to pursue some of the more promising branches in greater detail.

The team explored several of these concepts in more detail and built working proto-types of nailers incorporating two fundamentally different directions: (1) a motor winding a spring with energy released in a single blow, and (2) a motor with a rotating mass that repeatedly hit the nail at a rate of about 10 cycles per second until the nail was fully driven. Ultimately, the multiblow tool proved to be the most technically feasible approach and the final product (Exhibit 6-1) was based on this concept.

Summary

A product concept is an approximate description of the technology, working principles, and form of the product. The degree to which a product satisfies customers and can be successfully commercialized depends to a large measure on the quality of the underlying concept.

- The concept generation process begins with a set of customer needs and target specifications and results in a set of product concepts from which the team will make a final selection.
- In most cases, an effective development team will generate hundreds of concepts, of which 5 to 20 will merit serious consideration during the subsequent concept selection activity.
- The concept generation method presented in this chapter consists of five steps:
  1. *Clarify the problem.* Understand the problem and decompose it into simpler sub-problems.
  2. *Search externally.* Gather information from lead users, experts, patents, published literature, and related products.
  3. *Search internally.* Use individual and group methods to retrieve and adapt the knowledge of the team.
  4. *Explore systematically.* Use classification trees and combination tables to organize the thinking of the team and to synthesize solution fragments.
  5. *Reflect on the solutions and the process.* Identify opportunities for improvement in subsequent iterations or future projects.
- Although concept generation is an inherently creative process, teams can benefit from using a structured method. Such an approach allows full exploration of the design space and reduces the chance of oversight in the types of solution concepts considered. It also acts as a map for those team members who are less experienced in design problem solving.
- Despite the linear presentation of the concept generation process in this chapter, the team will likely return to each step of the process several times. Iteration is particularly common when the team is developing a radically new product.
- Professionals who are good at concept generation seem to always be in great demand as team members. Contrary to popular opinion, we believe concept generation is a skill that can be learned and developed.

References and Bibliography

Many current resources are available on the Internet via
www.ulrich-eppinger.net

Pahl and Beitz were the driving force behind structured design methods in Germany. We adapt many of their ideas for functional decomposition.

Pahl, Gerhard, and Wolfgang Beitz, *Engineering Design,* second edition, K. Wallace, editor, Springer-Verlag, New York, 1996.

Hubka and Eder have written in a detailed way about systematic concept generation for technical products.

Hubka, Vladimir, and W. Ernst Eder, *Theory of Technical Systems: A Total Concept Theory for Engineering Design,* Springer-Verlag, New York, 1988.

Von Hippel reports on his empirical research on the sources of new product concepts. His central argument is that lead users are the innovators in many markets.

von Hippel, Eric, *The Sources of Innovation,* Oxford University Press, New York, 1988.

VanGundy presents dozens of methods for problem solving, many of which are directly applicable to product concept generation.

VanGundy, Arthur B., Jr., *Techniques of Structured Problem Solving,* second edition, Van Nostrand Reinhold, New York, 1988.

Von Oech provides dozens of good ideas for improving individual and group creative performance.

von Oech, Roger, *A Whack on the Side of the Head: How You Can Be More Creative,* revised edition, Warner Books, New York, 1998.

McKim presents a holistic approach to developing creative thinking skills in individuals and groups.

McKim, Robert H., *Experiences in Visual Thinking,* second edition, Brooks/Cole Publishing, Monterey, CA, 1980.

Interesting research on a set of standard "templates" for identifying novel product concepts has been done by Goldenberg and Mazursky.

Goldenberg, Jacob, and David Mazursky, *Creativity in Product Innovation,* Cambridge University Press, Cambridge, 2002.

The following are two of the better English-language publications on TRIZ.

Altshuller, Genrich, *40 Principles: TRIZ Keys to Technical Innovation,* Technical Innovation Center, Worcester, MA, 1998.

Terninko, John, Alla Zusman, and Boris Zlotin, *Systematic Innovation: An Introduction to TRIZ,* St. Lucie Press, Boca Raton, FL, 1998.

McGrath presents studies comparing the relative performance of groups and individuals in generating new ideas.

McGrath Joseph E., *Groups: Interaction and Performance,* Prentice Hall, Englewood Cliffs, NJ, 1984.

Engineering handbooks are handy sources of information on standard technical solutions. Three good handbooks are:

Avallone, Eugene A., and Theodore Baumeister III (eds.), *Marks' Standard Handbook of Mechanical Engineering,* 10th edition, McGraw-Hill, New York, 1996.

Perry, Robert H., Don W. Green, and James O. Maloney (eds.), *Perry's Chemical Engineers' Handbook,* seventh edition, McGraw-Hill, New York, 1997.

Sclater, Neil, and Nicholas P. Chironis, *Mechanisms and Mechanical Devices Sourcebook,* third edition, McGraw-Hill, New York, 2001.

Exercises

1. Decompose the problem of designing a new barbecue grill. Try a functional decomposition as well as a decomposition based on the user interactions with the product.
2. Generate 20 concepts for the subproblem "prevent fraying of end of rope" as part of a system for cutting lengths of nylon rope from a spool.
3. Prepare an external-search plan for the problem of permanently applying serial numbers to plastic products.

Thought Questions

1. What are the prospects for computer support for concept generation activities? Can you think of any computer tools that would be especially helpful in this process?
2. What would be the relative advantages and disadvantages of involving actual customers in the concept generation process?
3. For what types of products would the initial focus of the concept generation activity be on the form and user interface of the product and not on the core technology? Describe specific examples.
4. Could you apply the five-step method to an everyday problem like choosing the food for a picnic?
5. Consider the task of generating new concepts for the problem of dealing with leaves on a lawn. How would a plastic-bag manufacturer's assumptions and problem decomposition differ from those of a manufacturer of lawn tools and equipment and from those of a company responsible for maintaining golf courses around the world? Should the context of the firm dictate the way concept generation is approached?

CHAPTER SEVEN

Concept Selection

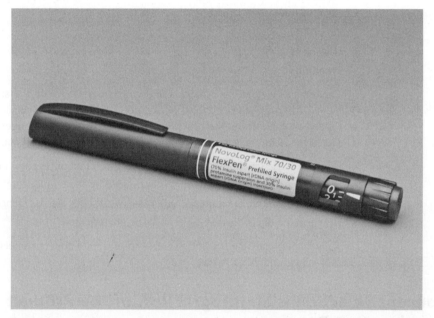

Courtesy of Novo Nordisk Pharmaceuticals Inc.

EXHIBIT 7-1
One of the existing outpatient syringes.

This chapter was developed in collaboration with Eric Howlett.

123

A medical supply company retained a product design firm to develop a reusable syringe with precise dosage control for outpatient use. One of the products sold by a competitor is shown in Exhibit 7-1. To focus the development effort, the medical supply company identified two major problems with its current product: cost (the existing model was made of stainless steel) and accuracy of dose metering. The company also requested that the product be tailored to the physical capabilities of the elderly, an important segment of the target market. To summarize the needs of its client and of the intended end users, the team established seven criteria on which the choice of a product concept would be based:

- Ease of handling.
- Ease of use.
- Readability of dose settings.
- Dose metering accuracy.
- Durability.
- Ease of manufacture.
- Portability.

The team described the concepts under consideration with the sketches shown in Exhibit 7-3. Although each concept nominally satisfied the key customer needs, the team was faced with choosing the best concept for further design, refinement, and production. The need to select one syringe concept from many raises several questions:

- How can the team choose the best concept, given that the designs are still quite abstract?
- How can a decision be made that is embraced by the whole team?
- How can desirable attributes of otherwise weak concepts be identified and used?
- How can the decision-making process be documented?

This chapter uses the syringe example to present a concept selection methodology addressing these and other issues.

Concept Selection Is an Integral Part of the Product Development Process

Early in the development process the product development team identifies a set of customer needs. By using a variety of methods, the team then generates alternative solution concepts in response to these needs. (See Chapter 4, Identifying Customer Needs, and Chapter 6, Concept Generation, for more detail on these activities.) *Concept selection* is the process of evaluating concepts with respect to customer needs and other criteria, comparing the relative strengths and weaknesses of the concepts, and selecting one or more concepts for further investigation, testing, or development. Exhibit 7-2 illustrates how the concept selection activity is related to the other activities that make up the concept development phase of the product development process. Although this chapter focuses on the selection of an overall product concept at the beginning of the development process, the method we present is also useful later in the development process when the team must select subsystem concepts, components, and production processes.

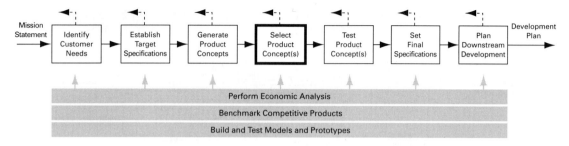

EXHIBIT 7-2   Concept selection is part of the overall concept development phase.

While many stages of the development process benefit from unbounded creativity and divergent thinking, concept selection is the process of narrowing the set of concept alternatives under consideration. Although concept selection is a convergent process, it is frequently iterative and may not produce a dominant concept immediately. A large set of concepts is initially winnowed down to a smaller set, but these concepts may subsequently be combined and improved to temporarily enlarge the set of concepts under consideration. Through several iterations a dominant concept is finally chosen. Exhibit 7-4 illustrates the successive narrowing and temporary widening of the set of options under consideration during the concept selection activity.

All Teams Use Some Method for Choosing a Concept

Whether or not the concept selection process is explicit, all teams use some method to choose among concepts. (Even those teams generating only one concept are using a method: choosing the first concept they think of.) The methods vary in their effectiveness and include the following:

- *External decision:* Concepts are turned over to the customer, client, or some other external entity for selection.
- *Product champion:* An influential member of the product development team chooses a concept based on personal preference.
- *Intuition:* The concept is chosen by its feel. Explicit criteria or trade-offs are not used. The concept just *seems* better.
- *Multivoting:* Each member of the team votes for several concepts. The concept with the most votes is selected.
- *Pros and cons:* The team lists the strengths and weaknesses of each concept and makes a choice based upon group opinion.
- *Prototype and test:* The organization builds and tests prototypes of each concept, making a selection based upon test data.
- *Decision matrices:* The team rates each concept against prespecified selection criteria, which may be weighted.

The concept selection method in this chapter is built around the use of decision matrices for evaluating each concept with respect to a set of selection criteria.

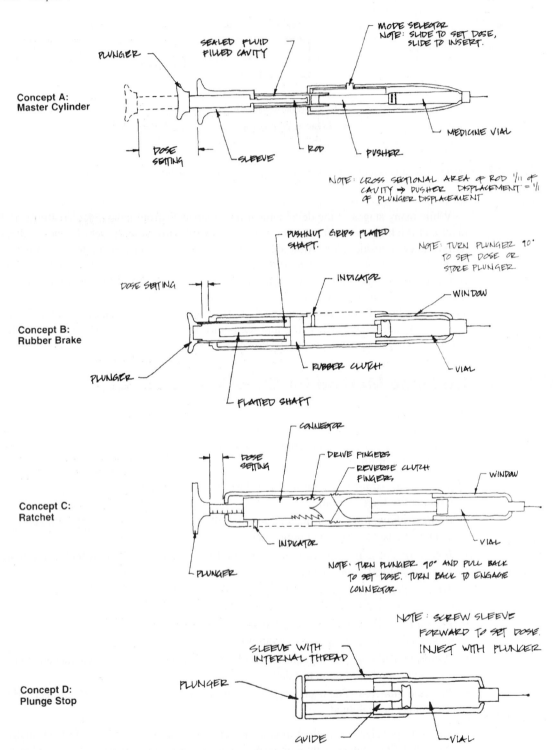

EXHIBIT 7-3 Seven concepts for the outpatient syringe. The product development team generated the seven sketches to describe the basic concepts under consideration.

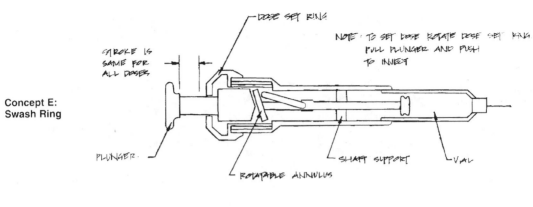

Concept E:
Swash Ring

DOSE SET RING

NOTE: TO SET DOSE ROTATE DOSE SET RING PULL PLUNGER AND PUSH TO INJECT

STROKE IS SAME FOR ALL DOSES

PLUNGER

ROTATABLE ANNULUS

SHAFT SUPPORT

VIAL

NOTE: SET DOSE BY PUSHING LEVER FORWARD. INJECT BY PUSHING LEVER BACK.

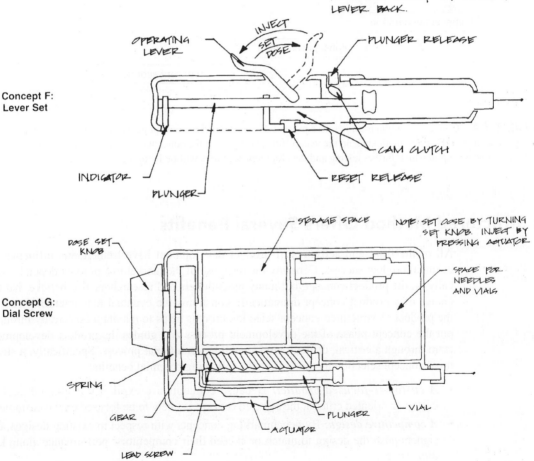

Concept F:
Lever Set

OPERATING LEVER

INJECT

SET DOSE

PLUNGER RELEASE

CAM CLUTCH

INDICATOR

PLUNGER

RESET RELEASE

Concept G:
Dial Screw

DOSE SET KNOB

STORAGE SPACE

NOTE: SET DOSE BY TURNING SET KNOB. INJECT BY PRESSING ACTUATOR

SPACE FOR NEEDLES AND VIALS

SPRING

GEAR

LEAD SCREW

ACTUATOR

PLUNGER

VIAL

EXHIBIT 7-3 *Continued*

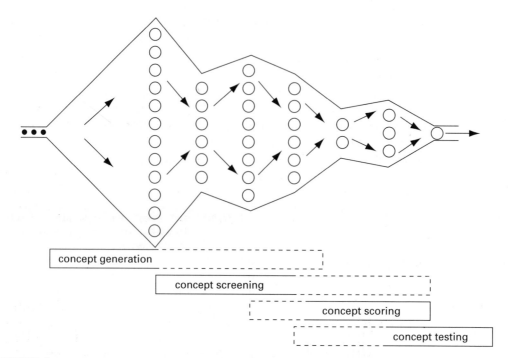

concept generation

concept screening

concept scoring

concept testing

EXHIBIT 7-4 Concept selection is an iterative process closely related to concept generation and testing. The concept screening and scoring methods help the team refine and improve the concepts, leading to one or more promising concepts upon which further testing and development activities will be focused.

A Structured Method Offers Several Benefits

All of the front-end activities of product development have tremendous influence on eventual product success. Certainly the response of the market to a product depends critically on the product concept, but many practitioners and researchers also believe that the choice of a product concept dramatically constrains the eventual manufacturing cost of the product. A structured concept selection process helps to maintain objectivity throughout the concept phase of the development process and guides the product development team through a critical, difficult, and sometimes emotional process. Specifically, a structured concept selection method offers the following potential benefits:

- *A customer-focused product:* Because concepts are explicitly evaluated against customer-oriented criteria, the selected concept is likely to be focused on the customer.
- *A competitive design:* By benchmarking concepts with respect to existing designs, designers push the design to match or exceed their competitors' performance along key dimensions.
- *Better product-process coordination:* Explicit evaluation of the product with respect to manufacturing criteria improves the product's manufacturability and helps to match the product with the process capabilities of the firm.

- *Reduced time to product introduction:* A structured method becomes a common language among design engineers, manufacturing engineers, industrial designers, marketers, and project managers, resulting in decreased ambiguity, faster communication, and fewer false starts.

- *Effective group decision making:* Within the development team, organizational philosophy and guidelines, willingness of members to participate, and team member experience may constrain the concept selection process. A structured method encourages decision making based on objective criteria and minimizes the likelihood that arbitrary or personal factors influence the product concept.

- *Documentation of the decision process:* A structured method results in a readily understood archive of the rationale behind concept decisions. This record is useful for assimilating new team members and for quickly assessing the impact of changes in the customer needs or in the available alternatives.

Overview of Methodology

We present a two-stage concept selection methodology, although the first stage may suffice for simple design decisions. The first stage is called *concept screening* and the second stage is called *concept scoring.* Each is supported by a decision matrix which is used by the team to rate, rank, and select the best concept(s). Although the method is structured, we emphasize the role of group insight to improve and combine concepts.

Concept selection is often performed in two stages as a way to manage the complexity of evaluating dozens of product concepts. The application of these two methods is illustrated in Exhibit 7-4. Screening is a quick, approximate evaluation aimed at producing a few viable alternatives. Scoring is a more careful analysis of these relatively few concepts in order to choose the single concept most likely to lead to product success.

During concept screening, rough initial concepts are evaluated relative to a common reference concept using the *screening matrix.* At this preliminary stage, detailed quantitative comparisons are difficult to obtain and may be misleading, so a coarse comparative rating system is used. After some alternatives are eliminated, the team may choose to move on to concept scoring and conduct more detailed analyses and finer quantitative evaluation of the remaining concepts using the *scoring matrix* as a guide. Throughout the screening and scoring process, several iterations may be performed, with new alternatives arising from the combination of the features of several concepts. Exhibits 7-5 and 7-7 illustrate the screening and scoring matrices, using the selection criteria and concepts from the syringe example.

Both stages, concept screening and concept scoring, follow a six-step process which leads the team through the concept selection activity. The steps are:

1. Prepare the selection matrix.
2. Rate the concepts.
3. Rank the concepts.
4. Combine and improve the concepts.
5. Select one or more concepts.
6. Reflect on the results and the process.

Although we present a well-defined process, the team, not the method, creates the concepts and makes the decisions that determine the quality of the product. Ideally, teams are made up of people from different functional groups within the organization. Each member brings unique views that increase the understanding of the problem and thus facilitate the development of a successful, customer-oriented product. The concept selection method exploits the matrices as visual guides for consensus building among team members. The matrices focus attention on the customer needs and other decision criteria and on the product concepts for explicit evaluation, improvement, and selection.

Concept Screening

Concept screening is based on a method developed by the late Stuart Pugh in the 1980s and is often called *Pugh concept selection* (Pugh, 1990). The purposes of this stage are to narrow the number of concepts quickly and to improve the concepts. Exhibit 7-5 illustrates the screening matrix used during this stage.

Step 1: Prepare the Selection Matrix

To prepare the matrix, the team selects a physical medium appropriate to the problem at hand. Individuals and small groups with a short list of criteria may use matrices on paper similar to Exhibit 7-5 or Appendix A for their selection process. For larger groups a chalkboard or flip chart is desirable to facilitate group discussion.

Next, the inputs (concepts and criteria) are entered on the matrix. Although possibly generated by different individuals, concepts should be presented at the same level of detail for meaningful comparison and unbiased selection. The concepts are best portrayed by both a

	Concepts						
Selection Criteria	**A** **Master** **Cylinder**	**B** **Rubber** **Brake**	**C** **Ratchet**	**D** **(Reference)** **Plunge Stop**	**E** **Swash** **Ring**	**F** **Lever** **Set**	**G** **Dial** **Screw**
Ease of handling	0	0	−	0	0	−	−
Ease of use	0	−	−	0	0	+	0
Readability of settings	0	0	+	0	+	0	+
Dose metering accuracy	0	0	0	0	−	0	0
Durability	0	0	0	0	0	+	0
Ease of manufacture	+	−	−	0	0	−	0
Portability	+	+	0	0	+	0	0
Sum +'s	2	1	1	0	2	2	1
Sum 0's	5	4	3	7	4	3	5
Sum −'s	0	2	3	0	1	2	1
Net Score	2	−1	−2	0	1	0	0
Rank	1	6	7	3	2	3	3
Continue?	Yes	No	No	Combine	Yes	Combine	Revise

EXHIBIT 7-5   The concept screening matrix. For the syringe example, the team rated the concepts against the reference concept using a simple code (+ for "better than," 0 for "same as," − for "worse than") in order to identify some concepts for further consideration. Note that the three concepts ranked "3" all received the same net score.

written description and a graphical representation. A simple one-page sketch of each concept greatly facilitates communication of the key features of the concept. The concepts are entered along the top of the matrix, using graphical or textual labels of some kind.

If the team is considering more than about 12 concepts, the *multivote* technique may be used to quickly choose the dozen or so concepts to be evaluated with the screening matrix. Multivoting is a technique in which members of the team simultaneously vote for three to five concepts by applying "dots" to the sheets describing their preferred concepts. The concepts with the most dots are chosen for concept screening. It is also possible to use the screening matrix method with a large number of concepts. This is facilitated by a spreadsheet and it is then useful to transpose the rows and columns. (Arrange the concepts in this case in the left column and the criteria along the top.)

The selection criteria are listed along the left-hand side of the screening matrix, as shown in Exhibit 7-5. These criteria are chosen based on the customer needs the team has identified, as well as on the needs of the enterprise, such as low manufacturing cost or minimal risk of product liability. The criteria at this stage are usually expressed at a fairly high level of abstraction and typically include from 5 to 10 dimensions. The selection criteria should be chosen to differentiate among the concepts. However, because each criterion is given equal weight in the concept screening method, the team should be careful not to list many relatively unimportant criteria in the screening matrix. Otherwise, the differences among the concepts relative to the more important criteria will not be clearly reflected in the outcome.

After careful consideration, the team chooses a concept to become the benchmark, or *reference concept,* against which all other concepts are rated. The reference is generally either an industry standard or a straightforward concept with which the team members are very familiar. It can be a commercially available product, a best-in-class benchmark product which the team has studied, an earlier generation of the product, any one of the concepts under consideration, or a combination of subsystems assembled to represent the best features of different products.

Step 2: Rate the Concepts

A relative score of "better than" (+), "same as" (0), or "worse than" (−) is placed in each cell of the matrix to represent how each concept rates in comparison to the reference concept relative to the particular criterion. It is generally advisable to rate every concept on one criterion before moving to the next criterion. However, with a large number of concepts, it is faster to use the opposite approach—to rate each concept completely before moving on to the next concept.

Some people find the coarse nature of the relative ratings difficult to work with. However, at this stage in the design process, each concept is only a general notion of the ultimate product, and more detailed ratings are largely meaningless. In fact, given the imprecision of the concept descriptions at this point, it is very difficult to consistently compare concepts to one another unless one concept (the reference) is consistently used as a basis for comparison.

When available, objective metrics can be used as the basis for rating a concept. For example, a good approximation of assembly cost is the number of parts in a design. Similarly, a good approximation of ease of use is the number of operations required to use the device. These objective metrics help to minimize the judgmental nature of the rating process. Some objective metrics suitable for concept selection may arise from the process of establishing target specifications for the product. (See Chapter 5, Product Specifications, for a discussion of metrics.) Absent objective metrics, ratings are established by team consensus, although secret ballot or other methods may also be useful. At this point the team may also wish to note which selection criteria need further investigation and analysis.

Step 3: Rank the Concepts

After rating all the concepts, the team sums the number of "better than," "same as," and "worse than" scores and enters the sum for each category in the lower rows of the matrix. From our example in Exhibit 7-5, concept A was rated to have two criteria better than, five the same as, and none worse than the reference concept. Next, a net score can be calculated by subtracting the number of "worse than" ratings from the "better than" ratings.

Once the summation is completed, the team rank-orders the concepts. Obviously, in general those concepts with more pluses and fewer minuses are ranked higher. Often at this point the team can identify one or two criteria which really seem to differentiate the concepts.

Step 4: Combine and Improve the Concepts

Having rated and ranked the concepts, the team should verify that the results make sense and then consider if there are ways to combine and improve certain concepts. Two issues to consider are:

- Is there a generally good concept which is degraded by one bad feature? Can a minor modification improve the overall concept and yet preserve a distinction from the other concepts?

- Are there two concepts which can be combined to preserve the "better than" qualities while annulling the "worse than" qualities?

Combined and improved concepts are then added to the matrix, rated by the team, and ranked along with the original concepts. In our example, the team noticed that concepts D and F could be combined to remove several of the "worse than" ratings to yield a new concept, DF, to be considered in the next round. Concept G was also considered for revision. The team decided that this concept was too bulky, so the excess storage space was removed while retaining the injection technique. These revised concepts are shown in Exhibit 7-6.

Step 5: Select One or More Concepts

Once the team members are satisfied with their understanding of each concept and its relative quality, they decide which concepts are to be selected for further refinement and analysis. Based upon previous steps, the team will likely develop a clear sense of which are the most promising concepts. The number of concepts selected for further review will be limited by team resources (personnel, money, and time). In our example, the team selected concepts A and E to be considered along with the revised concept G+ and the new concept DF. Having determined the concepts for further analysis, the team must clarify which issues need to be investigated further before a final selection can be made.

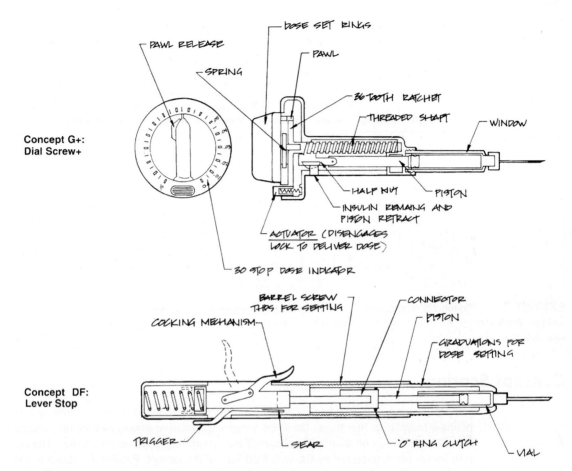

EXHIBIT 7-6 New and revised concepts for the syringe. During the selection process, the syringe team revised concept G and generated a new concept, DF, arising from the combination of concepts D and F.

The team must also decide whether another round of concept screening will be performed or whether concept scoring will be applied next. If the screening matrix is not seen to provide sufficient resolution for the next step of evaluation and selection, then the concept-scoring stage with its weighted selection criteria and more detailed rating scheme would be used.

Step 6: Reflect on the Results and the Process

All of the team members should be comfortable with the outcome. If an individual is not in agreement with the decision of the team, then perhaps one or more important criteria are missing from the screening matrix, or perhaps a particular rating is in error, or at least is not clear. An explicit consideration of whether the results make sense to everyone reduces the likelihood of making a mistake and increases the likelihood that the entire team will be solidly committed to the subsequent development activities.

		Concept							
		A **(Reference)** **Master Cylinder**		**DF** **Lever Stop**		**E** **Swash Ring**		**G+** **Dial Screw+**	
Selection Criteria	**Weight**	**Rating**	**Weighted Score**	**Rating**	**Weighted Score**	**Rating**	**Weighted Score**	**Rating**	**Weighted Score**
Ease of handling	5%	**3**	0.15	3	0.15	4	0.2	4	0.2
Ease of use	15%	**3**	0.45	4	0.6	4	0.6	3	0.45
Readability of settings	10%	2	0.2	**3**	0.3	5	0.5	5	0.5
Dose metering accuracy	25%	**3**	0.75	3	0.75	2	0.5	3	0.75
Durability	15%	2	0.3	5	0.75	4	0.6	**3**	0.45
Ease of manufacture	20%	**3**	0.6	3	0.6	2	0.4	2	0.4
Portability	10%	**3**	0.3	3	0.3	3	0.3	3	0.3
	Total Score	2.75		3.45		3.10		3.05	
	Rank	4		1		2		3	
	Continue?	No		Develop		No		No	

EXHIBIT 7-7 The concept scoring matrix. This method uses a weighted sum of the ratings to determine concept ranking. While concept A serves as the overall reference concept, the separate reference points for each criterion are signified by **bold** rating values.

Concept Scoring

Concept scoring is used when increased resolution will better differentiate among competing concepts. In this stage, the team weighs the relative importance of the selection criteria and focuses on more refined comparisons with respect to each criterion. The concept scores are determined by the weighted sum of the ratings. Exhibit 7-7 illustrates the scoring matrix used in this stage. In describing the concept scoring process, we focus on the differences relative to concept screening.

Step 1: Prepare the Selection Matrix

As in the screening stage, the team prepares a matrix and identifies a reference concept. In most cases a computer spreadsheet is the best format to facilitate ranking and sensitivity analysis. The concepts which have been identified for analysis are entered on the top of the matrix. The concepts have typically been refined to some extent since concept screening and may be expressed in more detail. In conjunction with more detailed concepts, the team may wish to add more detail to the selection criteria. The use of hierarchical relations is a useful way to illuminate the criteria. For the syringe example, suppose the team decided that the criterion "ease of use" did not provide sufficient detail to help distinguish among the remaining concepts. "Ease of use" could be broken down, as shown in Exhibit 7-8, to include "ease of injection," "ease of cleaning," and "ease of loading." The level of criteria detail will depend upon the needs of the team; it may not be necessary to expand the criteria at all. If the team has created a hierarchical list of customer needs, the secondary and tertiary needs are good candidates for more detailed selection criteria. (See Chapter 4, Identifying Customer Needs, for an explanation of pri-

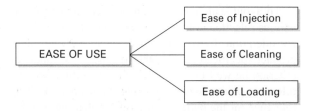

EXHIBIT 7-8   Hierarchical decomposition of selection criteria. In conjunction with more detailed concepts, the team may choose to break down criteria to the level of detail necessary for meaningful comparison.

mary, secondary, and tertiary needs, and see Appendixes A and B for examples of hierarchical selection criteria.)

After the criteria are entered, the team adds importance weights to the matrix. Several different schemes can be used to weight the criteria, such as assigning an importance value from 1 to 5, or allocating 100 percentage points among them, as the team has done in Exhibit 7-7. There are marketing techniques for empirically determining weights from customer data, and a thorough process of identifying customer needs may result in such weights (Urban and Hauser, 1993). However, for the purpose of concept selection the weights are often determined subjectively by team consensus.

Step 2: Rate the Concepts

As in the screening stage, it is generally easiest for the team to focus its discussion by rating all of the concepts with respect to one criterion at a time. Because of the need for additional resolution to distinguish among competing concepts, a finer scale is now used. We recommend a scale from 1 to 5:

Relative Performance	Rating
Much worse than reference	1
Worse than reference	2
Same as reference	3
Better than reference	4
Much better than reference	5

Another scale, such as 1 to 9, may certainly be used, but finer scales generally require more time and effort.

A single reference concept can be used for the comparative ratings, as in the screening stage; however, this is not always appropriate. Unless by pure coincidence the reference concept is of average performance relative to all of the criteria, the use of the same reference concept for the evaluation of each criterion will lead to "scale compression" for some of the criteria. For example, if the reference concept happens to be the easiest concept to manufacture, all of the remaining concepts will receive an evaluation of 1, 2, or 3 ("much worse than," "worse than," or "same as") for the ease-of-manufacture criterion, compressing the rating scale from five levels to three levels.

To avoid scale compression, we recommend using different reference points for the various selection criteria. Reference points may come from several of the concepts under consideration, from comparative benchmarking analysis, from the target values of the product specifications, or other means. It is important that the reference point for each criterion be well understood to facilitate direct one-to-one comparisons. Using multiple reference points does not prevent the team from designating one concept as the overall reference for the purposes of ensuring that the selected concept is competitive relative to this benchmark. Under such conditions the overall reference concept will simply not receive a neutral score.

Exhibit 7-7 shows the scoring matrix for the syringe example. The team believed that the master cylinder concept was not suitable as a reference point for two of the criteria, and other concepts were used as reference points in these cases.

Appendix B illustrates a more detailed scoring matrix for which the team rated the concepts on each criterion with no explicit reference points. These ratings were accomplished by discussing the merits of every concept with respect to one criterion at a time and arranging the scores on a 9-point scale.

Step 3: Rank the Concepts

Once the ratings are entered for each concept, weighted scores are calculated by multiplying the raw scores by the criteria weights. The total score for each concept is the sum of the weighted scores:

$$S_j = \sum_{i=1}^{n} r_{ij} w_i$$

where

r_{ij} = raw rating of concept *j* for the *i*th criterion
w_i = weighting for *i*th criterion
n = number of criteria
S_j = total score for concept *j*

Finally, each concept is given a rank corresponding to its total score, as shown in Exhibit 7-7.

Step 4: Combine and Improve the Concepts

As in the screening stage, the team looks for changes or combinations that improve concepts. Although the formal concept generation process is typically completed before concept selection begins, some of the most creative refinements and improvements occur during the concept selection process as the team realizes the inherent strengths and weaknesses of certain features of the product concepts.

Step 5: Select One or More Concepts

The final selection is not simply a question of choosing the concept that achieves the highest ranking after the first pass through the process. Rather, the team should explore this initial evaluation by conducting a sensitivity analysis. Using a computer spreadsheet, the team can vary weights and ratings to determine their effect on the ranking.

By investigating the sensitivity of the ranking to variations in a particular rating, the team members can assess whether uncertainty about a particular rating has a large impact on their choice. In some cases they may select a lower-scoring concept about which there is little uncertainty instead of a higher-scoring concept that may possibly prove unworkable or less desirable as they learn more about it.

Based on the selection matrix, the team may decide to select the top two or more concepts. These concepts may be further developed, prototyped, and tested to elicit customer feedback. See Chapter 8, Concept Testing, for a discussion of methods to assess customer response to product concepts.

The team may also create two or more scoring matrices with different weightings to yield the concept ranking for various market segments with different customer preferences. It may be that one concept is dominant for several segments. The team should also consider carefully the significance of differences in concept scores. Given the resolution of the scoring system, small differences are generally not significant.

For the syringe example, the team agreed that concept DF was the most promising and would be likely to result in a successful product.

Step 6: Reflect on the Results and the Process

As a final step the team reflects on the selected concept(s) and on the concept selection process. In some ways, this is the "point of no return" for the concept development process, so everyone on the team should feel comfortable that all of the relevant issues have been discussed and that the selected concept(s) have the greatest potential to satisfy customers and be economically successful.

After each stage of concept selection, it is a useful reality check for the team to review each of the concepts that are to be eliminated from further consideration. If the team agrees that any of the dropped concepts is better overall than some of those retained, then the source of this inconsistency should be identified. Perhaps an important criterion is missing, not weighted properly, or inconsistently applied.

The organization can also benefit from reflection on the process itself. Two questions are useful in improving the process for subsequent concept selection activities:

- In what way (if at all) did the concept selection method facilitate team decision making?
- How can the method be modified to improve team performance?

These questions focus the team on the strengths and weaknesses of the methodology in relation to the needs and capabilities of the organization.

Caveats

With experience, users of the concept selection methods will discover several subtleties. Here we discuss some of these subtleties and point out a few areas for caution.

- ***Decomposition of concept quality:*** The basic theory underlying the concept selection method is that selection criteria—and, by implication, customer needs—can be evaluated independently and that concept quality is the sum of the qualities of the concept relative to each criterion. The quality of some product concepts may not be easily decomposed into a set of independent criteria, or the performance of the concept relative to the different criteria may be difficult to relate to overall concept quality. For example, the overall appeal or performance of a tennis racquet design may arise in a highly complex way from its weight, ease of swinging, shock transmission, and energy absorption. Simply choosing a concept based on the sum of performance relative to each criterion may fail to capture complex relationships among these criteria. Keeney and Raiffa (1993) discuss the problem of multiattribute decision making, including the issue of nonlinear relationships among selection criteria.

- *Subjective criteria:* Some selection criteria, particularly those related to aesthetics, are highly subjective. Choices among alternatives based solely on subjective criteria must be made carefully. In general, the development team's collective judgment is not the best way to evaluate concepts on subjective dimensions. Rather, the team should narrow the alternatives to three or four and then solicit the opinions of representative customers from the target market for the product, perhaps using mock-ups or models to represent the concepts. (See Chapter 8, Concept Testing.)

- *To facilitate improvement of concepts:* While discussing each concept to determine its rating, the team may wish to make note of any outstanding (positive or negative) attributes of the concepts. It is useful to identify any features which could be applied to other concepts, as well as issues which could be addressed to improve the concept. Notes may be placed directly in the cells of the selection matrix. Such notes are particularly useful in step 4, when the team seeks to combine, refine, and improve the concepts before making a selection decision.

- *Where to include cost:* Most of the selection criteria are adaptations of the customer needs. However, "ease of manufacturing" and "manufacturing cost" are not customer needs. The only reason customers care about manufacturing cost is that it establishes the lower bound on sale price. Nevertheless, cost is an extremely important factor in choosing a concept, because it is one of the factors determining the economic success of the product. For this reason, we advocate the inclusion of some measure of cost or ease of manufacturing when evaluating concepts, even though these measures are not true customer needs. Similarly, there may be needs of other stakeholders that were not expressed by actual customers but are important for economic success of the product.

- *Selecting elements of aggregate concepts:* Some product concepts are really aggregations of several simpler concepts. If all of the concepts under consideration include choices from a set of simpler elements, then the simple elements can be evaluated first and in an independent fashion before the more complex concepts are evaluated. This sort of decomposition may follow partly from the structure used in concept generation. For example, if all of the syringes in our example could be used with all of several different needle types, then the selection of a needle concept could be conducted independently of the selection of an overall syringe concept.

- *Applying concept selection throughout the development process:* Although throughout this chapter we have emphasized the application of the method to the selection of a basic product concept, concept selection is used again and again at many levels of detail in the design and development process. For example, in the syringe example, concept selection could be used at the very beginning of the development project to decide between a single-use or multiple-use approach. Once the basic approach had been determined, concept selection could be used to choose the basic product concept, as illustrated in this chapter. Finally, concept selection could be used at the most detailed level of design for resolving decisions such as the choice of colors or materials.

Summary

Concept selection is the process of evaluating concepts with respect to customer needs and other criteria, comparing the relative strengths and weaknesses of the concepts, and selecting one or more concepts for further investigation or development.

- All teams use some method, implicit or explicit, for selecting concepts. Decision techniques employed for selecting concepts range from intuitive approaches to structured methods.
- Successful design is facilitated by structured concept selection. We recommend a two-stage process: concept screening and concept scoring.
- Concept screening uses a reference concept to evaluate concept variants against selection criteria. Concept scoring may use different reference points for each criterion.
- Concept screening uses a coarse comparison system to narrow the range of concepts under consideration.
- Concept scoring uses weighted selection criteria and a finer rating scale. Concept scoring may be skipped if concept screening produces a dominant concept.
- Both screening and scoring use a matrix as the basis of a six-step selection process. The six steps are:

  1. Prepare the selection matrix.
  2. Rate the concepts.
  3. Rank the concepts.
  4. Combine and improve the concepts.
  5. Select one or more concepts.
  6. Reflect on the results and the process.

- Concept selection is applied not only during concept development but throughout the subsequent design and development process.
- Concept selection is a group process that facilitates the selection of a winning concept, helps build team consensus, and creates a record of the decision-making process.

References and Bibliography

Many current resources are available on the Internet via
www.ulrich-eppinger.net

The concept selection methodology is a decision-making process. Souder outlines other decision techniques.

Souder, William E., *Management Decision Methods for Managers of Engineering and Research,* Van Nostrand Reinhold, New York, 1980.

For a more formal treatment of multiattribute decision making, illustrated with a set of eclectic and interesting case studies, see Keeney and Raiffa.

Keeney, Ralph L., and Howard Raiffa, *Decisions with Multiple Objectives: Preferences and Value Trade-Offs,* Cambridge University Press, New York, 1993.

Pahl and Beitz's influential engineering design textbook contains an excellent set of systematic methods. The book outlines two concept selection methods similar to concept scoring.

Pahl, Gerhard, and Wolfgang Beitz, *Engineering Design: A Systematic Approach,* second edition, Ken Wallace (ed.), Springer-Verlag, London, 1996.

Weighting alternatives for selection is not a new idea. The following is one of the earlier references for using selection matrices with weights:

Alger, J. R., and C. V. Hays, *Creative Synthesis in Design,* Prentice Hall, Englewood Cliffs, NJ, 1964.

The concept-screening method is based upon the concept selection process presented by Stuart Pugh. Pugh was known to criticize more quantitative methods, such as the concept-scoring method presented in this chapter. He cautioned that numbers can be misleading and can reduce the focus on creativity required to develop better concepts.

Pugh, Stuart, *Total Design,* Addison-Wesley, Reading, MA, 1990.

Concept scoring is similar to a method often called the Kepner-Tregoe method. It is described, along with other techniques for problem identification and solution, in their text.

Kepner, Charles H., and Benjamin B. Tregoe, *The Rational Manager,* McGraw-Hill, New York, 1965.

Urban and Hauser describe techniques for determining the relative importance of different product attributes.

Urban, Glen L., and John R. Hauser, *Design and Marketing of New Products,* second edition, Prentice Hall, Englewood Cliffs, NJ, 1993.

Otto and Wood present a method to include certainty bounds with the ratings given to concepts in concept scoring. These can be combined to derive an estimate of the error in selecting the highest-scoring concept and to compute a confidence interval for the results.

Otto, Kevin N., and Kristin L. Wood, "Estimating Errors in Concept Selection," *ASME Design Engineering Technical Conferences,* Vol. DE-83, 1995, pp. 397–412.

Exercises

1. How can the concept selection methods be used to benchmark or evaluate existing products? Perform such an evaluation for five automobiles you might consider purchasing.
2. Propose a set of selection criteria for the choice of a battery technology for use in a portable computer.
3. Perform concept screening for the four pencil holder concepts shown below. Assume the pencil holders are for a member of a product development team who is continually moving from site to site.
4. Repeat Exercise 3, but use concept scoring.

Zip Pouch   Screw Cap   Clam Shell   Slider

Thought Questions

1. How might you use the concept selection method to decide whether to offer a single product to the marketplace or to offer several different product options?

2. How might you use the method to determine which product features should be standard and which should be optional or add-ons?

3. Can you imagine an interactive computer tool that would allow a large group (say, 20 or more people) to participate in the concept selection process? How might such a tool work?

4. What could cause a situation in which a development team uses the concept selection method to agree on a concept that then results in commercial failure?

Appendix A

Concept-Screening Matrix Example

This matrix was created and used by a development team designing a collar to hold weights onto a barbell.

Selection Criteria	Handcuff	Master Lock	Velcro Belt	Rubber Belt	Alligator Clip	4-Part Latch (REF)	Torsional Spring	Screw Type	Wing Nut	Clothespin	Hose Clamp	C-Clamp	Spring-Loaded Bar	Magnetic Plates	Threaded Bar
Functionality															
Lightweight	+	0	+	+	+	0	+	−	−	+	0	0	+	+	0
Fits different bars	+	0	+	+	+	0	0	0	0	+	0	+	0	−	0
Weights secured laterally	0	0	−	−	0	0	0	−	+	−	0	0	−	0	+
Convenience															
Tighten from end/side	0	0	0	0	0	0	−	−	−	0	−	0	+	+	−
Does not roll	0	0	0	0	0	0	0	0	0	0	0	0	0	0	0
Change weights without removing collar	0	0	0	0	0	0	0	0	0	0	0	0	+	+	0
Convenience of placement when changing weights	0	0	+	+	0	0	−	−	−	0	−	0	+	+	−
Ergonomics															
Secure/release (one motion)	+	0	−	−	+	0	−	−	−	0	−	−	+	−	−
Low force to secure/release	0	0	0	0	−	0	−	0	−	0	0	0	+	−	0
RH/LH usage	0	0	0	0	0	0	0	−	−	0	−	−	0	0	−
Not slippery when wet	0	0	+	+	0	0	0	0	0	0	0	0	+	+	0
Use with one hand	+	0	0	0	+	0	0	+	0	0	0	0	+	+	0
Durability															
Longevity	−	−	−	−	0	0	0	+	0	−	+	+	−	−	+
Other															
Cost of raw materials	0	0	+	+	0	0	0	0	0	+	0	0	−	−	−
Manufacturability	0	−	+	+	0	0	0	0	−	+	+	0	−	−	−
Uses existing weight bars	0	0	0	0	0	0	0	0	0	0	0	0	−	0	−
Sum +'s	4	0	6	6	4	0	1	2	1	4	2	2	8	6	2
Sum 0's	11	14	7	7	11	16	11	8	8	11	10	12	3	4	7
Sum −'s	1	2	3	3	1	0	4	6	7	1	4	2	5	6	7
Net Score	3	−2	3	3	3	0	−3	−4	−6	3	−2	0	3	0	−5
Rank	1	10	1	1	1	7	12	13	15	1	10	7	1	7	15

Appendix B

Concept-Scoring Matrix Example

A development team generated this matrix while selecting a new concept for a spillproof beverage holder to be used on boats. Note that in this case the team chose not to define a single concept as the reference for all of the selection criteria.

Selection Criteria	Weight	Concept A Rating	Concept A Weighted Score	Concept C Rating	Concept C Weighted Score	Concept F Rating	Concept F Weighted Score	Concept I Rating	Concept I Weighted Score	Concept J Rating	Concept J Weighted Score	Concept K Rating	Concept K Weighted Score	Concept O Rating	Concept O Weighted Score
Flexible Use	20														
Use in different locations	15	7	105	7	105	8	120	6	90	6	90	5	75	7	105
Holds different beverages	5	5	25	5	25	3	15	4	20	5	25	3	15	3	15
Maintains Drink Condition	15														
Retains temperature of drink	13	5	65	5	65	5	65	1	13	5	65	5	65	5	65
Prevents water from getting in	2	5	10	7	14	5	10	5	10	5	10	5	10	5	10
Survives Boating Environment	5														
Doesn't break when dropped	1	6	6	6	6	9	9	7	7	5	5	9	9	6	6
Resists corrosion from sea spray	2	7	14	7	14	8	16	8	16	5	10	9	18	7	14
Floats when it falls in water	2	5	10	6	12	8	16	4	8	5	10	8	16	7	14
Keeps Drink Container Stable	20														
Prevents spilling	7	3	21	4	28	3	21	5	35	5	35	3	21	3	21
Prevents bouncing in waves	6	7	42	8	48	7	42	5	30	5	30	7	42	7	42
Will not slide during pitch/roll	7	5	35	5	35	5	35	5	35	5	35	5	35	5	35
Requires Little Maintenance	5														
Easily stored when not in use	1	7	7	6	6	8	8	9	9	4	4	8	8	7	7
Easy to maintain a clean appearance	2	6	12	6	12	3	6	4	8	5	10	5	10	6	12
Allows liquid to drain out bottom	2	5	10	5	10	5	10	5	10	5	10	5	10	5	10
Easy to Use	15														
Usable with one hand	5	7	35	7	35	7	35	6	30	5	25	7	35	7	35
Easy/comfortable to grip	5	8	40	8	40	6	30	5	25	5	25	6	30	8	40
Easy to exchange beverage containers	2	5	10	5	10	5	10	8	16	5	10	5	10	5	10
Works reliably	3	3	9	3	9	3	9	3	9	4	12	4	12	3	9
Attractive in Environment	10														
Doesn't damage boat surface	5	8	40	8	40	8	40	8	40	8	40	6	30	8	40
Attractive to look at	5	7	35	8	40	3	15	4	20	5	25	5	25	8	40
Manufacturing Ease	10														
Low-cost materials	4	5	20	4	16	7	28	8	32	4	16	8	32	6	24
Low complexity of parts	3	4	12	3	9	7	21	4	12	3	9	8	24	5	15
Low number of assembly steps	3	5	15	5	15	8	24	3	9	3	9	8	24	6	18
Total Score			578		594		585		484		510		556		587
Rank			4		1		3		7		6		5		2

CHAPTER EIGHT

Concept Testing

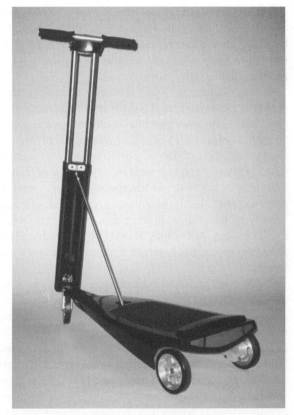

Courtesy of emPower Corporation

EXHIBIT 8-1
A prototype of emPower Corporation's electric scooter product concept.

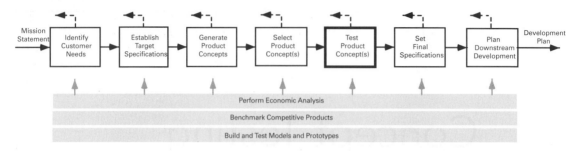

EXHIBIT 8-2   Concept testing in relation to other concept development activities.

The emPower Corporation, a start-up company, had developed a new product concept to address the personal transportation market. Exhibit 8-1 shows a photograph of a prototype of the product. The concept was a three-wheeled electric-powered scooter that could be folded up and carried easily. emPower wished to assess the customer response to this concept in order to decide whether to proceed with its development and to support the company's financing efforts.

In this chapter, we focus primarily on testing done during the concept development phase. In a concept test, the development team solicits a response to a description of the product concept from potential customers in the target market. This type of testing may be used to select which of two or more concepts should be pursued, to gather information from potential customers on how to improve a concept, and to estimate the sales potential of the product. Note that various other types of testing with potential customers may be completed at times other than during concept development. For example, some kind of customer test, usually based on only a verbal description of a concept, may be used in identifying the original product opportunity that forms the basis of the mission statement for the project. A test may also be used to refine the demand forecast after the development of a product is nearly complete, but before a firm commits to full production and launch.

Exhibit 8-2 shows concept testing relative to other concept development activities. Concept testing is closely related to concept selection (Chapter 7) in that both activities aim to further narrow the set of concepts under consideration. However, concept testing is distinct in that it is based on data gathered directly from potential customers and relies to a lesser degree on judgments made by the development team. The reason that concept testing generally follows concept selection is that a team cannot feasibly test more than a few concepts directly with potential customers. As a result, the team must first narrow the set of alternatives under consideration to very few. Concept testing is also closely related to prototyping (Chapter 12), because concept testing invariably involves some kind of representation of the product concept, often a prototype. One of the end results of a concept test may be an estimate of how many units of the product the company is likely to sell. This forecast is a key element of the information used in making an economic analysis of the product (Chapter 15).

A team may choose not to do any concept testing at all if the time required to test the concept is large relative to the product life cycles in the product category, or if the cost of testing is large relative to the cost of actually launching the product. For example, in the

Internet software business, some observers and practitioners argue that just launching a product and iteratively refining it with subsequent product generations is a better strategy than carefully testing a concept before developing it fully. While perhaps appropriate for some products, this strategy would be foolish in the development of, for example, a new commercial airplane, where development costs and time are huge and failure can be disastrous. Most product categories fall between these extremes, and in most cases some form of concept testing is warranted.

This chapter presents a seven-step method for testing product concepts:

1. Define the purpose of the concept test.
2. Choose a survey population.
3. Choose a survey format.
4. Communicate the concept.
5. Measure customer response.
6. Interpret the results.
7. Reflect on the results and the process.

We illustrate this method with the scooter example.

Step 1: Define the Purpose of the Concept Test

As a first step in concept testing, we recommend that the team explicitly articulate in writing the questions that the team wishes to answer with the test. Concept testing is essentially an experimental activity, and as with any experiment, knowing the purpose of the experiment is essential to designing an effective experimental method. This step is closely analogous to "defining the purpose" in prototyping. (See Chapter 12, Prototyping.) The primary questions addressed in concept testing are typically:

- Which of several alternative concepts should be pursued?
- How can the concept be improved to better meet customer needs?
- Approximately how many units are likely to be sold?
- Should development be continued?

Step 2: Choose a Survey Population

An assumption underlying the concept test is that the population of potential customers surveyed reflects that of the target market for the product. If the survey population is either more or less enthusiastic about the product than will be the eventual target audience for the product, then inferences based on the concept test will be biased. As a result, the team should choose a survey population that mirrors the target market in as many ways as possible. In the actual survey, the first few questions are called the *screener questions* and generally are used to verify that the respondent fits the definition of the target market for the product.

Often a product addresses multiple market segments. In such cases, an accurate concept test requires that potential customers from each target segment be surveyed. Surveying

EXHIBIT 8-3
Factors leading
to relatively
smaller or larger
survey sample
sizes.

Factors Favoring a Smaller Sample Size	Factors Favoring a Larger Sample Size
• Test occurs early in concept development process.	• Test occurs later in concept development process.
• Test is primarily intended to gather qualitative data.	• Test is primarily intended to assess demand quantitatively.
• Surveying potential customers is relatively costly in time or money.	• Surveying customers is relatively fast and inexpensive.
• Required investment to develop and launch the product is relatively small.	• Required investment to develop and launch the product is relatively high.
• A relatively large fraction of the target market is expected to value the product (i.e., many positively inclined respondents can be found without a large sample).	• A relatively small fraction of the target market is expected to value the product (i.e., many people have to be sampled to reliably estimate the fraction that value the product).

every possible segment may be prohibitively expensive in cost or time, and in such cases, the team may choose to survey potential customers from only the largest segment. However, when only one segment is sampled, inferences about the response of the entire market are likely to be biased.

For the scooter, there were two primary consumer segments: college students and urban commuters. The team decided to form a survey population from both segments. The team had also identified several smaller secondary segments, including transportation for factory and airport employees.

The sample size of the survey should be large enough that the team's confidence in the results is high enough to guide decision making. Sample sizes for concept testing are sometimes as small as 10 (e.g., when gathering qualitative feedback on a new surgical device for a highly specialized procedure) or as large as 1,000 (e.g., when trying to quantitatively assess the potential demand for a new portable telephone which is targeted at a market segment comprising 10 million households). Although there are no simple formulas for determining sample size, some of the factors driving sample size are shown in Exhibit 8-3.

Depending on the desired data to be collected from the concept-testing process, the team may actually structure multiple surveys with different objectives. Each of these surveys may involve a different sample population and a different sample size. The emPower team performed two different concept tests. In early concept testing, the team sampled only a dozen or so potential customers to solicit feedback on the attractiveness of the basic concept. Later, the team performed a purchase-intent survey of 1,000 customers. This survey was used to make a demand forecast on which financing decisions were based. Because of the importance of this objective, the team felt that the time and expense associated with such a large sample were justified.

Step 3: Choose a Survey Format

The following formats are commonly used in concept testing:

• *Face-to-face interaction:* In this format, an interviewer interacts directly with the respondent. Face-to-face interactions can take the form of *intercepts* (i.e., stopping people at a mall, in a park, or on a city street), interviews prearranged by telephone, interviews with potential customers at a trade-show booth, or focus groups (i.e., prearranged group discussions with 6–12 people).

- *Telephone:* Telephone interviews may be prearranged and targeted at very specific individuals (e.g., pediatric dentists) or may be "cold calls" of consumers from a target population.

- *Postal mail:* In mail surveys, concept-testing materials are sent and respondents are asked to return a completed form. Postal surveys are somewhat slower than other methods and suffer from relatively poor response rates. Some kind of incentive—often cash or a gift—is sometimes offered to increase response.

- *Electronic mail:* Electronic mail surveys are very similar to postal mail surveys, except that (as of this writing) respondents seem slightly more likely to reply than via postal mail. With the proliferation of unwanted e-mail, this tendency may not persist. Many electronic mail users react extremely negatively to unsolicited commercial correspondence. We therefore recommend that electronic mail surveys be used only when respondents are likely to perceive a benefit to their participation, or when the team has already established some kind of positive relationship with the target population.

- *Internet:* Using the Internet, a team may create a virtual concept-testing site in which survey participants can observe concepts and provide responses. An electronic mail message is usually used to recruit respondents to visit the test site.

Each of these formats presents risks of sample bias. For example, the use of electronic formats may bias the sample toward those who are technologically sophisticated. For some products, this sophistication is part of the profile of the target market (e.g., the target market for Internet software products is likely to be comfortable with electronic survey formats). Conversely, an Internet survey might be a particularly bad format for testing a television-based computer concept targeted at people without personal computers.

Exploratory testing, typical in the early phases of concept development, benefits from open-ended interactive formats. We recommend that the team use face-to-face formats when presenting multiple concept alternatives or when soliciting ideas for improving a concept. In these settings, the product developers themselves benefit from performing the interviews because they can directly observe reactions to the product in rich detail. As the purpose of the concept test becomes more focused, more structured formats such as mail and telephone become more appropriate. If the questions are very focused, the team can hire a market research firm to implement the concept test. When gathering data intended primarily for use in forecasting demand, third parties are generally used to collect the data in face-to-face formats. This helps to avoid a sympathy bias—respondents indicating that they like the concept in order to please an anxious product developer.

Step 4: Communicate the Concept

The choice of survey format is closely linked to the way in which the concept will be communicated. Concepts can be communicated in any of the following ways, listed in order of increasing richness of the description.

EXHIBIT 8-4
Sketch of
scooter concept.

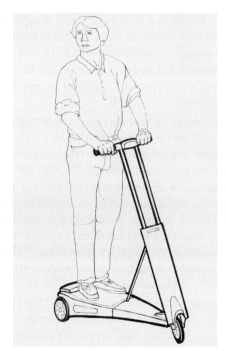

Sketch by David Wallace

• *Verbal description:* A verbal description is generally a short paragraph or a collection of bullet points summarizing the product concept. This description may be read by the respondent or may be read aloud by the person administering the survey. For example, the scooter concept might be described as follows:

> The product is a lightweight electric scooter that can be easily folded and taken with you inside a building or on public transportation. The scooter weighs about 25 pounds. It travels at speeds of up to 15 miles per hour and can go about 12 miles on a single charge. The scooter can be recharged in about two hours from a standard electric outlet. The scooter is easy to ride and has simple controls—just an accelerator button and a brake.

• *Sketch:* Sketches are usually line drawings showing the product in perspective, perhaps with annotations of key features. Exhibit 8-4 shows a sketch of the scooter concept.

• *Photos and renderings:* Photographs can be used to communicate the concept when appearance models exist for the product concept. Renderings are nearly photo-realistic illustrations of the concept. Renderings can be created with pens and markers or using computer-aided design tools. Exhibit 8-5 shows a rendering of the scooter created using computer-aided design software.

• *Storyboard:* A storyboard is a series of images that communicates a temporal sequence of actions involving the product. For example, one of the potential benefits of the scooter is that it can be easily stored and transported. This scenario is illustrated in the storyboard in Exhibit 8-6.

• *Video:* Video images allow even more dynamism than the storyboard. With video, the form of the product itself can be clearly communicated, as can the way in which the

EXHIBIT 8-5
Rendering of the
scooter from
computer-aided
design software.

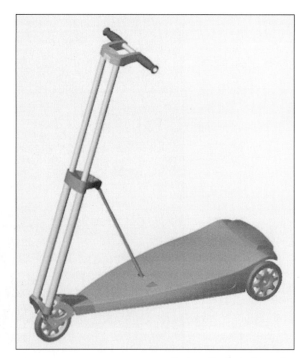

Courtesy of emPower Corporation

EXHIBIT 8-6
Storyboard
illustrating
storage,
transportation,
and use
scenarios.

Courtesy of emPower Corporation

EXHIBIT 8-7
Appearance
model of the
scooter concept.

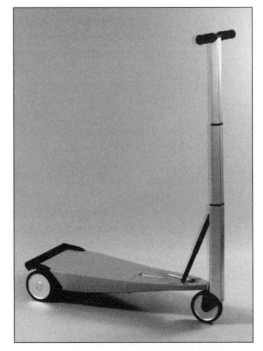

Courtesy of emPower Corporation

product is used. The scooter team used a video in its purchase-intent survey. The video
showed students and commuters riding prototypes of the product and showed an animation of the folding mechanism.

• *Simulation:* Simulation is generally implemented as software that mimics the function or interactive features of the product. Simulation would probably not be the ideal
way to communicate the key features of a scooter, but in some other cases simulation can
be effective. For example, in testing controls for electronic devices, a visual image of the
device can be created on the computer screen, and the user can control the simulated device via a touch screen or mouse clicks and can observe simulated displays and sounds.

• *Interactive multimedia:* Interactive multimedia combines the visual richness of
video with the interactivity of simulation. Using multimedia, you can display video and
still images of the product. The respondent can view verbal and graphical information
and can listen to audio information. Interaction allows the respondent to choose from
among several sources of available information on the product, and in some cases to experience the controls and displays of a simulated product. Unfortunately, the development of multimedia systems remains quite expensive and is therefore justified only for
very large product development efforts.

• *Physical appearance models:* Physical appearance models, also known as "looks-
like" models, vividly display the form and appearance of a product. They are often made
of wood or polymer foams and are painted to look like real products. In some cases, limited functionality is included in the model. The scooter team built several looks-like
models, one of which was articulated so that the folding feature could be demonstrated.
Exhibit 8-7 shows a photograph of this model.

EXHIBIT 8-8
Working
prototype of the
scooter concept.

Courtesy of emPower Corporation

• *Working prototypes:* When available, working prototypes, or works-like models, can be useful in concept testing. However, the use of working prototypes is also risky. The primary risk is that the respondents will equate the prototype with the finished product. In some cases, prototypes perform better than the ultimate product (e.g., because the prototype uses better, more expensive components such as motors or batteries). In most cases, the prototype performs worse than the ultimate product and is almost always less visually attractive than the ultimate product. Sometimes separate works-like and looks-like prototypes can be used, one to illustrate how the product will appear in production and the other to illustrate how it would work. Exhibit 8-8 shows a working prototype of the scooter, which was used in some early concept testing.

Matching the Survey Format with the Means of Communicating the Concept

The choice of survey format is tightly linked to the means of communicating the product concept. For example, the team obviously cannot demonstrate the scooter with a working model using a telephone survey. Exhibit 8-9 identifies which means of communicating concepts are appropriate for each survey format.

Issues in Communicating the Concept

When communicating the product concept, the team must decide how aggressively to promote the product and its benefits. The scooter could be described as an "electric-powered personal mobility device" or as an "exciting new electric scooter that provides freedom from gridlock." In our view, the description of the concept should closely mirror

EXHIBIT 8-9
Appropriate-
ness of different
survey formats
for different
ways of
communicating
the product
concept.

	Telephone	Electronic Mail	Postal Mail	Internet	Face-to-Face
Verbal description	•	•	•	•	•
Sketch		•	•	•	•
Photo or rendering		•	•	•	•
Storyboard		•	•	•	•
Video				•	•
Simulation				•	•
Interactive multimedia				•	•
Physical appearance model					•
Working prototype					•

the information that the user is likely to consider when making a purchase decision. If highly promotional information is used, it can be labeled as a "sample advertisement," perhaps supplemented by mock-ups of "magazine articles" or "comments by current owners" providing additional descriptions of the product.

Researchers and practitioners argue endlessly about whether the purchase price of the product should be included as part of the concept description. Price is a very powerful lever on customer response, and, therefore, pricing information can dramatically influence the results of a concept test. We recommend that price be omitted from the concept description unless the price of the product is expected to be unusually high or low. For example, the primary benefit of a concept may be that it provides basic functionality at a very low price. In this case, price must be included as part of the concept description. Conversely, a product may provide extremely high performance or unique features, but only at a relatively high price. In this case, price must also be included as part of the concept description. When the price of the product is likely to be quite similar to existing products and to customer expectations, price can be omitted from the concept description. Instead of including price in the concept description, we suggest that the respondent be asked explicitly what his or her expectation of price would be. If the resulting customer expectations differ substantially from the team's pricing plans, then the team may need to either consider modifications to the concept or repeat the concept test including price as a product attribute. Because the scooter was a new product category, for which customers had not developed clear pricing expectations, the emPower team chose to include their target price as part of the concept description.

Instead of showing a single concept, the team may choose to ask a respondent to select from several alternatives. This approach is attractive when the team is trying to decide among several concepts under consideration. A variant on this approach is to present the concept for the new product along with descriptions and pictures of the most successful existing products. This approach has the advantage of allowing respondents to directly assess attributes of the product concept in comparison to those of competitors. Assuming the products would be equally distributed and promoted, this approach also allows the team to estimate potential market share. Using a forced-choice survey technique is likely to be most effective in cases for which there is a narrowly defined product category with relatively few existing products.

Step 5: Measure Customer Response

Most concept test surveys first communicate the product concept and then measure customer response. When a concept test is performed early in the concept development phase, customer response is usually measured by asking the respondent to choose from two or more alternative concepts. Additional questions focus on why respondents react the way they do and on how the product concepts could be improved. Concept tests also generally attempt to measure *purchase intent.* The most commonly used purchase-intent scale has five response categories:

- Definitely would buy.
- Probably would buy.
- Might or might not buy.
- Probably would not buy.
- Definitely would not buy.

There are many alternatives to this scale, including providing seven or more response categories or asking respondents to indicate a numerical probability of purchase.

Exhibit 8-10 shows an example of a survey form for the scooter. This form was designed to be an interview guide for a face-to-face format in which both a brochure and a working prototype were used to communicate the product concept.

Step 6: Interpret the Results

If the team is simply interested in comparing two or more concepts, interpretation of the results is straightforward. If one concept dominates the others and the team is confident that the respondents understood the key differences among the concepts, then the team can simply choose the preferred concept. If the results are not conclusive, the team may decide to choose a concept based on cost or other considerations, or may decide to offer multiple versions of the product. Note that care must be applied in making this judgment for cases in which manufacturing costs are dramatically different among the concepts under comparison and in which no price information is communicated to the respondents. In such cases, respondents may be biased to select the most costly alternative.

In many cases the team is also interested in estimating the demand for a product in the period following launch, usually one year. Here we present a model for estimating the sales potential of *durables.* By *durables* we mean products that last several years, and for which there is, therefore, a negligible repeat-purchase rate. These products are in contrast to consumer packaged goods, like razor blades, toothpaste, or frozen food, for which forecasting models must consider rates of trial and subsequent repeat purchase.

Before proceeding with the model, we note that forecasting sales of new products is subject to a great deal of uncertainty and exhibits notoriously high errors. Nevertheless, forecasts do tend to be correlated with actual demand and so provide useful information to the team.

We estimate Q, the quantity of the product expected to be sold during a time period, as

$$Q = N \times A \times P$$

CONCEPT TEST SURVEY— Electric Powered Personal Transportation Device

I am gathering information for a new transportation product and am hoping that you would be willing to share your opinions with me.

Are you a college student?_____
<If the response is no, thank the respondent and end the survey.>

Do you live between one and three miles from campus?_____

Do you travel distances of one to three miles between classes or other activities during your day?_____
*<If the response is no to this **and** the previous question, thank the respondent and end the survey.>*

How do you currently get to campus from your home: _____

How do you currently get around campus during the day: _____

Here is a brochure for the product. *<Show the brochure.>*

The product is a lightweight electric scooter that can be easily folded and taken with you inside a building or on public transportation. The scooter weighs about 25 pounds. It travels at speeds of up to 15 miles per hour and can go about 12 miles on a single charge. The scooter can be recharged in about two hours from a standard electric outlet. The scooter is easy to ride and has simple controls— just an accelerator button and a brake.

If the product were priced at $489 and were available from a dealer on or near campus, how likely would you be to purchase the scooter within the next year?

❑	❑	❑	❑	❑
I would **definitely not** purchase the scooter	I would **probably not** purchase the scooter	I **might or might not** purchase the scooter	I would **probably** purchase the scooter	I would **definitely** purchase the scooter

---

Would you be interested in test riding a prototype of the product?

<Provide operating instructions and fit the helmet.>

Based on your experience with the product, how likely would you be to purchase the product within the next year?

❑	❑	❑	❑	❑
I would **definitely not** purchase the scooter	I would **probably not** purchase the scooter	I **might or might not** purchase the scooter	I would **probably** purchase the scooter	I would **definitely** purchase the scooter

---

How might this product be improved?

<Ask open-ended questions to elicit feedback on the concept.>

EXHIBIT 8-10   Example interview guide (abridged) for a concept test of the electric scooter.

N is the number of potential customers expected to make purchases during the time period. For an existing and stable product category (e.g., bicycles) *N* is the expected number of purchases to be made of existing products in the category over the time period.

A is the fraction of these potential customers or purchases for which the product is *available* and the customer is *aware* of the product. (In situations where awareness and availability are assumed to be separate independent factors, they are multiplied together to generate *A*.)

P is the probability that the product is purchased if available and if the customer is aware of it. *P* is estimated in turn by

$$P = C_{\text{definitely}} \times F_{\text{definitely}} + C_{\text{probably}} \times F_{\text{probably}}$$

$F_{\text{definitely}}$ is the fraction of survey respondents indicating in the concept test survey that they would *definitely purchase* (often called the "top box" score).

F_{probably} is the fraction of survey respondents indicating that they would *probably purchase* (often called the "second box" score).

$C_{\text{definitely}}$ and C_{probably} are calibration constants usually established based on the experience of a company with similar products in the past. Generally the values of $C_{\text{definitely}}$ and C_{probably} fall in these intervals: $0.10 < C_{\text{definitely}} < 0.50$, $0 < C_{\text{probably}} < 0.25$. Absent prior history, many teams use values of $C_{\text{definitely}} = 0.4$ and $C_{\text{probably}} = 0.2$. Note that these values reflect the typical bias of respondents to *overestimate* the probability that they would actually purchase the product.

Among other possible schemes for estimating P is a function that includes the fraction of respondents in all of the response categories, not just the top two.

For a product associated with an entirely new category (e.g., portable commuter scooters), the interpretation of these variables is slightly different. In this case, N is the number of customers in the target market for the new product, and P is the probability of a target-market customer purchasing the product within a given time period, often a year. Note that this interpretation is reflected in the survey questions in Exhibit 8-10, in which the respondent is asked to indicate the likelihood of purchase "within the next year."

To clarify the model, consider these two numerical examples corresponding to two different market segments and possible product positionings for the scooter concept.

Scooter Sold as Single-Person Transportation in Large Factories This is an existing category. Assume that scooters are currently sold into this market at a rate of 150,000 units per year ($N = 150,000$). Assume that the company sells the product through a single distributor that accounts for 25 percent of the sales in this category ($A = 0.25$). Assume that results from a concept test with factory managers responsible for purchasing transportation devices indicate a definitely-would-buy fraction of 0.30 and probably-would-buy fraction of 0.20. If we use a value of 0.4 for $C_{\text{definitely}}$ and 0.2 for C_{probably}, then

$$P = 0.4 \times 0.30 + 0.2 \times 0.20 = 0.16$$

and

$$Q = 150,000 \times 0.25 \times 0.16 = 6,000 \text{ units/year}$$

Scooter Sold to College Students This is a new category and therefore poses a much more difficult estimation challenge. First, what should be the value of N? Strictly speaking (as of this writing) there are very few existing sales of electric scooters to college students. However, we could define N several other ways. For example, how many students purchase bicycles or motor scooters intended for basic transportation of up to two miles. This number is approximately 1 million per year. Alternatively, how many students must travel distances of between one and three miles either in commuting from home or traveling between classes or other school activities. This number is approximately 2 million. Assume that we sample students in this second group, and that we obtain a definitely-would-buy fraction of 0.10 and a probably-would-buy fraction of 0.05. (Note that these numbers represent the fraction of respondents that indicate intent to purchase within one year.) Further assume that the company plans to sell the scooter through bicycle stores near campuses and advertise in campus newspapers, for the 100 largest college campuses in the United States. Based on this plan, the company expects that 30 percent of the students in the target market will be aware of the product and have convenient access to a dealer. If we use a value of 0.4 for $C_{\text{definitely}}$ and 0.2 for C_{probably}, then

$$P = 0.4 \times 0.10 + 0.2 \times 0.05 = 0.05$$

and

$$Q = 2,000,000 \times 0.30 \times 0.05 = 30,000 \text{ units in the first year}$$

The results of forecasts based on concept testing should be interpreted with caution. Some firms, mostly after repeated experience with similar products, have achieved impressive levels of accuracy in their forecasts. While forecasts do tend to be correlated with actual sales, most individual forecasts exhibit substantial errors. Some of the factors that can cause actual purchase patterns to differ from the purchase intentions expressed in surveys include:

- *Importance of word-of-mouth:* When the benefits of a product are not immediately obvious, the enthusiasm of existing users may be an important factor in generating demand. This factor is not generally captured in concept testing.
- *Fidelity of the concept description:* If the actual product differs substantially from the description of the product in the concept test, then actual sales are likely to differ from the forecast.
- *Pricing:* If the price of the product deviates substantially from the price indicated in the survey, or from the expectations of survey respondents, then forecasts are likely to be inaccurate.
- *Level of promotion:* Spending on advertising and other forms of promotion can increase demand for most products. The influence of promotion is accounted for only weakly in the forecasting model via the awareness/availability term and via the materials used to present the concept(s).

Step 7: Reflect on the Results and the Process

The primary benefit of the concept test is in getting feedback from real potential customers. The qualitative insights gathered through open-ended discussions with respondents about the proposed concepts may be the most important result of concept testing, especially early in the development process. The team should reflect on this evidence as well as on the numerical outcome of its forecast.

The team benefits from thinking about the impact of the three key variables in the forecasting model: (1) the overall size of the market, (2) the availability and awareness of the product, and (3) the fraction of customers who are likely to purchase. Considering alternative markets for the product can sometimes increase the first factor. The second factor can be increased through distribution arrangements and promotion plans. The third factor can be increased through changes to the product design (and possibly advertising) that improve the attractiveness of the product. In considering these factors, a sensitivity analysis can yield useful insights and aids in decision making. For example, what would be the impact on sales if the team were able to secure a partnership with a retailer and therefore increase A by 20 percent?

In reflecting on the results of the concept test, the team should ask two key diagnostic questions. First, was the concept communicated in a way that is likely to elicit customer response that reflects true intent? For example, if one of the primary benefits of the concept is its aesthetic appeal, was the concept presented in a way that this aspect of the product was clear to respondents? Second, is the resulting forecast consistent with observed sales rates of similar products? For example, if only 1,000 gasoline-powered Go-Ped scooters (a competing product) are currently sold to college students each year, why does the emPower team believe it will sell 30 times as many of its product?

Finally, we note that experience with a new product is likely to be applicable to future, similar products. The team can benefit from its experience by documenting the results of its concept testing and by attempting to reconcile these results with subsequent observations of product success.

Summary

A concept test solicits a direct response to a description of the product concept from potential customers in the target market. Concept testing is distinct from concept selection in that it is based on data gathered directly from potential customers and relies to a lesser degree on judgments made by the development team.

- Concept testing can verify that customer needs have been adequately met by the product concept, assess the sales potential of a product concept, and/or gather customer information for refining the product concept.
- Concept testing is appropriate at several different points in the development process: when identifying the original product opportunity, when selecting which of two or more concepts should be pursued, when assessing the sales potential of a product concept, and/or when deciding whether to continue further development and commercialization of the product.
- We recommend a seven-step method for testing product concepts:

  1. Define the purpose of the concept test.
  2. Choose a survey population.
  3. Choose a survey format.
  4. Communicate the concept.
  5. Measure customer response.
  6. Interpret the results.
  7. Reflect on the results and the process.

References and Bibliography

Many current resources are available on the Internet via
www.ulrich-eppinger.net

Crawford and Di Benedetto examine some forecasting models of frequently purchased goods.
Crawford, C. Merle, and C. Anthony Di Benedetto, *New Products Marketing*, seventh edition, McGraw-Hill, New York, 2003.

Jamieson and Bass review methods for interpreting purchase intent data and discuss the factors that may explain differences between stated intention and behavior.

Jamieson, Linda F., and Frank M. Bass, "Adjusting Stated Intention Measures to Predict Trial Purchase of New Products: A Comparison of Models and Methods," *Journal of Marketing Research,* Vol. 26, August 1989, pp. 336–345.

When forecasting the growth of a new product category, diffusion models, which are discussed by Mahajan et al., may be useful.

Mahajan, Vijay, Eitan Muller, and Frank M. Bass, "Diffusion of New Products: Empirical Generalizations and Managerial Uses," *Marketing Science,* Vol. 14, No. 3, Part 2 of 2, 1995, pp. G79–G88.

Vriens and his colleagues report on a study of the differences in concept testing results using verbal descriptions and pictorial descriptions of products.

Vriens, Marco, Gerard H. Loosschilder, Edward Rosbergen, and Dick R. Wittink, "Verbal versus Realistic Pictorial Representations in Conjoint Analysis with Design Attributes," *Journal of Product Innovation Management,* Vol. 15, No. 5, 1998, pp. 455–467.

Dahan and Srinivasan show that the results of concept testing using the Internet are very similar to those using physical models of the product concepts.

Dahan, Ely, and V. Srinivasan, "The Predictive Power of Internet-Based Product Concept Testing Using Visual Depiction and Animation," *Journal of Product Innovation Management,* Vol. 17, No. 2, March 2000, pp. 99–109.

Urban et al. report on the use of multimedia systems for describing product concepts and simulating sources of consumer information.

Urban, Glen L., John R. Hauser, William J. Qualls, Bruce D. Weinberg, Jonathan D. Bohlmann, and Roberta A. Chicos, "Information Acceleration: Validation and Lessons from the Field," *Journal of Marketing Research,* Vol. 34, February 1997, pp. 143–153.

Exercises

1. What are some different ways you could communicate a concept for a new user interface for an automotive audio system? What are the strengths and weaknesses of each approach?
2. Roughly estimate N for the following products. List your assumptions.
   a. A sleeping pillow for air travelers.
   b. An electronic weather station (monitoring temperature, pressure, humidity, etc.) for homes.

Thought Questions

1. Why do you think respondents typically overestimate the likelihood that they will purchase a product?
2. When might it not be advantageous to communicate the product concept to potential customers using a working prototype? Under what circumstances is it better to use some other format?

Appendix

Estimating Market Sizes

Rough estimates of market size can often be made through comparisons with similar products or with known sizes of demographic groups. Exhibits 8-11 and 8-12 contain some numbers that may be useful.

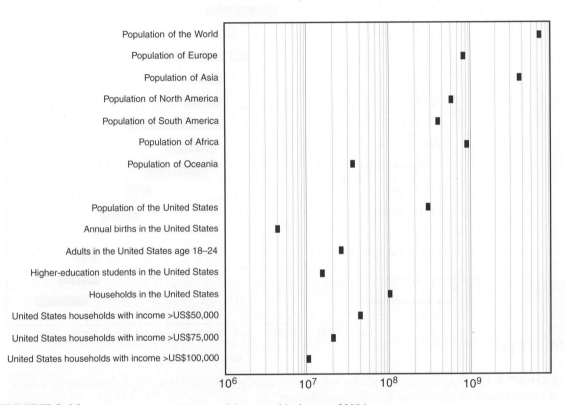

EXHIBIT 8-11 Approximate population and demographic data as of 2004.

Source: U.S. Government Statistics

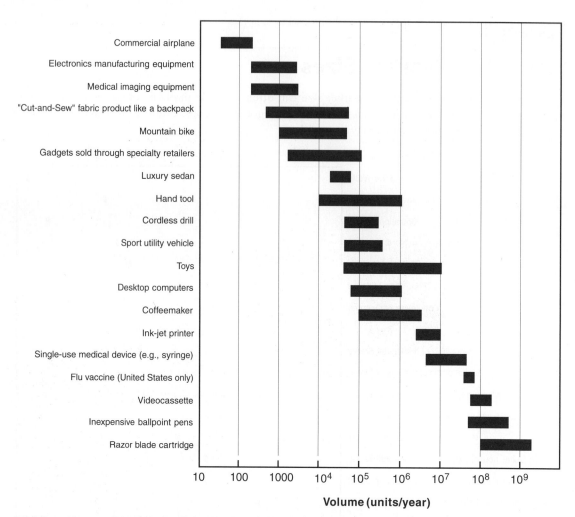

EXHIBIT 8-12   Approximate annual sales volume of miscellaneous products. These figures represent the volume of a typical single model produced by a single manufacturer.

Source: Various

Product Architecture

EXHIBIT 9-1
Three Hewlett-Packard printers from the same product platform: an office model, a photo model, and a model including scanning capability.

A product development team within Hewlett-Packard's home printing division was considering how to respond to the simultaneous pressures to increase product variety and to reduce manufacturing costs. Several of the division's printer products are shown in Exhibit 9-1. Ink jet printing had become the dominant technology for consumer and small-office printing involving color. Excellent black and white print quality and near-photographic color print quality could be obtained using a printer costing less than $200. Driven by the increasing value of color ink jet printers, sales of the three leading competitors together were millions of units per year. However, as the market matured, commercial success required that printers be tuned to the subtle needs of more focused market segments and that the manufacturing costs of these products be continually reduced.

In considering their next steps, the team members asked:

- How would the architecture of the product impact their ability to offer product variety?
- What would be the cost implications of different product architectures?
- How would the architecture of the product impact their ability to complete the design within 12 months?
- How would the architecture of the product influence their ability to manage the development process?

Product architecture is the assignment of the functional elements of a product to the physical building blocks of the product. We focus this chapter on the task of establishing the product architecture. The purpose of the product architecture is to define the basic physical building blocks of the product in terms of what they do and what their interfaces are to the rest of the device. Architectural decisions allow the detailed design and testing of these building blocks to be assigned to teams, individuals, and/or suppliers, such that development of different portions of the product can be carried out simultaneously.

In the next two sections of this chapter, we define product architecture and illustrate the profound implications of architectural decisions using, as examples, the Hewlett-Packard printer and several other products. We then present a method for establishing the product architecture and focus on the printer example for illustration. (Note that the details of the printer example have been somewhat disguised to preserve Hewlett-Packard's proprietary product information.) After presenting the method, we discuss the relationships among product architecture, product variety, and supply chain performance, and we provide guidance for platform planning, an activity closely linked to the product architecture.

What Is Product Architecture?

A product can be thought of in both functional and physical terms. The *functional elements* of a product are the individual operations and transformations that contribute to the overall performance of the product. For a printer, some of the functional elements are "store paper" and "communicate with host computer." Functional elements are usually described in schematic form before they are reduced to specific technologies, components, or physical working principles.

The *physical elements* of a product are the parts, components, and subassemblies that ultimately implement the product's functions. The physical elements become more defined as development progresses. Some physical elements are dictated by the product concept, and others become defined during the detail design phase. For example, the DeskJet embodies a product concept involving a thermal ink delivery device, implemented by a print cartridge. This physical element is inextricably linked to the product concept and was essentially an assumption of the development project.

The physical elements of a product are typically organized into several major physical building blocks, which we call *chunks*. (This term has attained some popularity within major U.S. manufacturing firms.) Each chunk is then made up of a collection of components that implement the functions of the product. The *architecture* of a product is the scheme by which the functional elements of the product are arranged into physical chunks and by which the chunks interact.

Perhaps the most important characteristic of a product's architecture is its modularity. Consider the two different designs for bicycle braking and shifting controls shown in Exhibit 9-2. In the traditional design (left), the shift control function and the brake control function are allocated to separate chunks, which in fact are mounted in separate locations on the bicycle. This design exhibits a modular architecture. In the design on the right, the shift and brake control functions are allocated to the same chunk. This design exhibits an integral architecture—in this case motivated by aerodynamic and ergonomic concerns.

A *modular architecture* has the following two properties:

- Chunks implement one or a few functional elements in their entirety.
- The interactions between chunks are well defined and are generally fundamental to the primary functions of the product.

The most modular architecture is one in which each functional element of the product is implemented by exactly one physical chunk and in which there are a few well-defined interactions between the chunks. Such a modular architecture allows a design change to be made to one chunk without requiring a change to other chunks for the product to function correctly. The chunks may also be designed quite independently of one another.

EXHIBIT 9-2
Two models of bicycle brake and shifting controls. The product on the left exemplifies a modular architecture; the product on the right has a more integral architecture.

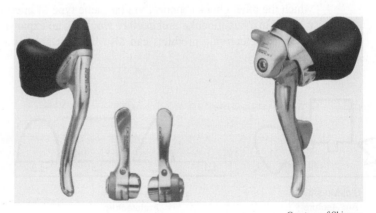

Courtesy of Shimano

The opposite of a modular architecture is an *integral architecture.* An integral architecture exhibits one or more of the following properties:

- Functional elements of the product are implemented using more than one chunk.
- A single chunk implements many functional elements.
- The interactions between chunks are ill defined and may be incidental to the primary functions of the products.

A product embodying an integral architecture will often be designed with the highest possible performance in mind. Implementation of functional elements may be distributed across multiple chunks. Boundaries between the chunks may be difficult to identify or may be nonexistent. Many functional elements may be combined into a few physical components to optimize certain dimensions of performance; however, modifications to any one particular component or feature may require extensive redesign of the product.

Modularity is a relative property of a product architecture. Products are rarely strictly modular or integral. Rather, we can say that they exhibit either more or less modularity than a comparative product, as in the brake and shift controls example in Exhibit 9-2.

Types of Modularity

Modular architectures comprise three types: slot, bus, and sectional (Ulrich, 1995). Each type embodies a one-to-one mapping from functional elements to chunks and well-defined interfaces. The differences between these types lie in the way the interactions between chunks are organized. Exhibit 9-3 illustrates the conceptual differences among these types of architectures.

- *Slot-modular architecture:* Each of the interfaces between chunks in a slot-modular architecture is of a different type from the others, so that the various chunks in the product cannot be interchanged. An automobile radio is an example of a chunk in a slot-modular architecture. The radio implements exactly one function, but its interface is different from any of the other components in the vehicle (e.g., radios and speedometers have different types of interfaces to the instrument panel).
- *Bus-modular architecture:* In a bus-modular architecture, there is a common *bus* to which the other chunks connect via the same type of interface. A common example of a chunk in a bus-modular architecture would be an expansion card for a personal computer. Nonelectronic products can also be built around a bus-modular architecture.

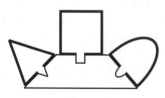

Slot-Modular
Architecture

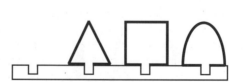

Bus-Modular
Architecture

Sectional-Modular
Architecture

EXHIBIT 9-3   Three types of modular architectures.

Track lighting, shelving systems with rails, and adjustable roof racks for automobiles all embody a bus-modular architecture.

- *Sectional-modular architecture:* In a sectional-modular architecture, all interfaces are of the same type, but there is no single element to which all the other chunks attach. The assembly is built up by connecting the chunks to each other via identical interfaces. Many piping systems adhere to a sectional-modular architecture, as do sectional sofas, office partitions, and some computer systems.

Slot-modular architectures are the most common of the modular architectures because for most products each chunk requires a different interface to accommodate unique interactions between that chunk and the rest of the product. Bus-modular and sectional-modular architectures are particularly useful for situations in which the overall product must vary widely in configuration, but whose chunks can interact in standard ways with the rest of the product. These situations can arise when all of the chunks can use the same type of power, fluid connection, structural attachment, or exchanges of signals.

When Is the Product Architecture Defined?

A product's architecture begins to emerge during concept development. This happens informally—in the sketches, function diagrams, and early prototypes of the concept development phase. Generally, the maturity of the basic product technology dictates whether the product architecture is fully defined during concept development or during system-level design. When the new product is an incremental improvement on an existing product concept, then the product architecture is defined within the product concept. This is for two reasons. First, the basic technologies and working principles of the product are predefined, and so conceptual-design efforts are generally focused on better ways to embody the given concept. Second, as a product category matures, supply chain (i.e., production and distribution) considerations and issues of product variety begin to become more prominent. Product architecture is one of the development decisions that most impacts a firm's ability to efficiently deliver high product variety. Architecture therefore becomes a central element of the product concept. However, when the new product is the first of its kind, concept development is generally concerned with the basic working principles and technology on which the product will be based. In this case, the product architecture is often the initial focus of the system-level design phase of development.

Implications of the Architecture

Decisions about how to divide the product into chunks and about how much modularity to impose on the architecture are tightly linked to several issues of importance to the entire enterprise: product change, product variety, component standardization, product performance, manufacturability, and product development management. The architecture of the product therefore is closely linked to decisions about marketing strategy, manufacturing capabilities, and product development management.

Product Change

Chunks are the physical building blocks of the product, but the architecture of the product defines how these blocks relate to the function of the product. The architecture therefore also defines how the product can be changed. Modular chunks allow changes to be made to a few isolated functional elements of the product without necessarily affecting the design of other chunks. Changing an integral chunk may influence many functional elements and require changes to several related chunks.

Some of the motives for product change are:

- *Upgrade:* As technological capabilities or user needs evolve, some products can accommodate this evolution through upgrades. Examples include changing the processor board in a computer printer or replacing a pump in a cooling system with a more powerful model.
- *Add-ons:* Many products are sold by a manufacturer as a basic unit, to which the user adds components, often produced by third parties, as needed. This type of change is common in the personal computer industry (e.g., third-party mass storage devices may be added to a basic computer).
- *Adaptation:* Some long-lived products may be used in several different use environments, requiring adaptation. For example, machine tools may need to be converted from 220-volt to 110-volt power. Some engines can be converted from a gasoline to a propane fuel supply.
- *Wear:* Physical elements of a product may deteriorate with use, necessitating replacement of the worn components to extend the useful life of the product. For example, many razors allow dull blades to be replaced, tires on vehicles can usually be replaced, most rotational bearings can be replaced, and many appliance motors can be replaced.
- *Consumption:* Some products consume materials, which can then be easily replenished. For example, copiers and printers frequently contain print cartridges, cameras take film cartridges, glue guns consume glue sticks, torches have gas cartridges, and watches contain batteries, all of which are generally replaceable.
- *Flexibility in use:* Some products can be configured by the user to provide different capabilities. For example, many cameras can be used with different lens and flash options, some boats can be used with several awning options, and fishing rods may accommodate several rod-reel configurations.
- *Reuse:* In creating subsequent products, the firm may wish to change only a few functional elements while retaining the rest of the product intact. For example, consumer electronics manufacturers may wish to update a product line by changing only the user interface and enclosure while retaining the inner workings from a previous model.

In each of these cases, a modular architecture allows the firm to minimize the *physical* changes required to achieve a *functional* change.

Product Variety

Variety refers to the range of product models the firm can produce within a particular time period in response to market demand. Products built around modular product architectures can be more easily varied without adding tremendous complexity to the manufacturing system. For example, Swatch produces hundreds of different watch models, but can achieve this variety at relatively low cost by assembling the variants from different combinations of standard chunks (Exhibit 9-4). A large number of different hands, faces, and wristbands can be combined with a relatively small selection of movements and cases to create seemingly endless combinations.

EXHIBIT 9-4
Swatch uses a modular architecture to enable high-variety manufacturing.

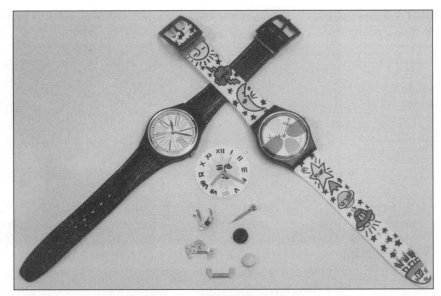

Photo by Stuart Cohen

Component Standardization

Component standardization is the use of the same component or chunk in multiple products. If a chunk implements only one or a few widely useful functional elements, then the chunk can be standardized and used in several different products. Such standardization allows the firm to manufacture the chunk in higher volumes than would otherwise be possible. This in turn may lead to lower costs and increased quality. For example, the watch movement shown in Exhibit 9-4 is identical for many Swatch models. Component standardization may also occur outside the firm when several manufacturers' products all use a chunk or component from the same supplier. For example, the watch battery shown in Exhibit 9-4 is made by a supplier and standardized across several manufacturers' product lines.

Product Performance

We define *product performance* as how well a product implements its intended functions. Typical product performance characteristics are speed, efficiency, life, accuracy, and noise. An integral architecture facilitates the optimization of holistic performance characteristics and those that are driven by the size, shape, and mass of a product. Such characteristics include acceleration, energy consumption, aerodynamic drag, noise, and aesthetics. Consider, for example, a motorcycle. A conventional motorcycle architecture assigns the structural-support functional element to a frame chunk and the power-conversion functional element to a transmission chunk. Exhibit 9-5 shows a photograph of the BMW R1100RS. The architecture of this motorcycle assigns both the structural-support function and the power-conversion function to the transmission chunk. This integral architecture allows the motorcycle designers to exploit the secondary structural properties of the transmission casing in order to eliminate the extra size and mass of a separate frame. The practice of implementing multiple functions using a single physical element is called *function sharing*. An integral architecture allows for redundancy to be eliminated through

Courtesy of BMW Motorcycle Group

function sharing (as in the case of the motorcycle) and allows for geometric nesting of components to minimize the volume a product occupies. Such function sharing and nesting also allows materials use to be minimized, potentially reducing the cost of manufacturing the product.

Manufacturability

In addition to the cost implications of product variety and component standardization described above, the product architecture also directly affects the ability of the team to design each chunk to be produced at low cost. One important design-for-manufacturing (DFM) strategy involves the minimization of the number of parts in a product through *component integration*. However, to maintain a given architecture, the integration of physical components can only be easily considered within each of the chunks. Component integration across several chunks is difficult, if not impossible, and would alter the architecture dramatically. Since the product architecture constrains subsequent detail design decisions in this way, the team must consider the manufacturing implications of the architecture. For this reason DFM begins during the system-level design phase while the layout of the chunks is being planned. For details about the implementation of DFM, see Chapter 11, Design for Manufacturing.

Product Development Management

Responsibility for the detail design of each chunk is usually assigned to a relatively small group within the firm or to an outside supplier. Chunks are assigned to a single individual or group because their design requires careful resolution of interactions, geometric and otherwise, among components within the chunk. With a modular architecture, the group assigned to design a chunk deals with known, and relatively limited, functional interactions with other chunks. If a functional element is implemented by two or more chunks, as in some integral architectures, detail design will require close coordination among different groups. This coordination is likely to be substantially more involved and challenging than the limited coordination required among groups designing different chunks in a modular design. For this reason, teams relying on outside suppliers or on a geographically dispersed team often opt for a modular architecture in which development responsibilities can be split according to the chunk boundaries. Another possibility is to have several functional elements allocated to the same chunk. In this case, the work of the group assigned to that chunk involves a great deal of internal coordination across a larger group.

Modular and integral architectures also demand different project management styles. Modular approaches require very careful planning during the system-level design phase, but detail design is largely concerned with ensuring that the teams assigned to chunks are meeting the performance, cost, and schedule requirements for their chunks. An integral architecture may require less planning and specification during system-level design, but such an architecture requires substantially more integration, conflict resolution, and coordination during the detail design phase.

Establishing the Architecture

Because the product architecture will have profound implications for subsequent product development activities and for the manufacturing and marketing of the completed product, it should be established in a cross-functional effort by the development team. The end result of this activity is an approximate geometric layout of the product, descriptions of the major chunks, and documentation of the key interactions among the chunks. We recommend a four-step method to structure the decision process, which is illustrated using the DeskJet printer example. The steps are:

1. Create a schematic of the product.
2. Cluster the elements of the schematic.
3. Create a rough geometric layout.
4. Identify the fundamental and incidental interactions.

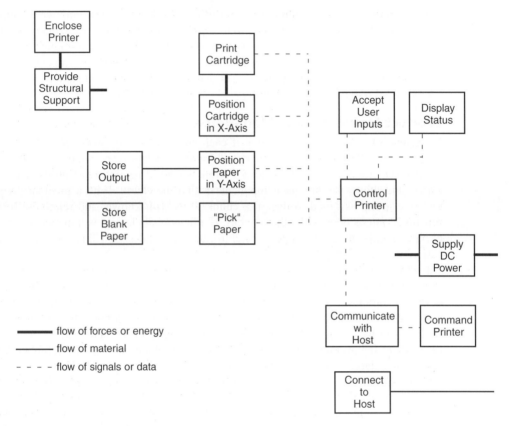

EXHIBIT 9-6 Schematic of the DeskJet printer. Note the presence of both functional elements (e.g., "Store Output") and physical elements (e.g., "Print Cartridge"). For clarity, not all connections among elements are shown.

Step 1: Create a Schematic of the Product

A *schematic* is a diagram representing the team's understanding of the constituent elements of the product. A schematic for the DeskJet is shown in Exhibit 9-6. At the end of the concept development phase, some of the elements in the schematic are physical concepts, such as the front-in/front-out paper path. Some of the elements correspond to critical components, such as the print cartridge the team expects to use. However, some of the elements remain described only functionally. These are the functional elements of the product that have not yet been reduced to physical concepts or components. For example, "display status" is a functional element required for the printer, but the particular approach of the display has not yet been decided. Those elements that have been reduced to physical concepts or components are usually central to the basic product concept the team has generated and selected. Those elements that remain unspecified in physical terms are usually ancillary functions of the product.

The schematic should reflect the team's best understanding of the state of the product, but it does not have to contain every imaginable detail, such as "sense out-of-paper condition" or "shield radio frequency emissions." These and other more detailed functional elements are deferred to a later step. A good rule of thumb is to aim for fewer than 30 elements in the schematic, for the purpose of establishing the product architecture. If the product is a complex system, involving hundreds of functional elements, then it is useful to omit some of the minor ones and to group some others into higher-level functions to be decomposed later. (See Defining Secondary Systems, later in this chapter.)

The schematic created will not be unique. The specific choices made in creating the schematic, such as the choice of functional elements and their arrangement, partly define the product architecture. For example, the functional element "control printer" is represented as a single centralized element in Exhibit 9-6. An alternative would be to distribute the control of each of the other elements of the product throughout the system and have coordination done by the host computer. Because there is usually substantial latitude in the schematic, the team should generate several alternatives and select an approach that will facilitate the consideration of several architectural options.

Step 2: Cluster the Elements of the Schematic

The challenge of step 2 is to assign each of the elements of the schematic to a chunk. One possible assignment of elements to chunks is shown in Exhibit 9-7, where nine chunks are used. Although this was the approximate approach taken by the DeskJet team, there are several other viable alternatives. At one extreme, each element could be assigned to its own chunk, yielding 15 chunks. At the other extreme, the team could decide that the product would have only one major chunk and then attempt to physically integrate all of the elements of the product. In fact, consideration of all possible clusterings of elements would yield thousands of alternatives. One procedure for managing the complexity of the alternatives is to begin with the assumption that each element of the schematic will be assigned to its own chunk, and then to successively cluster elements where advantageous. To determine when there are advantages to clustering, consider these factors, which echo the implications discussed in the previous section:

- *Geometric integration and precision:* Assigning elements to the same chunk allows a single individual or group to control the physical relationships among the elements. Elements requiring precise location or close geometric integration can often be best designed if they are part of the same chunk. For the DeskJet printer, this would suggest clustering the elements associated with positioning the cartridge in the x-axis and positioning the paper in the y-axis.

- *Function sharing:* When a single physical component can implement several functional elements of the product, these functional elements are best clustered together. This is the situation exemplified by the BMW motorcycle transmission (Exhibit 9-5). For the DeskJet printer, the team believed that the status display and the user controls could be incorporated into the same component, and so clustered these two elements together.

- *Capabilities of vendors:* A trusted vendor may have specific capabilities related to a project, and in order to best take advantage of such capabilities a team may choose to cluster those elements about which the vendor has expertise into one chunk. In the case of the DeskJet printer, an internal team did the majority of the engineering design work, and so this was not a major consideration.

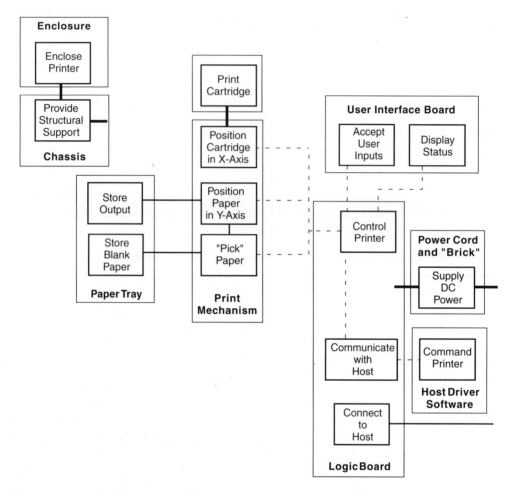

EXHIBIT 9-7   Clustering the elements into chunks. Nine chunks make up this proposed architecture for the DeskJet printer.

- ***Similarity of design or production technology:*** When two or more functional elements are likely to be implemented using the same design and/or production technology, then incorporating these elements into the same chunk may allow for more economical design and/or production. A common strategy, for example, is to combine all functions that are likely to involve electronics in the same chunk. This allows the possibility of implementing all of these functions with a single circuit board.

- ***Localization of change:*** When a team anticipates a great deal of change in some element, it makes sense to isolate that element into its own modular chunk, so that required changes to the element can be carried out without disrupting any of the other chunks. The Hewlett-Packard team anticipated changing the physical appearance of the product over its life cycle, and so chose to isolate the enclosure element into its own chunk.

- ***Accommodating variety:*** Elements should be clustered together to enable the firm to vary the product in ways that will have value for customers. The printer was to be sold

EXHIBIT 9-8
Geometric
layout of the
printer.

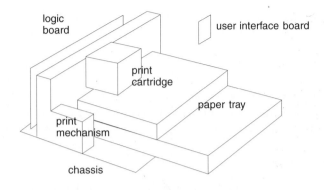

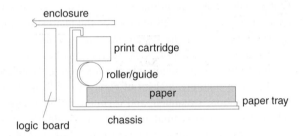

EXHIBIT 9-8
Geometric layout of the printer.

around the world in regions with different electrical power standards. As a result, the team created a separate chunk for the element associated with supplying DC power.

- *Enabling standardization:* If a set of elements will be useful in other products, they should be clustered together into a single chunk. This allows the physical elements of the chunk to be produced in higher quantities. Hewlett-Packard's internal standardization was a key motive for using an existing print cartridge, and so this element is preserved as its own chunk.

- *Portability of the interfaces:* Some interactions are easily transmitted over large distances. For example, electrical signals are much more portable than are mechanical forces and motions. As a result, elements with electronic interactions can be easily separated from one another. This is also true, but to a lesser extent, for fluid connections. The flexibility of electrical interactions allowed the Hewlett-Packard team to cluster the control and communication functions into the same chunk. Conversely, the elements related to paper handling are much more geometrically constrained by their necessary mechanical interactions.

Step 3: Create a Rough Geometric Layout

A geometric layout can be created in two or three dimensions, using drawings, computer models, or physical models (of cardboard or foam, for example). Exhibit 9-8 shows a geometric layout of the DeskJet printer, positioning the major chunks. Creating a geometric layout forces the team to consider whether the geometric interfaces among the chunks are feasible and to work out the basic dimensional relationships among the chunks. By considering a cross section of the printer, the team realized that there was a fundamental trade-off between how much paper could be stored in the paper tray and the height of the machine. In this step, as in the previous step, the team benefits from generating several alternative layouts and selecting the best one. Layout decision criteria are closely related to the clustering issues in step 2. In some cases, the team may discover that the clustering derived in step 2 is not geometrically feasible and thus some of the elements would have to be reassigned to other chunks. Creating the rough layout should be coordinated with the industrial designers on the team in cases where the aesthetic and human interface issues of the product are important and strongly related to the geometric arrangement of the chunks.

Step 4: Identify the Fundamental and Incidental Interactions

Most likely a different person or group will be assigned to design each chunk. Because the chunks interact with one another in both planned and unintended ways, these different groups will have to coordinate their activities and exchange information. In order to better manage this coordination process, the team should identify the known interactions between chunks during the system-level design phase.

There are two categories of interactions between chunks. First, *fundamental interactions* are those corresponding to the lines on the schematic that connect the chunks to one another. For example, a sheet of paper flows from the paper tray to the print mechanism. This interaction is planned, and it should be well understood, even from the very earliest schematic, since it is fundamental to the system's operation. Second, *incidental interactions* are those that arise because of the particular physical implementation of functional elements or because of the geometric arrangement of the chunks. For example, vibrations induced by the actuators in the paper tray could interfere with the precise location of the print cartridge in the x-axis.

While the fundamental interactions are explicitly represented by the schematic showing the clustering of elements into chunks, the incidental interactions must be documented in some other way. For a small number of interacting chunks (fewer than about 10), an *interaction graph* is a convenient way to represent the incidental interactions. Exhibit 9-9 shows a possible interaction graph for the DeskJet printer, representing the known incidental interactions. For larger systems this type of graph becomes confusing, and an *interac-*

EXHIBIT 9-9
Incidental interaction graph.

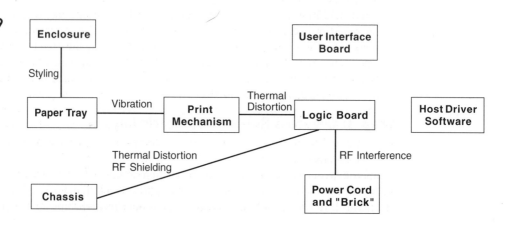

tion matrix is useful instead and can be used to display both fundamental and incidental interactions. See Eppinger (1997) for an example of using such a matrix, which is also used to cluster the functional elements into chunks based on quantification of their interactions.

The interaction graph in Exhibit 9-9 suggests that vibration and thermal distortion are incidental interactions among the chunks that create heat and involve positioning motions. These interactions represent challenges in the development of the system, requiring focused coordination efforts within the team.

We can use the mapping of the interactions between the chunks to provide guidance for structuring and managing the remaining development activities. Chunks with important interactions should be designed by groups with strong communication and coordination between them. Conversely, chunks with little interaction can be designed by groups with less coordination. Eppinger (1997) describes a matrix-based method for prescribing such system-level coordination needs in larger projects.

It is also possible, through careful advance coordination, to develop two interacting chunks in a completely independent fashion. This is facilitated when the interactions between the two chunks can be reduced in advance to a completely specified interface that will be implemented by both chunks. It is relatively straightforward to specify interfaces to handle the fundamental interactions, while it can be difficult to do so for incidental interactions.

Knowledge of the incidental interactions (and sometimes of the fundamental interactions as well) develops as system-level and detail design progress. The schematic and the interaction graph or matrix can be used for documenting this information as it evolves. The network of interactions among subsystems, modules, and components is sometimes called the *system architecture.*

Delayed Differentiation

When a firm offers several variants of a product, the product architecture is a key determinant of the performance of the *supply chain*—the sequence of production and distribution activities that links raw materials and components to finished products in the hands of customers.

Imagine three different versions of the printer, each adapted to a different electrical power standard in three different geographic regions. Consider at what point along the supply chain the product is uniquely defined as one of these three variants. Assume that the supply chain consists of three basic activities: assembly, transportation, and packaging. Exhibit 9-10 illustrates how the number of distinct variants of the product evolves as the product moves through the supply chain. In Scenario A, the three versions of the printer are defined during assembly, then transported, and finally packaged. In Scenario B, the assembly activity is divided into two stages, most of the product is assembled in the first stage, the product is then transported, assembly is completed, and finally the product is packaged. In Scenario B, the components associated with power conversion are assembled after transportation, and so the product is not differentiated until near the end of the supply chain.

Postponing the differentiation of a product until late in the supply chain is called *delayed differentiation* or simply *postponement,* and may offer substantial reductions in the costs of operating the supply chain, primarily through reductions in inventory requirements. For

Scenario A: Early Differentiation

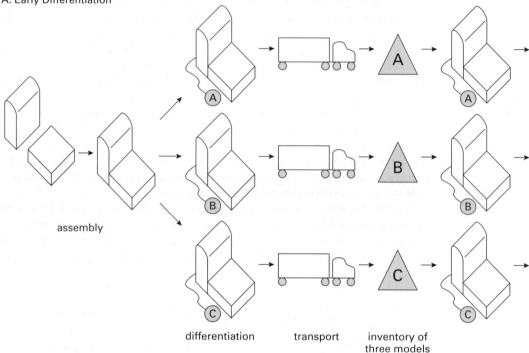

Scenario B: Postponement

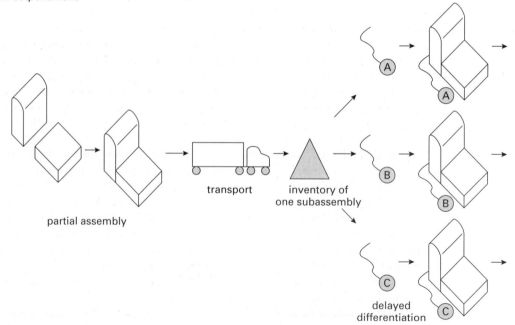

EXHIBIT 9-10 Postponement involves delaying differentiation of the product until late in the supply chain. In Scenario A, three versions of the product are created during assembly and before transportation. In Scenario B, the three versions of the product are not created until after transportation.

most products, and especially for innovative products, demand for each version of a product is unpredictable. That is, there is a component of demand that varies randomly from one time period to the next. To offer consistently high product availability in the presence of such demand uncertainty requires that inventory be held somewhere near the end of the supply chain. (To understand why this is so, imagine a McDonald's restaurant trying to respond to minute-to-minute fluctuations in demand for french fries if it peeled, cut, and fried potatoes only after an order was placed. Instead, it maintains an inventory of cooked french fries which can be quickly scooped into a package and delivered.) For printers, transportation by ship between production and distribution sites may require several weeks. So to be responsive to fluctuations in demand, substantial inventories must be held after transportation. The amount of inventory required for a given target level of availability is a function of the magnitude of the variability in demand.

Postponement enables substantial reductions in the cost of inventories because there is substantially less randomness in the demand for the basic elements of the product (e.g., the platform) than there is for the differentiating components of the variants of the product. This is because in most cases demand for different versions of a product is somewhat uncorrelated, so that when demand for one version is high, it is possible that demand for some other version of the product will be low.

Two design principles are necessary conditions for postponement.

1. The differentiating elements of the product must be concentrated in one or a few chunks. In order to differentiate the product through one or a few simple process steps, the differentiating attributes of the product must be defined by one or a few components of the product. Consider the case of the different electrical power requirements for printers in different geographical regions. If the differences between a product adapted for 120VAC power in the United States and 220VAC power in Europe were associated with several components distributed throughout the product (e.g., power cord, power switch, transformer, rectifier, etc., all in different chunks), there would be no way to delay differentiation of the product without also delaying the assembly of these several chunks. (See Exhibit 9-11, top.) If, however, the only difference between these two models is a single chunk containing a cord and a power supply "brick," then the difference between the two versions of the product requires differences in only one chunk and one assembly operation. (See Exhibit 9-11, bottom.)

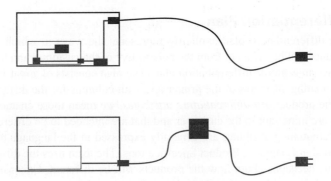

EXHIBIT 9-11   To enable postponement, the differentiating attributes of the product must be concentrated in one or a few chunks. In the top case, the power supply is distributed across the cord, enclosure, chassis, and logic board. In the bottom case, the power supply is confined to the cord and a power supply "brick."

2. The product and production process must be designed so that the differentiating chunk(s) can be added to the product near the end of the supply chain. Even if the differentiating attributes of the product correspond to a single chunk, postponement may not be possible. This is because the constraints of the assembly process or product design may require that this chunk be assembled early in the supply chain. For example, one could envision the consumer packaging of the printer (i.e., the printed carton) being a primary differentiating chunk because of different language requirements for different markets. If transporting the product from the factory to the distribution center required that the printer be assembled into its carton, then it would be impossible to postpone the differentiation of the product with respect to packaging type. To avoid this problem, Hewlett-Packard devised a clever packaging scheme in which molded trays are used to position several dozen bare assembled printers on each of several layers of a large shipping pallet, which can then be wrapped with plastic film and loaded directly into a shipping container. This approach allows differentiation of the carton to occur after the printers have been transported to the distribution center and the appropriate power supply installed.

Platform Planning

Hewlett-Packard provides DeskJet products to customers with different needs. For illustrative purposes, think of these customers as belonging to three market segments: *family, student,* and *small-office/home-office (SOHO)*. To serve these customers, Hewlett-Packard could develop three entirely different products, it could offer only one product to all three segments, or it could differentiate these products through differences in only a subset of the printer components. (See Chapter 3, Product Planning, for discussion of related decisions.)

A desirable property of the product architecture is that it enables a company to offer two or more products that are highly differentiated yet share a substantial fraction of their components. The collection of assets, including component designs, shared by these products is called a product *platform.* Planning the product platform involves managing a basic trade-off between distinctiveness and commonality. On the one hand, there are market benefits to offering several very distinctive versions of a product. On the other hand, there are design and manufacturing benefits to maximizing the extent to which these different products share common components. Two simple information systems allow the team to manage this trade-off: the *differentiation plan* and the *commonality plan.*

Differentiation Plan

The differentiation plan explicitly represents the ways in which multiple versions of a product will be different from the perspective of the customer and the market. Exhibit 9-12 shows an example differentiation plan. The plan consists of a matrix with rows for the differentiating attributes of the printer and with columns for the different versions or models of the product. By *differentiating attributes,* we mean those characteristics of the product that are important to the customer and that are intended to be different across the products. Differentiating attributes are generally expressed in the language of specifications, as described in Chapter 5, Product Specifications. The team uses the differentiation plan to codify its decisions about how the products will be different. Unconstrained, the differentiation plan would exactly match the preferences of the customers in the market segments targeted by each different product. Unfortunately, such plans generally imply products that are prohibitively costly.

Differentiating Attributes	Family	Student	SOHO (Small Office, Home Office)
Black print quality	"Near Laser" quality 300dpi	"Laser" quality 600dpi	"Laser" quality 600dpi
Color print quality	"Near photo" quality	Equivalent to DJ600	Equivalent to DJ600
Print speed	6 pages/minute	8 pages/minute	10 pages/minute
Footprint	360mm deep × 400mm wide	340mm deep × 360mm wide	400mm deep × 450mm wide
Paper storage	100 sheets	100 sheets	150 sheets
Style	"Consumer"	"Youth Consumer"	"Commercial"
Connectivity to computer	USB and Parallel Port	USB	USB
Operating system compatibility	Macintosh and Windows	Macintosh and Windows	Windows

EXHIBIT 9-12   An example differentiation plan for a family of three printers.

Chunks	Number of Types	Family	Student	SOHO (Small Office, Home Office)
Print cartridge	2	"Manet" Cartridge	"Picasso" Cartidge	"Picasso" Cartridge
Print mechanism	2	"Aurora" series	Narrow "Aurora" series	"Aurora" series
Paper tray	2	Front-in Front-out	Front-in Front-out	Tall Front-in Front-out
Logic board	2	"Next gen" board with parallel port	"Next gen" board	"Next gen" board
Enclosure	3	Home style	Youth style	Soft office style
Driver software	5	Version A-PC, Version A-Mac	Version B-PC, Version B-Mac	Version C

EXHIBIT 9-13   An example commonality plan for a family of three printers.

Commonality Plan

The commonality plan explicitly represents the ways in which the different versions of the product are the same physically. Exhibit 9-13 shows a commonality plan for the printer example. The plan consists of a matrix with rows representing the chunks of the product. The third, fourth, and fifth columns correspond to the three different versions of the product. The second column indicates the number of different types of each chunk that are implied by the plan. The team fills each cell in the remaining columns with a label for each different version of a chunk that will be used to make up the product. Unconstrained, most manufacturing engineers would probably choose to use only one version of each chunk in all variants of the product. Unfortunately, this strategy would result in products that are undifferentiated.

Managing the Trade-Off between Differentiation and Commonality

The challenge in platform planning is to resolve the tension between the desire to differentiate the products and the desire for these products to share a substantial fraction of their components. Examination of the differentiation plan and the commonality plan reveals several trade-offs. For example, the student printer has the potential to offer the benefit of a small footprint, which is likely to be important to space-conscious college students. However, this differentiating attribute implies that the student printer would require a different print mechanism chunk, which is likely to add substantially to the investment required to design and produce the printer. This tension between a desire to tailor the benefits of a product to the target market segment and the desire to minimize investment is highlighted when the team attempts to make the differentiation plan and the commonality plan consistent. We offer several guidelines for managing this tension.

- *Platform planning decisions should be informed by quantitative estimates of cost and revenue implications:* Estimating the profit contribution from a one-percentage-point increase in market share is a useful benchmark against which to measure the potential increase in manufacturing and supply-chain costs of additional versions of a chunk. In estimating supply chain costs, the team must consider the extent to which the differentiation implied by the differentiation plan can be postponed or whether it must be created early in the supply chain.
- *Iteration is beneficial:* In our experience, teams make better decisions when they make several iterations based on approximate information than when they agonize over the details during relatively fewer iterations.
- *The product architecture dictates the nature of the trade-off between differentiation and commonality:* The nature of the trade-off between differentiation and commonality is not fixed. Generally, modular architectures enable a higher proportion of components to be shared than integral architectures. This implies that when confronted with a seemingly intractable conflict between differentiation and commonality, the team should consider alternative architectural approaches, which may provide opportunities to enhance both differentiation and commonality.

For the printer example, the tension between differentiation and commonality might be resolved by a compromise. The revenue benefits of a slightly narrower student printer are not likely to exceed the costs associated with creating an entirely different, and narrower, print mechanism. The costs of different print mechanisms are likely to be especially high given that the print mechanism involves substantial tooling investments. Also, because the print mechanism is created early in the supply chain, postponement of differentiation would be substantially less feasible if it required different print mechanisms. For these reasons, the team would most likely choose to use a single, common print mechanism and forgo the possible revenue benefits of a narrower footprint for the student printer.

Related System-Level Design Issues

The four-step method for establishing the product architecture guides the early system-level design activities, but many more detailed activities remain. Here we discuss some of the issues that frequently arise during subsequent system-level design activities and their implications for the product architecture.

Defining Secondary Systems

The schematic in Exhibit 9-6 shows only the key elements of the product. There are many other functional and physical elements not shown, some of which will only be conceived and detailed as the system-level design evolves. These additional elements make up the secondary systems of the product. Examples include safety systems, power systems, status monitors, and structural supports. Some of these systems, such as safety systems, will span several chunks. Fortunately, secondary systems usually involve flexible connections such as wiring and tubing and can be considered after the major architectural decisions have been made. Secondary systems cutting across the boundaries of chunks present a special management challenge: Should a single group or individual be assigned to design a secondary system even though the system will be made up of components residing in several different chunks? Or should the group or individuals responsible for the chunks be responsible for coordinating among themselves to ensure that the secondary systems will work as needed? The former approach is more typical, where specific individuals or subteams are assigned to focus on the secondary systems.

Establishing the Architecture of the Chunks

Some of the chunks of a complex product may be very complex systems in their own right. For example, many of the chunks in the DeskJet printer involve dozens of parts. Each of these chunks may have its own architecture—the scheme by which it is divided into smaller chunks. This problem is essentially identical to the architectural challenge posed at the level of the entire product. Careful consideration of the architecture of the chunks is nearly as important as the creation of the architecture of the overall product. For example, the print cartridge consists of the subfunctions *store ink* and *deliver ink* for each of four colors of ink. Several architectural approaches are possible for this chunk, including, for example, the use of independently replaceable reservoirs for each ink color.

Creating Detailed Interface Specifications

As the system-level design progresses, the fundamental interactions indicated by lines on the schematic in Exhibit 9-6 are specified as much more detailed collections of signals, material flows, and exchanges of energy. As this refinement occurs, the specification of the interfaces between chunks should also be clarified. For example, Exhibit 9-14 shows an overview of a possible specification of an interface between a black print cartridge and a logic board for a printer. Such interfaces represent the "contracts" between chunks and are often detailed in formal specification documents.

EXHIBIT 9-14
Specification
of interface
between black
print cartridge
and logic board.

Line	Name	Properties
1	PWR-A	+12VDC, 5mA
2	PWR-B	+5VDC, 10mA
3	STAT	TTL
4	LVL	100KΩ-1MΩ
5	PRNT1	TTL
6	PRNT2	TTL
7	PRNT3	TTL
8	PRNT4	TTL
9	PRNT5	TTL
10	PRNT6	TTL
11	GND	

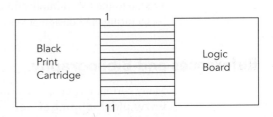

Summary

Product architecture is the scheme by which the functional elements of the product are arranged into physical chunks. The architecture of the product is established during the concept development and system-level design phases of development.

- Product architecture decisions have far-reaching implications, affecting such things as product change, product variety, component standardization, product performance, manufacturability, and product development management.
- A key characteristic of a product architecture is the degree to which it is modular or integral.
- Modular architectures are those in which each physical chunk implements a specific set of functional elements and has well-defined interactions with the other chunks.
- There are three types of modular architectures: slot-modular, bus-modular, and sectional-modular.
- Integral architectures are those in which the implementation of functional elements is spread across chunks, resulting in ill-defined interactions between the chunks.
- We recommend a four-step method for establishing the product architecture:

  1. Create a schematic of the product.
  2. Cluster the elements of the schematic.
  3. Create a rough geometric layout.
  4. Identify the fundamental and incidental interactions.

- This method leads the team through the preliminary architectural decisions. Subsequent system-level and detail design activities will contribute to a continuing evolution of the architectural details.
- The product architecture can enable postponement, the delayed differentiation of the product, which offers substantial potential cost savings.
- Architectural choices are closely linked to platform planning, the balancing of differentiation and commonality when addressing different market segments with different versions of a product.
- Due to the broad implications of architectural decisions, inputs from marketing, manufacturing, and design are essential in this aspect of product development.

References and Bibliography

Many current resources are available on the Internet via
www.ulrich-eppinger.net

The basic concepts of product architecture and its implications are developed and discussed in this article.

Ulrich, Karl, "The Role of Product Architecture in the Manufacturing Firm," *Research Policy,* Vol. 24, 1995, pp. 419–440.

Many of the issues involved in establishing a product architecture are treated from a slightly different perspective in the systems engineering literature. Hall provides an overview along with many relevant references. Maier and Rechtin discuss the architecture of complex systems.

Hall, Arthur D., III, *Metasystems Methodology: A New Synthesis and Unification,* Pergamon Press, Elmsford, NY, 1989.

Maier, Mark W. and Eberhardt Rechtin, *The Art of Systems Architecting,* second edition, CRC Press, Boca Raton, FL, 2000.

The linkage between product variety and product architecture is discussed by Pine in the context of *mass customization,* or very high variety manufacturing.

Pine, B. Joseph, II, *Mass Customization: The New Frontier in Business Competition,* Harvard Business School Press, Boston, 1992.

Clark and Fujimoto discuss the practice of "black box" supplier interactions in their book on product development in the automobile industry. In this situation, the manufacturer specifies only the function and interface of a chunk or component and the supplier handles the detailed implementation issues.

Clark, Kim B., and Takahiro Fujimoto, *Product Development Performance: Strategy, Organization, and Management in the World Auto Industry,* Harvard Business School Press, Boston, 1991.

Alexander and Simon are among the earliest authors to discuss the partitioning of a system into minimally interacting chunks.

Alexander, Christopher, *Notes on the Synthesis of Form,* Harvard University Press, Cambridge, MA, 1964.

Simon, Herbert, "The Architecture of Complexity," in *The Sciences of the Artificial,* third edition, MIT Press, Cambridge, MA, 1996. (Based on an article that appeared originally in 1965.)

Eppinger has developed matrix-based methods to help analyze system architectures based on documentation of the interactions between chunks and the teams that implement the chunks.

Eppinger, Steven D., "A Planning Method for Integration of Large-Scale Engineering Systems," International Conference on Engineering Design, ICED 97, Tampere, Finland, August 1997, pp. 199–204.

Further detail on delayed differentiation and supply chain performance may be found in the work of Lee and colleagues.

Lee, Hau L., "Effective Inventory and Service Management through Product and Process Re-Design," *Operations Research,* Vol. 44, No. 1, 1996, pp. 151–159.

Lee, Hau L., and C. Tang, "Modelling the Costs and Benefits of Delayed Product Differentiation," *Management Science,* Vol. 43, No. 1, January 1997, pp. 40–53.

Lee, Hau L., Cory Billington, and Brent Carter, "Hewlett-Packard Gains Control of Inventory and Service through Design for Localization," *Interfaces,* August 1993, pp. 1–11.

The platform planning method presented in this chapter is derived in part from Robertson and Ulrich's more comprehensive discussion.

Robertson, David, and Karl Ulrich, "Planning for Product Platforms," *Sloan Management Review,* Vol. 39, No. 4, Summer 1998, pp. 19–31.

Exercises

1. Draw a schematic for a wristwatch, using only functional elements (without assuming any particular physical working principles or components).
2. Describe the architecture of a Swiss army knife. What advantages and disadvantages does this architecture provide?
3. Take apart a small electromechanical product (which you are willing to sacrifice if necessary). Draw a schematic including the essential functional elements. Identify two or three possible clusterings of these elements into chunks. Is there any evidence to suggest which architecture was chosen by the development team?

Thought Questions

1. Do service products, such as bank accounts or insurance policies, have architectures?
2. Can a firm achieve high product variety without a modular product architecture? How (or why not)?
3. The argument for the motorcycle architecture shown in Exhibit 9-5 is that it allows for a lighter motorcycle than the more modular alternative. What are the other advantages and disadvantages? Which approach is likely to cost less to manufacture?
4. There are thousands of architectural decisions to be made in the development of an automobile. Consider all of the likely fundamental and incidental interactions that any one functional element (say, safety restraints) would have with the others. How would you use the documentation of such interactions to guide the decision about what chunk to place this functional element in?
5. The schematic shown in Exhibit 9-6 includes 15 elements. Consider the possibility of assigning each element to its own chunk. What are the strengths and weaknesses of such an architecture?

CHAPTER TEN

Industrial Design

Courtesy of Motorola, Inc.

EXHIBIT 10-1
Motorola portable cellular telephones. These photos show the original MicroTAC model from 1989 (left) and the newer StarTAC phone introduced in 1996 (right).

This chapter was developed in collaboration with Paul Brody.

In the mid-1990s, Motorola launched a product development effort to augment its very successful MicroTAC line of cellular telephones with an exciting new product. The MicroTAC Flip Phone brand platform had seen five generations of the product released since 1989, with each one thinner and lighter than the last. Each MicroTAC model used the same configuration of battery packs, adapters, and accessories, allowing users to easily upgrade to the newer phones.

The new StarTAC design emerged from a product vision to be "more wearable" than previous cellular telephones. This required a new architecture, abandoning the existing MicroTAC platform of batteries and accessories which had set certain width and height constraints. Upon its introduction in 1996, customers judged the StarTAC design, shown in Exhibit 10-1, to be just as radical as its predecessor.

Sales to early adopters came quickly after a successful market introduction in which Hollywood celebrities were shown with the product. Within three years, StarTAC sales had reached millions of units. This success can be attributed to several factors:

- *Small size and weight:* With its slim lithium ion battery installed, the StarTAC has a volume of 82 cubic centimeters and a weight of 88 grams, making it the smallest and lightest cellular phone on the market at the time. The StarTAC can be folded in the palm of the hand, like a small wallet, and placed into a pocket or purse. It can be worn like a pager or even as a necklace.

- *Performance features:* The StarTAC incorporates all the features of its predecessor, yet in a much smaller package. Important features include a continuous talk time of 60 minutes with the slim battery (110 minutes with an extra-capacity battery), an alphanumeric memory to store numbers and names, a stack to recall any of the last 10 numbers dialed, caller identification, voice messaging, silent vibration to alert the user to incoming calls, and a full range of accessories.

- *Superior ergonomics:* The StarTAC's sleek, ergonomic design complements the human face. The shape of the handset, particularly the angled position of the earpiece with respect to the mouthpiece, conforms to the user for superior comfort. The spacing and position of the buttons on the keypad are based on accepted standards allowing for fast and accurate dialing. The folding design allows the user to answer or end calls by opening or closing the keypad cover.

- *Durability:* As with all Motorola products, the StarTAC was designed to meet rigorous specifications. It can be dropped from a height of four feet onto a cement floor or sat upon in the open position without sustaining any visible or operational damage. The StarTAC can also withstand temperature extremes, humidity, shock, dust, and vibration.

- *Ease of manufacture:* The StarTAC benefits from many earlier generations of Motorola's cellular telephone technology. The single circuit board consists entirely of electronic components which can be assembled using automated equipment. StarTAC production cells can be replicated at Motorola factories around the world to meet global capacity demands.

- *Appearance:* The sleek appearance and black color give the StarTAC a futuristic look associated with innovation. Because of its aesthetic appeal the StarTAC became somewhat of a status symbol that evokes strong feelings of pride among owners.

The StarTAC development team included electrical, mechanical, software, and manufacturing engineers, whose contributions were instrumental in developing the technologies and manufacturing processes that allowed the product to achieve its performance, size, and weight. However, without the contributions of industrial designers, who helped to define the size, shape, and human factors, the StarTAC would never have taken its innovative form. In fact, the Motorola team could easily have developed "just another phone," smaller and lighter than the last MicroTAC model. Instead, a revolutionary concept generated by the industrial designers on the team turned the project into a dramatic success.

Industrial designers are primarily responsible for the aspects of a product that relate to the user—the product's aesthetic appeal (how it looks, sounds, feels, smells) and its functional interfaces (how it is used). For many U.S. manufacturers, industrial design has historically been an afterthought. Managers used industrial designers to style, or "gift wrap," a product after its technical features were determined. Companies would then market the product on the merits of its technology alone, although customers certainly evaluate a product using more holistic judgments, including ergonomics and style.

Today, a product's core technology is generally not enough to ensure commercial success. The globalization of markets has resulted in the design and manufacture of a wide array of consumer products. Fierce competition makes it unlikely that a company will enjoy a sustainable competitive advantage through technology alone. Accordingly, companies such as Motorola are increasingly using industrial design as an important tool for both satisfying customer needs and differentiating their products from those of their competition.

This chapter introduces engineers and managers to industrial design (ID) and explains how the ID process takes place in relation to other product development activities. We refer to the StarTAC example throughout this chapter to explain critical ideas. Specifically, this chapter presents:

- A historical perspective on ID and a working definition of ID.
- Statistics on typical investments in ID.
- A method for determining the importance of ID to a particular product.
- The costs and benefits of investing in ID.
- How ID helps to establish a corporation's identity.
- Specific steps industrial designers follow while designing a product.
- A description of how the ID process changes according to product type.
- A method for assessing the quality of the ID effort for a completed product.

What Is Industrial Design?

The birth of ID is often traced to western Europe in the early 1900s. (See Lorenz, 1986, for an account of the history of ID, which is summarized here.) Several German companies, including AEG, a large electrical manufacturer, commissioned a multitude of craftsmen and architects to design various products for manufacture. Initially, these early European designers had little direct impact on industry; however, their work resulted in lasting theories that have influenced and shaped what is today known as industrial design. Early European theories on ID, such as the Bauhaus movement, went beyond mere functionalism; they emphasized the importance of geometry, precision, simplicity, and economy in the design of products. In short, early European designers believed that a product should be designed "from the inside out." Form should follow function.

In the United States, however, early concepts of ID were distinctly different. While early European industrial designers were architects and engineers, most industrial designers in America were actually theater designers and artist-illustrators. Not surprisingly, ID in the United States was often at the service of sales and advertising, where a product's exterior was all important and its insides mattered little. Pioneers in U.S. industrial design, including Walter Dorwin Teague, Norman Bel Geddes, and Raymond Loewy, emphasized streamlining in product design. This trend is best evidenced in U.S. products of the 1930s. From fountain pens to baby buggies, products were designed with nonfunctional aerodynamic shapes in an attempt to create product appeal. The auto industry provides another example. The shapes of European automobiles of the thirties were fairly simple and smooth, while U.S. cars of the same era were decorated with such nonfunctional features as tailfins and chrome teeth.

By the 1970s, however, European design had strongly influenced American ID thinking, largely through the works of Henry Dreyfuss and Eliot Noyes. Heightened competition in the marketplace forced companies to search for ways to improve and differentiate their products. Increasingly, companies accepted the notion that the role of ID needed to go beyond mere shape and appearance. Success stories such as Bell, Deere, Ford, and IBM, all of which effectively integrated ID into their product development process, helped further this thinking. Today, industrial design is practiced in the United States by professionals in many diverse settings ranging from small design consulting firms to in-house design offices within large manufacturing companies.

The Industrial Designers Society of America (IDSA) defines industrial design as "the professional service of creating and developing concepts and specifications that optimize the function, value, and appearance of products and systems for the mutual benefit of both user and manufacturer." This definition is broad enough to include the activities of the entire product development team. In fact, industrial designers focus their attention upon the form and user interaction of products. Dreyfuss (1967) lists five critical goals that industrial designers can help a team to achieve when developing new products:

- *Utility:* The product's human interfaces should be safe, easy to use, and intuitive. Each feature should be shaped so that it communicates its function to the user.
- *Appearance:* Form, line, proportion, and color are used to integrate the product into a pleasing whole.
- *Ease of maintenance:* Products must also be designed to communicate how they are to be maintained and repaired.
- *Low costs:* Form and features have a large impact on tooling and production costs, so these must be considered jointly by the team.
- *Communication:* Product designs should communicate the corporate design philosophy and mission through the visual qualities of the products.

Industrial designers are typically educated in four-year university programs where they study sculpture and form; develop drawing, presentation, and model-making skills; and gain a basic understanding of materials, manufacturing techniques, and finishes. In industrial practice, designers receive additional exposure to basic engineering, advanced manufacturing/fabrication processes, and common marketing practices. Their ability to express ideas visually can facilitate the process of concept development for the team. Industrial designers may create most of the concept sketches, models, and renderings used by the team throughout the development process, even though the ideas come from the entire team.

Assessing the Need for Industrial Design

To assess the importance of ID to a particular product, we first review some investment statistics and then define the dimensions of a product which are dependent upon good ID.

Expenditures for Industrial Design

Exhibit 10-2 shows approximate values of investment in ID for a variety of products. Both the total expenditures on ID and the percentage of the product development budget invested in ID are shown for consumer and industrial products spanning various industries. These statistics should give design teams a rough idea of how much ID investment will be required for a new product.

The exhibit shows that the range of expenditures on ID is tremendous. For products with relatively little user interaction such as some types of industrial equipment, the cost of ID is only in the tens of thousands of dollars. On the other hand, the development of an intensely visual and interactive product such as an automobile requires millions of dollars of ID effort. The relative cost of ID as a fraction of the overall development budget also shows a wide range. For a technically sophisticated product, such as a new aircraft, the ID cost can be insignificant relative to the engineering and other development expenditures. This does not suggest, however, that ID is unimportant for such products; it suggests only that the other development functions are more costly. Certainly the success of a new automobile design is highly dependent on its aesthetic appeal and the quality of the user interfaces, two dimensions largely determined by ID; yet the ID expense of $10 million is modest, relative to the entire development budget.

How Important Is Industrial Design to a Product?

Most products on the market can be improved in some way or another by good ID, and all products that are used, operated, or seen by people depend critically on ID for commercial success.

With this in mind, a convenient means for assessing the importance of ID to a particular product is to characterize importance along two dimensions: ergonomics and aesthetics. (Note that we use the term *ergonomics* to encompass all aspects of a product that relate to its human interfaces.) The more important each dimension is to the product's success, the more dependent the product is on ID. Therefore, by answering a series of questions along each dimension we can qualitatively assess the importance of ID.

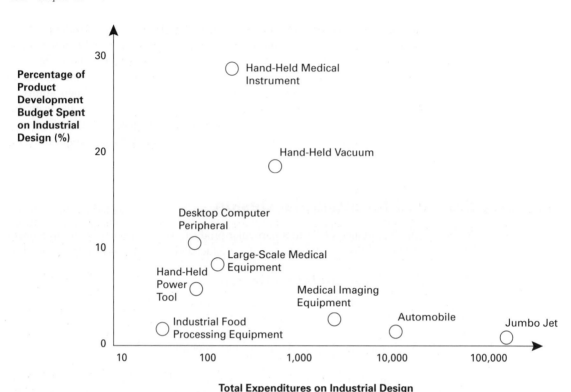

EXHIBIT 10-2 Industrial design expenditures for some consumer and industrial products.

Ergonomic Needs

- *How important is ease of use?* Ease of use may be extremely important both for frequently used products, such as an office photocopier, and for infrequently used products, such as a fire extinguisher. Ease of use is more challenging if the product has multiple features and/or modes of operation which may confuse or frustrate the user. When ease of use is an important criterion, industrial designers will need to ensure that the features of the product effectively communicate their function.

- *How important is ease of maintenance?* If the product needs to be serviced or repaired frequently, then ease of maintenance is crucial. For example, a user should be able to clear a paper jam in a printer or photocopier easily. Again, it is critical that the features of the product communicate maintenance/repair procedures to the user. However, in many cases, a more desirable solution is to eliminate the need for maintenance entirely.

- *How many user interactions are required for the product's functions?* In general, the more interactions users have with the product, the more the product will depend on ID. For example, a doorknob typically requires only one interaction, whereas a portable computer may require a dozen or more, all of which the industrial designer must understand in depth. Furthermore, each interaction may require a different design approach and/or additional research.

- *How novel are the user interaction needs?* A user interface requiring incremental improvements to an existing design will be relatively straightforward to design, such as the buttons on a next-generation desktop computer mouse. A more novel user interface may require substantial research and feasibility studies, such as the built-in trackball in the first Macintosh PowerBook notebook computer.
- *What are the safety issues?* All products have safety considerations. For some products, these can present significant challenges to the design team. For example, the safety concerns in the design of a child's toy are much more prominent than those for a new computer mouse.

Aesthetic Needs

- *Is visual product differentiation required?* Products with stable markets and technology are highly dependent upon ID to create aesthetic appeal and, hence, visual differentiation. In contrast, a product such as a computer's internal disk drive, which is differentiated by its technological performance, is less dependent on ID.
- *How important are pride of ownership, image, and fashion?* A customer's perception of a product is in part based upon its aesthetic appeal. An attractive product may be associated with high fashion and image and will likely create a strong sense of pride among its owners. This may similarly be true for a product that looks and feels rugged or conservative. When such characteristics are important, ID will play a critical role in determining the product's ultimate success.
- *Will an aesthetic product motivate the team?* A product that is aesthetically appealing can generate a sense of team pride among the design and manufacturing staff. Team pride helps to motivate and unify everyone associated with the project. An early ID concept gives the team a concrete vision of the end result of the development effort.

To demonstrate this method, we can use the above questions to assess the importance of industrial design in the development of the Motorola StarTAC. Exhibit 10-3 displays the results of such analysis. We find that both ergonomics and aesthetics were extremely important for the StarTAC. Accordingly, ID did indeed play a large role in determining many of the product's critical success factors.

The Impact of Industrial Design

The previous section focused primarily upon the importance of ID in satisfying customer needs. Next we explore both the direct economic impact of investing in ID as well as the impact ID has upon corporate identity.

Is Industrial Design Worth the Investment?

Managers will often want to know, for a specific product or for a business operation in general, how much effort should be invested in industrial design. While it is difficult to answer this question precisely, we can offer several insights by considering the costs and benefits.

The costs of ID include direct cost, manufacturing cost, and time cost, described below:

- *Direct cost* is the cost of the ID services. This quantity is determined by the number of designers used, duration of the project, and number of models required, plus material costs and other related expenses. In 2004, consulting services cost $75 to $175 per

Needs	Level of Importance			Explanation of Rating
	Low	Medium	High	
Ergonomics				
Ease of use			⊶○⊣	Critical for a portable telephone since it may be used frequently, may be needed in emergency situations, and can be operated by motorists while driving. The product's function must be communicated through its design.
Ease of maintenance	⊢○⊸			As with many integrated electronics products there is very little maintenance required.
Quantity of user interactions			○⊸	There are many important user interactions such as: changing the battery, dialing, programming the features, sending and receiving calls.
Novelty of user interactions			○	Design solutions associated with some of the customer interactions are straightforward, such as the numeric keypad, since there is a wealth of human factors data which dictate the basic dimensions. However, other interfaces, such as the hinged cover which folds and opens the phone, were quite different from earlier models and therefore required careful study.
Safety	⊢○⊸			There were few safety issues for ID to consider on the StarTAC itself. However, since many customers use cellular telephones in automobiles, a line of accessories needed to be designed for safe, convenient, hands-free operation.
Aesthetics				
Product differentiation			⊶○⊣	There were hundreds of models of cellular phones on the market when the StarTAC was introduced. Its appearance (including its size and shape) was essential for differentiation.
Pride of ownership, fashion, or image			⊶○⊣	The StarTAC was intended to be a highly visible product used by people for business and personal communication in public areas. It had to be physically attractive for everyday use.
Team motivation			○	The StarTAC's novel form turned out to be an important inspiration to the development team and selling point for senior management.

EXHIBIT 10-3 Assessing the importance of industrial design for the StarTAC.

hour plus costs for models, photos, and the like. The true cost of internal corporate design services is generally about the same.

- *Manufacturing cost* is the expense incurred to implement the product details created through ID. Surface finishes, stylized shapes, rich colors, and many other design details can increase tooling cost and/or production cost. Note, however, that many ID details can be implemented at practically no cost, particularly if ID is involved early enough in the process (see below). In fact, some ID inputs can actually reduce manufacturing costs—particularly when the industrial designer works closely with the manufacturing engineers.

- *Time cost* is the penalty associated with extended lead time. As industrial designers attempt to refine the ergonomics and aesthetics of a product, multiple design iterations and/or prototypes will be necessary. This may result in a delay in the product's introduction, which will likely have an economic cost.

The benefits of using ID include increased product appeal and greater customer satisfaction through additional or better features, strong brand identity, and product differentiation. These benefits usually translate into a price premium and/or increased market share (as compared to marketing the product without the ID efforts).

These costs and benefits of ID were estimated as part of a study conducted at MIT that assessed the impact of detail design decisions on product success factors for a set of competing products in the market (automatic drip coffee makers). Although the relation is difficult to quantify precisely, this study found a significant correlation between product aesthetics (as rated by practicing industrial designers) and the retail price for each product, but no correlation between aesthetics and manufacturing cost. The researchers could not conclude whether the manufacturers had priced their products optimally, nor could they determine unequivocally if aesthetics of the products enabled manufacturers to garner higher prices. However, the study suggests that an increase in price of $1 per unit for typical sales volumes would be worth several million dollars in profits over the life of these products. Industrial designers asked to price design services for such products gave a range from $75,000 to $250,000, suggesting that if ID could add even one dollar's worth of perceived benefit to the consumer, it would pay back handsomely (Pearson, 1992).

A second study, conducted at the Open University in England, also suggests that investing in ID yields a positive return. This study tracked the commercial impact of investing in engineering and ID for 221 design projects at small and medium-sized manufacturing firms. The study claims that investing in industrial design consultants led to profits in over 90 percent of all implemented projects, and when comparisons were possible with previous, less ID-oriented products, sales increased by an average of 41 percent (Roy and Potter, 1993).

For a specific decision, performing simple calculations and sensitivity analyses can help quantify the likely economic returns from ID. For example, if investing in ID will likely result in a price premium of $10 per unit, what will be the net economic benefit when summed over the original market sales projections? Similarly, if investing in ID will likely result in a greater demand for the product—by, say, 1,000 units per year—what will be the net economic benefit when summed at the original unit price? The rough estimates of these benefits can be compared to the expected cost of the ID effort. Spreadsheet models are commonly used for this kind of financial decision making and can easily be applied to estimate the expected payback of ID for a project. (Chapter 15, Product Development Economics, describes a method for developing such a financial model.)

How Does Industrial Design Establish a Corporate Identity?

Corporate identity is derived from "the visual style of an organization," a factor that affects the firm's positioning in the market (Olins, 1989). A company's identity emerges primarily through what people see. Advertising, logos, signage, uniforms, buildings, packaging, and product designs all contribute to creating corporate identity.

In product-based companies, ID plays an important role in determining the company's identity. Industrial design determines a product's style, which is directly related to the public perception of the firm. When a company's products maintain a consistent and recognizable appearance, *visual equity* is established. A consistent look and feel may be associated with the product's color, form, style, or even its features. When a firm enjoys a positive reputation, such visual equity is valuable, as it can create a positive association with quality for future products. Some companies that have effectively used ID to establish visual equity and corporate identity through their product lines include:

- *Apple Computer, Inc.:* The original Macintosh had a small, upright shape and a benign buff coloring. This design purposely gave the product a nonthreatening, user-friendly look that has since been associated with all of Apple's products.
- *Rolex Watch Co.:* The Rolex line of watches maintains a classic look and solid feel that signifies quality and prestige.
- *Braun AG:* Braun kitchen appliances and shavers have clean lines and basic colors. The Braun name has long been associated with simplicity and quality.
- *Bang & Olufsen:* B&O high-fidelity consumer electronics systems are designed to have sleek lines and impressive visual displays, providing an image of technological innovation.
- *Motorola, Inc.:* The original MicroTAC Flip Phone design is instantly recognizable as Motorola's innovation. The newer StarTAC model also uses a folding concept in a much smaller package, emphasizing Motorola's leadership in a rapidly evolving industry.

The Industrial Design Process

Many large companies have internal industrial design departments. Small companies tend to use contract ID services provided by consulting firms. In either case, industrial designers should participate fully on cross-functional product development teams. Within these teams, engineers will generally follow a process to generate and evaluate concepts for the technical features of a product. In a similar manner, most industrial designers follow a process for designing the aesthetics and ergonomics of a product. Although this approach may vary depending on the firm and the nature of the project, industrial designers also generate multiple concepts and then work with engineers to narrow these options down through a series of evaluation steps.

Specifically, the ID process can be thought of as consisting of the following phases:

1. Investigation of customer needs.
2. Conceptualization.

3. Preliminary refinement.
4. Further refinement and final concept selection.
5. Control drawings.
6. Coordination with engineering, manufacturing, and vendors.

This section discusses each of these phases in order, and the following section will discuss the timing of these phases within the overall product development process.

Investigation of Customer Needs

The product development team begins by documenting customer needs as described in Chapter 4, Identifying Customer Needs. Since industrial designers are skilled at recognizing issues involving user interactions, ID involvement is crucial in the needs process. For example, in researching customer needs for a new medical instrument, the team would study an operating room, interview physicians, and conduct focus groups. While involvement of marketing, engineering, and ID certainly leads to a common, comprehensive understanding of customer needs for the whole team, it particularly allows the industrial designer to gain an intimate understanding of the interactions between the user and the product.

Unlike many development efforts, the StarTAC project did not rely heavily upon focus groups or formal market research. Motorola believed that the high level of secrecy surrounding the project, and the difficulty in gaining customer input for next-generation products, made these techniques impractical. Instead, the team used extensive input from Motorola employees to understand the evolution of user needs. Marketing personnel stressed the importance of Motorola's leadership in size, weight, and style. Engineering supplied information on technical limitations involving geometry and materials.

Conceptualization

Once the customer needs and constraints are understood, the industrial designers help the team to conceptualize the product. During the concept generation stage engineers naturally focus their attention upon finding solutions to the technical subfunctions of the product. (See Chapter 6, Concept Generation.) At this time, the industrial designers concentrate upon creating the product's form and user interfaces. Industrial designers make simple sketches, known as *thumbnail sketches,* of each concept. These sketches are a fast and inexpensive medium for expressing ideas and evaluating possibilities. Exhibit 10-4 shows one such sketch from the StarTAC project.

The proposed concepts may then be matched and combined with the technical solutions under exploration. Concepts are grouped and evaluated by the team according to the customer needs, technical feasibility, cost, and manufacturing considerations. (See Chapter 7, Concept Selection.)

It is unfortunate that in some companies, industrial designers work quite independently from engineering. When this happens, ID is likely to propose concepts involving strictly form and style, and there are usually numerous iterations when engineering finds the concepts technically infeasible. Firms have therefore found it beneficial to tightly coordinate the efforts of industrial designers and engineers throughout the concept development phase so that these iterations can be accomplished more quickly—even in sketch form.

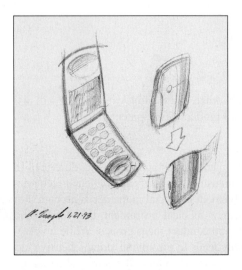

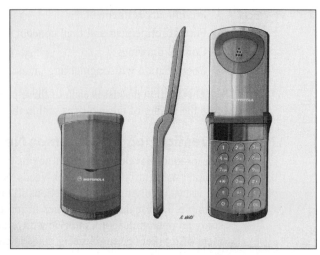

Courtesy of Motorola, Inc.

EXHIBIT 10-4 A quick thumbnail concept sketch (left) and more detailed rendering (right) showing some of the early concepts in the StarTAC development project.

Preliminary Refinement

In the preliminary refinement phase, industrial designers build models of the most promising concepts. *Soft models* are typically made in full scale using foam or foam-core board. They are the second fastest method—only slightly slower than sketches—used to evaluate concepts.

Although generally quite rough, these models are invaluable because they allow the development team to express and visualize product concepts in three dimensions. Concepts are evaluated by industrial designers, engineers, marketing personnel, and (at times) potential customers through the process of touching, feeling, and modifying the models. Typically, designers will build as many models as possible depending on time and financial constraints. Concepts that are particularly difficult to visualize require more models than do simpler ones.

The StarTAC industrial designers used numerous soft models to assess the overall size, proportion, and shape of many proposed concepts. Of particular concern was the feel of the product in the hand and against the face. These attributes can only be assessed using physical models. A soft model from the StarTAC project is shown in Exhibit 10-5.

Further Refinement and Final Concept Selection

At this stage, industrial designers often switch from soft models and sketches to hard models and information-intensive drawings known as *renderings.* Renderings show the details of the design and often depict the product in use. Drawn in two or three dimensions, they convey a great deal of information about the product. Renderings are often used for color studies and for testing customers' reception to the proposed product's features and functionality. A rendering from the StarTAC project is shown in Exhibit 10-4.

The final refinement step before selecting a concept is to create *hard models.* These models are still technically nonfunctional yet are close replicas of the final design with a

EXHIBIT 10-5
A soft model (left) and a hard model (right) used by the StarTAC industrial designers to study alternative forms.

Courtesy of Motorola, Inc.

very realistic look and feel. They are made from wood, dense foam, plastic, or metal; are painted and textured; and have some "working" features such as buttons that push or sliders that move. Because a hard model can cost thousands of dollars, a product development team usually has the budget to make only a few. Exhibit 10-5 shows one of the several hard models built during the StarTAC development process.

Hard models can be used to gain additional customer feedback in focus groups, to advertise and promote the product at trade shows, to sell the concept to senior management within an organization, and to further refine the final concept.

Control Drawings

Industrial designers complete their development process by making *control drawings* of the final concept. Control drawings document functionality, features, sizes, colors, surface finishes, and key dimensions. Although they are not detailed part drawings (known as engineering drawings), they can be used to fabricate final design models and other prototypes. Typically, these drawings are given to the detailed part designers for completion. Exhibit 10-6 shows a control drawing of the StarTAC design.

Coordination with Engineering, Manufacturing, and External Vendors

The industrial designers must continue to work closely with engineering and manufacturing personnel throughout the subsequent product development process. Some industrial design consulting firms offer quite comprehensive product development services, including detailed engineering design and the selection and management of outside vendors of materials, tooling, components, and assembly services.

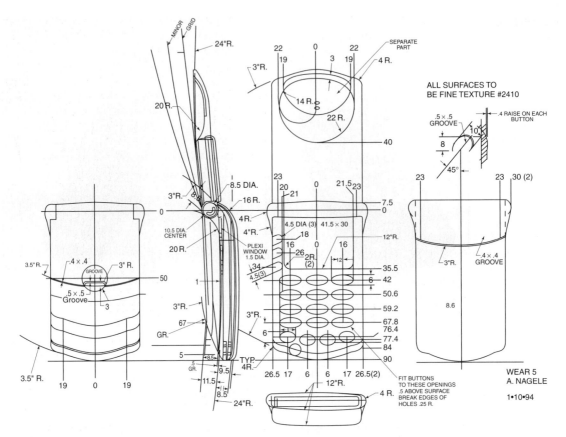

EXHIBIT 10-6   StarTAC control drawing showing the final StarTAC shape and dimensions.

Courtesy of Motorola, Inc.

The Impact of Computer-Based Tools on the ID Process

Since the 1990s, computer-aided industrial design (CAID) tools have had a significant impact on industrial designers and their work. Using CAID tools, industrial designers can generate, display, and rapidly modify three-dimensional designs on a high-resolution monitor. In this manner, ID can potentially generate a greater number of detailed concepts more quickly, which may lead to more innovative design solutions. The visual realism of CAID can enhance communication within the product development team and eliminate much of the inaccuracy of the manually generated sketches historically provided by industrial designers (Cardaci, 1992). Furthermore, CAID systems may be used to generate control drawings, and data can be transferred into engineering design (3D solid modeling) systems, allowing the entire development process to be more easily integrated. Exhibit 10-7 shows a CAID rendering of the StarTAC.

EXHIBIT 10-7
Computer-
generated
concept image.
This design was
created using
CAID software
by Alias
Research Inc.

Courtesy of Motorola, Inc.

Management of the Industrial Design Process

Industrial design may be involved in the overall product development process during several different phases. The timing of the ID effort depends upon the nature of the product being designed. To explain the timing of the ID effort it is convenient to classify products as technology-driven products and user-driven products.

- *Technology-driven products:* The primary characteristic of a technology-driven product is that its core benefit is based on its technology, or its ability to accomplish a specific technical task. While such a product may still have important aesthetic or ergonomic requirements, consumers will most likely purchase the product primarily for its technical performance. For example, a hard disk drive for a computer is largely technology driven. It follows that for the development team of a technology-driven product, the engineering or technical requirements will be paramount and will dominate development efforts. Accordingly, the role of ID is often limited to packaging the core technology. This entails determining the product's external appearance and ensuring that the product communicates its technological capabilities and modes of interaction to the user.

- *User-driven products:* The core benefit of a user-driven product is derived from the functionality of its interface and/or its aesthetic appeal. Typically there is a high degree of user interaction for these products. Accordingly, the user interfaces must be safe, easy to use, and easy to maintain. The product's external appearance is often important

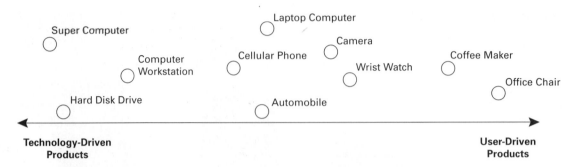

EXHIBIT 10-8   Classification of some common products on the continuum from technology-driven product to user-driven product.

to differentiate the product and to create pride of ownership. For example, an office chair is largely user driven. While these products may be technically sophisticated, the technology does not differentiate the product; thus, for the product development team, the ID considerations will be more important than the technical requirements. The role of engineering may still be important to determine any technical features of the product; however, since the technology is often already established, the development team focuses on the user aspects of the product.

Exhibit 10-8 classifies a variety of familiar products. Rarely does a product belong at one of the two extremes. Instead, most products fall somewhere along the continuum.

These classifications can be dynamic. For example, when a company develops a product based on a new core technology, the company is often interested in bringing the product to market as quickly as possible. Since little emphasis is placed on how the product looks or is used, the initial role of ID is small. However, as competitors enter the market, the product may need to compete more along user or aesthetic dimensions. The product's original classification shifts, and ID assumes an extremely important role in the development process. One example is the Sony Walkman. The core benefit of the first Walkman model was its technology (miniature tape player). As competition entered this market, however, Sony relied heavily on ID to create aesthetic appeal and enhanced utility, adding to the technical advantages of subsequent models.

Timing of Industrial Design Involvement

Typically, ID is incorporated into the product development process during the later phases for a technology-driven product and throughout the entire product development process for a user-driven product. Exhibit 10-9 illustrates these timing differences. Note that the ID process is a subprocess of the product development process; it is parallel but not separate. As shown in the exhibit, the ID process described above may be rapid relative to the overall development process. The technical nature of the problems that confront engineers in their design activities typically demands substantially more development effort than do the issues considered by ID.

Exhibit 10-9 shows that for a technology-driven product, ID activities may begin fairly late in the program. This is because ID for such products is focused primarily on packaging issues. For a user-driven product, ID is involved much more fully. In fact, the ID process may dominate the overall product development process for many user-driven products.

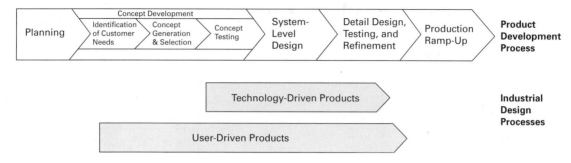

EXHIBIT 10-9 Relative timing of the industrial design process for two types of products.

Product Development Activity	Type of Product	
	Technology-Driven	**User-Driven**
Identification of Customer Needs	ID typically has no involvement.	ID works closely with marketing to identify customer needs. Industrial designers participate in focus groups or one-on-one customer interviews.
Concept Generation and Selection	ID works with marketing and engineering to ensure that human factors and user-interface issues are addressed. Safety and maintenance issues are often of primary importance.	ID generates multiple concepts according to the industrial design process flow described earlier.
Concept Testing	ID helps engineering to create prototypes, which are shown to customers for feedback.	ID leads in the creation of models to be tested with customers by marketing.
System-Level Design	ID typically has little involvement.	ID narrows down the concepts and refines the most promising approaches.
Detail Design, Testing, and Refinement	ID is responsible for packaging the product once most of the engineering details have been addressed. ID receives product specifications and constraints from engineering and marketing.	ID selects a final concept, then coordinates with engineering, manufacturing, and marketing to finalize the design.

EXHIBIT 10-10 The role of industrial design according to product type.

Exhibit 10-10 describes the responsibilities of ID during each phase of the product development process and how they relate to the other activities of the development team. As with the timing of ID involvement, the responsibilities of ID may also change according to product type.

Assessing the Quality of Industrial Design

Assessing the quality of ID for a finished product is an inherently subjective task. However, we can qualitatively determine whether ID has accomplished its goals by considering each aspect of the product that is influenced by ID. Below are five categories for

Assessment Category	Level of Importance			Explanation of Rating
	Low	Medium	High	
1. Quality of the User Interfaces	├──────────────○──────┤			In general, the StarTAC is both easy to use and comfortable. For example: calls can be answered by simply opening the keypad cover, numbers can be easily dialed into the keypad, and the functions are readily accessible. The StarTAC's drawbacks include that the visual display can be difficult to interpret because it mixes upper- and lowercase alphabetic characters, and that some users inadvertently remove the battery when attempting to open the phone for the first time.
2. Emotional Appeal	├──────────────────○─┤			The StarTAC has a high emotional appeal which stems from its sleek appearance and tiny size.
3. Ability to Maintain and Repair the Product	├────────────○──────┤			Although maintenance and repair are not of primary importance to the customer, the StarTAC rates high in this category. The battery can be removed and replaced easily. Customers can install various batteries depending on their preference for size, weight, and talk time.
4. Appropriate Use of Resources	├─────────────○──────┤			The final design includes only those features that satisfy real customer needs. Materials were selected to satisfy manufacturing constraints, to withstand extreme environmental conditions, and to meet strict appearance criteria.
5. Product Differentiation	├──────────────────○─┤			The StarTAC's appearance is clearly unique. It is easily identified when viewed in a public area or next to a competitor's product.

EXHIBIT 10-11   Assessment of industrial design's role in the StarTAC.

evaluating a product. These categories roughly match Dreyfuss's five critical goals for ID, presented above. We use these categories to develop specific questions, allowing the product to be rated along five dimensions. Exhibit 10-11 demonstrates this method by showing results for the StarTAC.

1. Quality of the User Interfaces

This is a rating of how easy the product is to use. Interface quality is related to the product's appearance, feel, and modes of interaction.

- Do the features of the product effectively communicate their operation to the user?
- Is the product's use intuitive?

- Are all features safe?
- Have all potential users and uses of the product been identified?

Examples of product-specific questions include:

- Is the grip comfortable?
- Does the tuning knob turn easily and smoothly?
- Is the power switch easy to locate?
- Is the display easy to read and understand?

2. Emotional Appeal

This is a rating of the overall consumer appeal of the product. Appeal is achieved in part through appearance, feel, sound, and smell.

- Is the product attractive? Is it exciting?
- Does the product express quality?
- What images come to mind when viewing it?
- Does the product inspire pride of ownership?
- Does the product evoke feelings of pride among development team and sales staff?

Examples of product-specific questions include:

- How does the car door sound when slammed?
- Does the hand tool feel solid and sturdy?
- Does the coffee maker look good on the kitchen counter?

3. Ability to Maintain and Repair the Product

This is a rating of the ease of product maintenance and repair. Maintenance and repair should be considered along with the other user interactions.

- Is the maintenance of the product obvious? Is it easy?
- Do product features effectively communicate disassembly and assembly procedures?

Examples of product-specific questions include:

- How easy and obvious is it to clear a paper jam in the copier?
- How difficult is it to disassemble and clean the food processor?
- How difficult is it to change the batteries in the Walkman, a remote controller, a wristwatch?

4. Appropriate Use of Resources

This is a rating of how well resources were used in satisfying the customer needs. Resources typically refer to the dollar expenditures on ID and other functions. These factors tend to drive costs such as manufacturing. A poorly designed product, one with unnecessary features, or a product made from an exotic material will affect tooling, manufacturing processes, assembly processes, and the like. This category asks whether these investments were well spent.

- How well were resources used to satisfy the customer requirements?
- Is the material selection appropriate (in terms of cost and quality)?

- Is the product over- or underdesigned (does it have features that are unnecessary or neglected)?
- Were environmental/ecological factors considered?

5. Product Differentiation

This is a rating of a product's uniqueness and consistency with the corporate identity. This differentiation arises predominantly from appearance.

- Will a customer who sees the product in a store be able to single it out because of its appearance?
- Will it be remembered by a consumer who has seen it in an advertisement?
- Will it be recognized when seen on the street?
- Does the product fit with or enhance the corporate identity?

From an ID perspective, as shown in Exhibit 10-11, the StarTAC is an excellent product. It is recognizable, durable, easy to fabricate, and has strong customer appeal. Since these features are extremely important to the consumer, ID played a critical role in determining the immediate market success of the product.

Summary

This chapter introduces the topic of industrial design, explains its benefits to product quality, and illustrates how the ID process takes place.

- The primary mission of ID is to design the aspects of a product that relate to the user: aesthetics and ergonomics.
- Most products can benefit in some way or another from ID. The more a product is looked at or used by people, the more it will depend on good ID for its success.
- For products that are characterized by a high degree of user interaction and the need for aesthetic appeal, ID should be involved throughout the product development process. Early involvement of industrial designers will ensure that critical aesthetic and user requirements will not be overlooked or ignored by the technical staff.
- When a product's success relies more on technology, ID can be integrated into the development process later.
- Active involvement of ID on the product development team can help to promote good communication between functional groups. Such communication facilitates coordination and ultimately translates into higher-quality products.

References and Bibliography

Many current resources are available on the Internet via
www.ulrich-eppinger.net

For more information about industrial design—its history, impact, future, and practice—the following books and articles are recommended. The brief history of ID presented in this chapter was adapted from Lorenz's book.

Lorenz, Christopher, *The Design Dimension: Product Strategy and the Challenge of Global Marketing,* Basil Blackwell, Oxford, UK, 1986.

Caplan, Ralph, *By Design: Why There Are No Locks on the Bathroom Doors in the Hotel Louis XIV, and Other Object Lessons,* St. Martin's Press, New York, 1982.

Lucie-Smith, Edward, *A History of Industrial Design,* Van Nostrand Reinhold, New York, 1983.

Dreyfuss, Henry, *Designing for People,* Paragraphic Books, New York, 1967.

Dreyfuss, Henry, "The Industrial Designer and the Businessman," *Harvard Business Review,* November 1950, pp. 77–85.

Harkins, Jack, "The Role of Industrial Design in Developing Medical Devices," *Medical Device and Diagnostic Industry,* September 1992, pp. 51–54, 94–97.

Norman discusses good and bad examples of product design and provides principles and guidelines for good design practice.

Norman, Donald A., *The Design of Everyday Things,* Doubleday, New York, 1990.

Thackara presents case studies of the winners of the 1997 European Design Prize, providing contemporary examples of best practice in design of consumer, industrial, and high-technology products.

Thackara, John, *Winners! How Today's Successful Companies Innovate by Design,* BIS, Amsterdam, 1977.

Computer-aided industrial design (CAID), introduced in this article by Cardaci, has become an important part of ID practice today, replacing traditional rendering in many situations.

Cardaci, Kitty, "CAID: A Tool for the Flexible Organization," *Design Management Journal,* Design Management Institute, Boston, Vol. 3, No. 2, Spring 1992, pp. 72–75.

The following studies are among the very few which have critically assessed the value of ID to products and their manufacturers. An issue of *Design Management Journal* was devoted to this topic.

Pearson, Scott, "Using Product Archaeology to Understand the Dimensions of Design Decision Making," S.M. Thesis, MIT Sloan School of Management, May 1992.

Roy, Robin, and Stephen Potter, "The Commercial Impacts of Investment in Design," *Design Studies,* Vol. 14, No. 2, April 1993, pp. 171–193.

Design Management Journal, Vol. 5, No. 2, Spring 1994.

Olins describes how a firm develops a corporate identity through design and communication.

Olins, Wally, *Corporate Identity: Making Business Strategy Visible through Design,* Harvard Business School Press, Boston, 1989.

Several excellent case studies involving the ID process and product development issues surrounding ID have been written by the Design Management Institute. Also the publications @ *Issue* (quarterly) and *I.D.* (monthly) include case studies, examples, and discussion of ID practices.

Publications Catalogue, Design Management Institute, 29 Temple Place, Boston, MA, 02111.

I.D.: The International Design Magazine, Magazine Publications, L.P., New York. Published monthly.

@ Issue: The Journal of Business and Design, Corporate Design Foundation, Boston. Published quarterly.

While industrial designers are best found through personal referral, IDSA can provide a list of ID consultants in any geographic region.

Directory of Industrial Designers, Industrial Designers Society of America, 1142 Walker Road, Great Falls, VA, 22066. Published annually.

Exercises

1. Visit a local specialty store (e.g., kitchen supplies, tools, office supply, gifts) and photograph (or purchase) a set of competing products. Assess each one in terms of the five ID quality categories as shown in Exhibit 10-11. Which product would you purchase? Would you be willing to pay more for it than for the others?

2. Develop several concept sketches for a common product. Try designing the product form both "from the inside out" and "from the outside in." Which is easier for you? Possible simple products include a stapler, a garlic press, an alarm clock, a reading light, and a telephone.

3. List some firms that you feel have a strong corporate identity. What aspects of their products helped to develop this identity?

Thought Questions

1. By what cause-and-effect mechanism does ID affect a product's manufacturing cost? Under what conditions would ID increase or decrease manufacturing cost?

2. What types of products might not benefit from ID involvement in the development process?

3. The term *visual equity* is sometimes used to refer to the value of the distinctive appearance of a firm's products. How is such equity obtained? Can it be "purchased" over a short time period, or does it accrue slowly?

CHAPTER ELEVEN

Design for Manufacturing

Courtesy of General Motors Corp.

EXHIBIT 11-1
The General Motors 3.8-liter V6 engine.

General Motors Powertrain Division manufactures about 3,500 3.8-liter V6 engines every day (Exhibit 11-1). Facing such high production volumes, the company had a strong interest in reducing the cost of the engine while simultaneously enhancing its quality. A team was formed to improve one of the most expensive subassemblies in the engine: the air intake manifold. (The intake manifold's primary function is to route air from the throttle to the intake valves at the cylinders.) The original and redesigned intake manifold assemblies are shown in Exhibit 11-2. This chapter presents a method of design for manufacturing using the GM V6 intake manifold as an example.

EXHIBIT 11-2

The original and redesigned air intake manifolds. The body of the original manifold (top) is made of cast aluminum. The redesigned manifold (bottom) is made of molded thermoplastic composite.

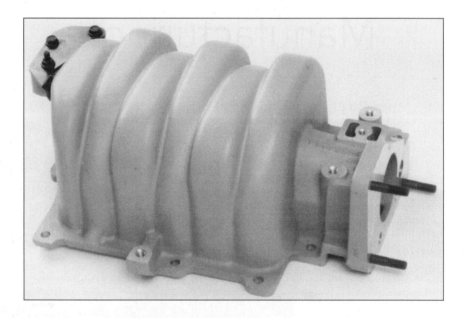

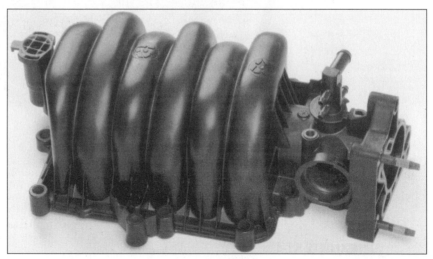

Photos by Stuart Cohen

Design for Manufacturing Defined

Customer needs and product specifications are useful for guiding the concept phase of product development; however, during the later development activities teams often have difficulty linking needs and specifications to the specific design issues they face. For this reason, many teams practice "design for X" (DFX) methodologies, where X may correspond to one of dozens of quality criteria such as reliability, robustness, serviceability, environmental impact, or manufacturability. The most common of these methodologies is *design for manufacturing* (DFM), which is of universal importance because it directly addresses manufacturing costs.

This chapter is primarily about DFM, but it is also intended to illustrate, by example, these general principles which apply to methodologies for achieving any of the Xs in DFX:

- Detail design decisions can have substantial impact on product quality and cost.
- Development teams face multiple, and often conflicting, goals.
- It is important to have metrics with which to compare alternative designs.
- Dramatic improvements often require substantial creative efforts early in the process.
- A well-defined method assists the decision-making process.

Manufacturing cost is a key determinant of the economic success of a product. In simple terms, economic success depends on the profit margin earned on each sale of the product and on how many units of the product the firm can sell. Profit margin is the difference between the manufacturer's selling price and the cost of making the product. The number of units sold and the sales price are to a large degree determined by the overall quality of the product. Economically successful design is therefore about ensuring high product quality while minimizing manufacturing cost. DFM is one method for achieving this goal; effective DFM practice leads to low manufacturing costs without sacrificing product quality. (See Chapter 15, Product Development Economics, for a more detailed discussion of models relating manufacturing costs to economic success.)

DFM Requires a Cross-Functional Team

Design for manufacturing is one of the most integrative practices involved in product development. DFM utilizes information of several types, including (1) sketches, drawings, product specifications, and design alternatives; (2) a detailed understanding of production and assembly processes; and (3) estimates of manufacturing costs, production volumes, and ramp-up timing. DFM therefore requires the contributions of most members of the development team as well as outside experts. DFM efforts commonly draw upon expertise from manufacturing engineers, cost accountants, and production personnel, in addition to product designers. Many companies use structured, team-based workshops to facilitate the integration and sharing of views required for DFM.

DFM Is Performed throughout the Development Process

DFM begins during the concept development phase, when the product's functions and specifications are being determined. When choosing a product concept, cost is almost always one of the criteria on which the decision is made—even though cost estimates at this phase are highly subjective and approximate. When product specifications are finalized, the team makes trade-offs between desired performance characteristics. For example, weight reduction may increase manufacturing costs. At this point, the team may have an approximate bill of materials (a list of parts) with estimates of costs. During the system-level design phase of development, the team makes decisions about how to break up the product into individual components, based in large measure on the expected cost and manufacturing complexity implications. Accurate cost estimates finally become available during the detail design phase of development, when many more decisions are driven by manufacturing concerns.

Overview of the DFM Process

Our DFM method is illustrated in Exhibit 11-3. It consists of five steps plus iteration:

1. Estimate the manufacturing costs.
2. Reduce the costs of components.
3. Reduce the costs of assembly.
4. Reduce the costs of supporting production.
5. Consider the impact of DFM decisions on other factors.

As shown in Exhibit 11-3, the DFM method begins with the estimation of the manufacturing cost of the proposed design. This helps the team to determine at a general level which aspects of the design—components, assembly, or support—are most costly. The team then directs its attention to the appropriate areas in the subsequent steps. This process is iterative. It is not unusual to recompute the manufacturing cost estimate and to improve the design of the product dozens of times before agreeing that it is good enough. As long as the product design is improving, these DFM iterations may continue even until pilot production begins. At some point, the design is frozen (or "released"), and any further modifications are considered formal "engineering changes" or become part of the next generation of the product.

In the next section, we use the original GM V6 intake manifold as an example and explain how manufacturing costs are determined. Then, recognizing that accurate cost estimates are difficult (if not impossible) to obtain, we present several useful methods for reducing the costs of components, assembly, and production support. We use the redesigned intake manifold and other products as examples to illustrate these DFM principles. Finally, we discuss the results achieved through DFM and some of the broader implications of DFM decisions.

Step 1: Estimate the Manufacturing Costs

Exhibit 11-4 shows a simple input-output model of a manufacturing system. The inputs include raw materials, purchased components, employees' efforts, energy, and equipment. The outputs include finished goods and waste. Manufacturing cost is the sum of all of the expenditures for the inputs of the system and for disposal of the wastes produced by the system. As the metric of cost for a product, firms generally use *unit manufacturing cost,* which is computed by dividing the total manufacturing costs for some period (usu-

EXHIBIT 11-3
The design for
manufacturing
(DFM) method.

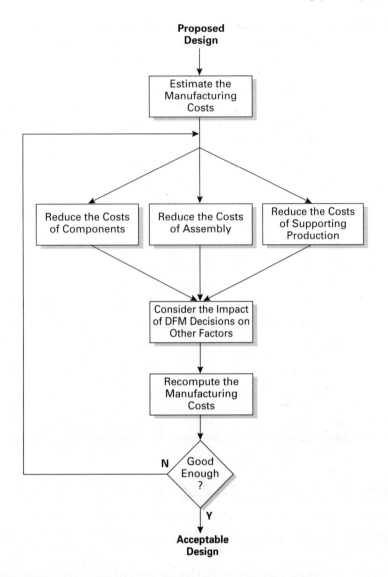

ally a quarter or a year) by the number of units of the product manufactured during that period. This simple concept is complicated in practice by several issues:

- What are the boundaries of the manufacturing system? Should the field service operations be included? What about product development activities?
- How do we "charge" the product for the use of expensive general-purpose equipment that lasts for many years?
- How are costs allocated among more than one product line in large, multiproduct manufacturing systems?

EXHIBIT 11-4
A simple input-output model of a manufacturing system.

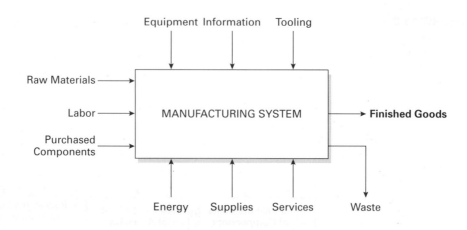

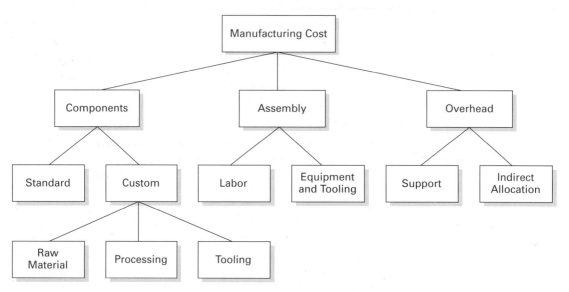

EXHIBIT 11-5   Elements of the manufacturing cost of a product.

These are issues around which much of the field of managerial accounting is built, and we do not treat them in depth here. Nevertheless, we will be mindful of these complications as we discuss cost and DFM in this chapter.

Exhibit 11-5 shows one way of categorizing the elements of manufacturing cost. Under this scheme, the unit manufacturing cost of a product consists of costs in three categories:

1. *Component costs:* The *components* of a product (also simply called parts of the product) may include *standard parts* purchased from suppliers. Examples of standard components include motors, switches, electronic chips, and screws. Other components are *custom parts,* made according to the manufacturer's design from raw materials, such as sheet steel, plastic pellets, or aluminum bars. Some custom components are made in the manufacturer's own plant, while others may be produced by suppliers according to the manufacturer's design specifications.

2. *Assembly costs:* Discrete goods are generally assembled from parts. The process of assembling almost always incurs labor costs and may also incur costs for equipment and tooling.

3. *Overhead costs:* Overhead is the category used to encompass all of the other costs. We find it useful to distinguish between two types of overhead: *support costs* and other *indirect allocations.* Support costs are the costs associated with materials handling, quality assurance, purchasing, shipping, receiving, facilities, and equipment/tooling maintenance (among others). These are the support systems required to manufacture the product, and these costs do greatly depend upon the product design. Nevertheless, since these costs are often shared by more than one product line, they are lumped together in the category of overhead. Indirect allocations are the costs of manufacturing that cannot be directly linked to a particular product but which must be paid for to be in business. For example, the salary of the security guard and the cost of maintenance to the building and grounds are indirect costs because these activities are shared among several different products and are difficult to allocate directly to a specific product. Because indirect costs are not specifically linked to the design of the product, they are not relevant to DFM, even though they do contribute to the cost of the product.

Fixed Costs versus Variable Costs

Another way to divide manufacturing costs is between *fixed costs* and *variable costs.* Fixed costs are those that are incurred in a predetermined amount, regardless of how many units of the product are manufactured. Purchasing the injection mold required for the new intake manifold is an example of a fixed cost. Whether 1,000 or 1 million units are produced, the fixed cost of the mold is incurred and does not change. Another example is the cost of setting up the factory work area for the intake manifold assembly line. This cost is also fixed, regardless of how many units are produced. Despite the terminology, however, no cost is truly fixed. If we quadruple the production quantity, we may have to build another production line. Conversely, we may be able to consolidate two assembly cells if we cannot use all the capacity due to dramatically lower production quantities. When considering a cost as fixed, ranges of production quantities and the assumed time horizon should be specified.

Variable costs are those incurred in direct proportion to the number of units produced. For example, the cost of raw materials is directly proportional to how many intake manifolds are produced, and therefore to how many 3.8-liter V6 engines are made. Assembly labor is sometimes considered a variable cost as well because many firms can adjust the staffing of assembly operations by shifting workers to other areas on short notice.

The Bill of Materials

Since manufacturing cost estimation is fundamental to DFM, it is useful to keep this information well organized. Exhibit 11-6 shows an information system for recording manufacturing cost estimates. It basically consists of a bill of materials (BOM) augmented with cost information. The BOM (usually pronounced *bomb*) is a list of each individual component in the product. Frequently the BOM is created using an indented format in which

Component	Purchased Materials	Processing (Machine + Labor)	Assembly (Labor)	Total Unit Variable Cost	Tooling and Other NRE, K$	Tooling Lifetime, K units	Total Unit Fixed Cost	Total Cost
Manifold machined casting	12.83	5.23		18.06	1960	500+	0.50	18.56
EGR return pipe	1.30		0.15	1.45				1.45
PCV assembly								
Valve	1.35		0.14	1.49				1.49
Gasket	0.05		0.13	0.18				0.18
Cover	0.76		0.13	0.89				0.89
Screws (3)	0.06		0.15	0.21				0.21
Vacuum source block assembly								
Block	0.95		0.13	1.08				1.08
Gasket	0.03		0.05	0.08				0.08
Screw	0.02		0.09	0.11				0.11
Total Direct Costs	17.35	5.23	0.95	23.53	1960		0.50	24.03
Overhead Charges	2.60	9.42	1.71				0.75	14.48
Total Cost								**38.51**

EXHIBIT 11-6   Indented bill of materials showing cost estimates for the original intake manifold and related components. The EGR (exhaust gas recirculation), PCV (positive crankcase ventilation), and vacuum block components are included here to facilitate comparison with the redesigned manifold assembly.

the assembly "tree structure" is illustrated by the indentation of components and sub-assembly names.

The columns of the BOM show the cost estimates broken down into fixed and variable costs. The variable costs may include materials, machine time, and labor. Fixed costs consist of tooling and other nonrecurring expenses (NRE) such as specialized equipment and one-time setup costs. The tooling lifetime is used to compute the unit fixed cost (unless the tool's expected lifetime exceeds the product's lifetime volume, in which case the lower product volume is used). To compute total cost, overhead is added according to the firm's accepted cost accounting scheme. Note that additional fixed costs, such as depreciation of capital equipment used for several products, are often also included in the overhead charge.

Estimating the Costs of Standard Components

The costs of standard components are estimated by either (1) comparing each part to a substantially similar part the firm is already producing or purchasing in comparable volumes or (2) soliciting price quotes from vendors or suppliers. The costs of minor components (e.g., bolts, springs, and inserts) are usually obtained from the firm's experience with similar components, while the costs of major components are usually obtained from vendor quotes.

In obtaining price quotes, the estimated production quantities are extremely important. For example, the unit price on a purchase of a dozen screws or inserts may be 10 times higher than the unit prices paid by GM when purchasing 100,000 of these parts every month. If the anticipated production quantities are high enough, an application engineer or sales engineer is usually quite willing to work with the development team to specify a component properly. For internally fabricated standard components, if the required quantities are high, there may not be available production capacity, necessitating the purchase of additional equipment or the use of outside suppliers.

Some suppliers will design and fabricate a custom variation to a standard component if production quantities are high enough. For example, small electric motors, such as those found in powered hand tools, are often designed and built specifically for the product application. If the production quantities are high enough (say, 100,000 per year in this case), these custom motors are quite economical ($1 to $5 per unit, depending on the performance characteristics). For the intake manifold, the volumes are sufficiently high that custom studs, bushings, and other parts may not cost much more than standard components. However, as we discuss later, introducing new parts can add substantial cost and complexity to the production system and field service operations, which increases the support costs.

Vendors for most standard components can be found in the *Thomas Register of American Manufacturers* or by looking for company names on components used in related products. To obtain a price quote, first request a catalog or product literature (now generally available on the Internet). Then, either choose a part number or, if a custom component will be used, write a one-page description of the requirements of the component. Next, telephone the vendor, ask to speak to someone in "sales," and request price information. Make sure to inform vendors that the information is for estimation purposes only; otherwise, they may claim they do not have enough information to determine exact prices.

Estimating the Costs of Custom Components

Custom components, which are parts designed especially for the product, are made by the manufacturer or by a supplier. Most custom components are produced using the same types of production processes as standard components (e.g., injection molding, stamping, machining); however, custom parts are typically special-purpose parts, useful only in a particular manufacturer's products.

When the custom component is a single part, we estimate its cost by adding up the costs of raw materials, processing, and tooling. In cases where the custom component is actually an assembly of several parts, then we consider it a "product" in and of itself; to arrive at the cost of this "product" we estimate the cost of each subcomponent and then add assembly and overhead costs (these costs are described below). For the purposes of this explanation, we assume the component is a single part.

The raw materials costs can be estimated by computing the mass of the part, allowing for some scrap (e.g., 5 percent to 50 percent for an injection molded part, and 25 percent to 100 percent for a sheet metal part), and multiplying by the cost (per unit mass) of the raw material. A table of raw material costs is given in Appendix A (Exhibit 11-17).

Processing costs include costs for the operator(s) of the processing machinery as well as the cost of using the equipment itself. Most standard processing equipment costs between $25 per hour (a simple stamping press) and $75 per hour (a medium-sized, computer-controlled milling machine) to operate, including depreciation, maintenance, utilities, and labor costs. Estimating the processing time generally requires experience with the type of equipment to be used. However, it is useful to understand the range of

EXHIBIT 11-7
Cost estimate
for the original
intake manifold.
Note that the
processing costs
for casting and
machining
reflect the costs
for a complete
casting line and
several
machining
stations.

Variable Cost		
Materials	5.7 kg aluminum at $2.25/kg	$12.83
Processing (casting)	150 units/hr at $530/hr	3.53
Processing (machining)	200 units/hr at $340/hr	1.70
Fixed Cost		
Tooling for casting	$160,000/tool at 500K units/tool (lifetime)	0.32
Machine tools and fixtures	$1,800,000/line at 10M units (lifetime)	0.18
Total Direct Cost		**$18.56**
Overhead charges		$12.09
Total Unit Cost		**$30.65**

typical costs for common production processes. For this purpose, tables of approximate processing times and costs are given in Appendix B for a variety of stampings, castings, injection moldings, and machined parts.

Tooling costs are incurred for the design and fabrication of the cutters, molds, dies, or fixtures required to use certain machinery to fabricate parts. For example, an injection molding machine requires a custom injection mold for every different type of part it produces. These molds generally range in cost from $10,000 to $500,000. Approximate tooling costs are also given for the parts listed in Appendix B. The unit tooling cost is simply the cost of the tooling divided by the number of units to be made over the life of the tool. A high-quality injection mold or stamping die can usually be used for a few million parts.

The cost of the original intake manifold's machined casting is estimated as shown in Exhibit 11-7. Note that the estimate reveals that the cost is dominated by the expense of the aluminum material. We will see that the redesign using a composite material not only reduced the material costs but also eliminated machining and allowed many features to be formed into the molded body.

Estimating the Cost of Assembly

Products made of more than one part require assembly. For products made in quantities of less than several hundred thousand units per year, this assembly is almost always performed manually. One exception to this generalization is the assembly of electronic circuit boards, which is now almost always done automatically, even at relatively low volumes. There will likely be more exceptions in the coming years, as flexible, precision automation becomes more common.

Manual assembly costs can be estimated by summing the estimated time of each assembly operation and multiplying by a labor rate. Assembly operations require from about 4 seconds to about 60 seconds each, depending upon the size of the parts, the difficulty of the operation, and the production quantities. At high volumes, workers can specialize in a particular set of operations, and special fixtures and tools can assist the assembly. Appendix C contains a table of approximate times for manual assembly of various products, which is helpful in estimating the range of times required for assembly operations. A popular method for estimating assembly times has been developed over the past 30 years by Boothroyd Dewhurst Inc. and is now available as a software tool. This system involves a tabular information system for keeping track of the estimated assembly

EXHIBIT 11-8

Assembly cost estimation for the PCV valve assembly of the redesigned intake manifold.

Source: Manual assembly tables in Boothroyd and Dewhurst, 1989

Component	Quantity	Handling Time	Insertion Time	Total Time
Valve	1	1.50	1.50	3.00
O-rings	2	2.25	4.00	12.50
Spring	1	2.25	6.00	8.25
Cover	1	1.95	6.00	7.95
Total Time (seconds)				**31.70**
Assembly Cost at $45/hour				**$0.40**

times for each part. The system is supported by a comprehensive database of standard handling and insertion times for a wide range of situations. Special software is also available for estimating the assembly cost of electronic circuit boards.

Assembly labor can cost from less than $1 per hour in low-wage countries to more than $40 per hour in some industrialized nations. In the United States, assembly labor is likely to cost between $10 and $20 per hour. (Each firm has different assembly labor cost structures, and some industries, such as the automobile and aircraft industries, have substantially higher cost structures.) These figures include an allowance for benefits and other worker-related expenses and are meant to reflect the true cost to the firm of assembly labor.

Consider the redesigned intake manifold. The assembly cost of the PCV (positive crankcase ventilation) valve assembly is estimated as shown in Exhibit 11-8.

Estimating the Overhead Costs

Accurately estimating overhead costs for a new product is difficult, and the industry practices are not very satisfying. Nevertheless, we will describe the standard industry practice here and identify some of its problems. Applying the overhead estimation schemes used by most firms is simple. Estimating the actual overhead costs incurred by the firm due to a particular product is not. The indirect costs of supporting production are very difficult to track and assign to particular product lines. The future costs of supporting production are even more difficult to predict for a new product.

Most firms assign overhead charges by using *overhead rates* (also called burden rates). Overhead rates are typically applied to one or two *cost drivers*. Cost drivers are parameters of the product which are directly measurable. Overhead charges are added to direct costs in proportion to the drivers. Common cost drivers are the cost of any purchased materials, the cost of assembly labor, and the number of hours of equipment time the product consumes. For example, the overhead rate for purchased materials might be 10 percent and the overhead rate for assembly labor might be 80 percent. (Of course, purchased components already have the vendor's overhead included in the price; we only add the purchasing overhead.) Under these conditions, a product containing $100 of purchased components and $10 of assembly labor would incur $18 of overhead costs (10 percent of $100 plus 80 percent of $10). Some typical overhead structures are given in Appendix D for different types of products and firms.

The problem with this scheme is that it implies that overhead costs are directly proportional to the cost drivers. A thought experiment reveals that this cannot always be so: Most firms use "cost of purchased materials" as one cost driver, yet why would any of their overhead costs actually change if a vendor of a $50 component raises its price to

$60? The answer is that they would not change at all. Overhead rates are used as a convenient way to account for overhead costs, but this scheme can yield inaccurate estimates of the true costs experienced by the manufacturer to support production.

This problem is partially addressed by activity-based costing (ABC) methods (Kaplan, 1990). Under the ABC approach, a firm utilizes more and different cost drivers and allocates all indirect costs to the associated cost drivers where they fit best. As a result, the firm may have overhead rates applied to various dimensions of product complexity (such as the number of different machining operations required or the number of different components or suppliers needed), in addition to overhead on tooling, materials, machine time, and direct labor. For the purposes of estimating manufacturing costs, the use of more cost drivers not only allows more accurate overhead cost estimates to be made but also provides important insights for reducing overhead costs by focusing attention on the cost drivers.

Step 2: Reduce the Costs of Components

For most highly engineered discrete goods the cost of purchased components will be the most significant element of the manufacturing cost. This section presents several strategies for minimizing these costs. Many of these strategies can be followed even without the benefit of accurate cost estimates. In this case, these strategies become *design rules,* or rules of thumb, to guide DFM cost reduction decisions.

Understand the Process Constraints and Cost Drivers

Some component parts may be costly simply because the designers did not understand the capabilities, cost drivers, and constraints of the production process. For example, a designer may specify a small internal corner radius on a machined part without realizing that physically creating such a feature requires an expensive electro-discharge machining (EDM) operation. A designer may specify dimensions with excessively tight tolerances, without understanding the difficulty of achieving such accuracy in production. Sometimes these costly part features are not even necessary for the component's intended function; they arise out of lack of knowledge. It is often possible to redesign the part to achieve the same performance while avoiding costly manufacturing steps; however, to do this the design engineer needs to know what types of operations are difficult in production and what drives their costs.

In some cases, the constraints of a process can be concisely communicated to designers in the form of design rules. For example, the capabilities of an automatic laser cutting machine for sheet metal can be concisely communicated in terms of allowable material types, material thicknesses, maximum part dimensions, minimum slot widths, and cutting accuracy. When this is possible, part designers can avoid exceeding the normal capabilities of a process and thereby avoid incurring unusually high costs.

For some processes, the cost of producing a part is a simple mathematical function of some attributes of the part, which would be the cost drivers for the process. For example, a welding process could have a cost directly proportional to two attributes of the product: (1) the number of welds and (2) the total length of welds the machine creates.

For processes whose capabilities are not easily described, the best strategy is to work closely with the people who deeply understand the part production process. These manufacturing experts will generally have plenty of ideas about how to redesign components to reduce production costs.

EXHIBIT 11-9
Cost estimate
for the
redesigned
intake manifold
(two moldings).

Variable Cost		
Materials (manifold housing)	1.4 kg glass-filled nylon at $2.75/kg	$ 3.85
Materials (intake runner insert)	0.3 kg glass-filled nylon at $2.75/kg	0.83
Molding (manifold housing)	80 units/hr at $125/hr	1.56
Molding (intake runner insert)	100 units/hr at $110/hr	1.10
Fixed Cost		
Mold tooling (manifold housing)	$350,000/tool at 1.5M units/tool	$ 0.23
Mold tooling (intake runner insert)	$150,000/tool at 1.5M units/tool	0.10
Total Direct Cost		**$ 7.67**
Overhead charges		**$ 5.99**
Total Unit Cost		**$13.66**

Redesign Components to Eliminate Processing Steps

Careful scrutiny of the proposed design may lead to suggestions for redesign that can result in simplification of the production process. Reducing the number of steps in the part fabrication process generally results in reduced costs as well. Some process steps may simply not be necessary. For example, aluminum parts may not need to be painted, especially if they will not be visible to the user of the product. In some cases, several steps may be eliminated through substitution of an alternative process step. A common example of this strategy is "net-shape" fabrication. A net-shape process is one that produces a part with the final intended geometry in a single manufacturing step. Typical examples include molding, casting, forging, and extrusion. Frequently designers are able to use one of the net-shape processes to create a part that is very close to the final requirement (near net shape) and may demand only minor additional processing (e.g., drilling and tapping a hole, cutting to length).

The original intake manifold required an expensive casting, followed by several machining operations. The redesigned manifold is molded in two parts to net shape. The cost estimate for these two moldings is shown in Exhibit 11-9. (Compare with Exhibit 11-7.)

Choose the Appropriate Economic Scale for the Part Process

The manufacturing cost of a product usually drops as the production volume increases. This phenomenon is labeled *economies of scale*. Economies of scale for a fabricated component occur for two basic reasons: (1) fixed costs are divided among more units and (2) variable costs become lower because the firm can justify the use of larger and more efficient processes and equipment. For example, consider an injection-molded plastic part. The part may require a mold which costs $50,000. If the firm produces 50,000 units of the part over the product's lifetime, each part will have to assume $1 of the cost of the mold. If, however, 100,000 units are produced, each part will assume only $0.50 of the cost of mold. As production volumes increase further, the firm may be able to justify a four-cavity mold, for which each cycle of the molding machine produces four parts instead of one. As shown in Exhibit 11-9, the tooling costs for the redesigned intake manifold are quite high; however, spread over the life of the tool, the unit fixed cost is small.

Processes can be thought of as incurring fixed and variable costs. Fixed costs are incurred once per part type regardless of how many parts are produced. Variable costs are incurred each time a part is made. Processes with inherently low fixed costs and high

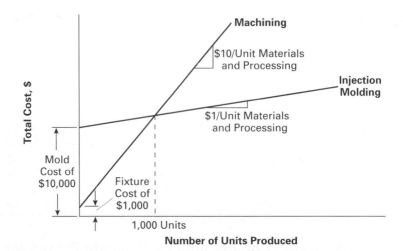

EXHIBIT 11-10
Total cost of a hypothetical part as a function of the number of units produced for injection molding versus machining.

variable costs, such as machining, are appropriate when few parts will be made, while processes with inherently high fixed costs and low variable costs, such as injection molding, are appropriate when many parts will be made. This concept is illustrated by the graph in Exhibit 11-10. As shown in the exhibit, if production volume is expected to be below 1,000 units, machining would be more economical; otherwise, injection molding would incur lower total costs.

Standardize Components and Processes

The principle of economies of scale also applies to the selection of components and processes. As the production volume of a component increases, the unit cost of the component decreases. Quality and performance often increase as well with increasing production quantities because the producer of the component can invest in learning and improvement of the component's design and its production process. For a given expected product volume, the benefits of substantially higher component volumes can be achieved through the use of standard components.

Standard components are those common to more than one product. This standardization may occur within the product line of a single firm or may occur, via an outside supplier, across the product lines of several firms. For example, the use of the 3.8-liter V6 engine in several GM cars is an example of *internal standardization.* The use of a common 10-millimeter socket head cap screw across several auto manufacturers is an example of *external standardization.* In either case, all other things being equal, the component unit cost is lower than if the component were used in only a single product.

The redesigned intake manifold is used on all of GM's 3.8-liter V6 engines, even though each particular vehicle application requires different EGR (exhaust gas recirculation) return and vacuum hose routings. To accommodate this, the new intake manifold has two standard interfaces, a vacuum port and an EGR port. For each vehicle model, a custom vacuum block and EGR adapter are used. This allows the major component, the intake manifold, to be standardized internally, rather than using a different manifold for each vehicle.

**EXHIBIT
11-11**
An example of
standardization
within a model.
Wheels of the
Ford Explorer
are the same on
the right and left
sides of the car.

Courtesy of Ford Motor Co.

Components may also be standardized within the same model. For example, most auto manufacturers use the same type of wheel on the right and left side of their cars, even though this causes directional "spokes" to have different orientations on different sides (Exhibit 11-11).

Adhere to "Black Box" Component Procurement

A component cost reduction strategy used effectively in the Japanese auto industry is called *black box* supplier design. Under this approach, the team provides a supplier with only a black box description of the component—a description of what the component has to do, not how to achieve it (Clark and Fujimoto, 1991). This kind of specification leaves the vendor with the widest possible latitude to design or select the component for minimum cost. An additional advantage of this approach is that it relieves the internal team of the responsibility to engineer and design the component. Successful black box development efforts require careful system-level design and extremely clear definitions of the functions, interfaces, and interactions of each component. (See Chapter 9, Product Architecture.)

For the redesigned intake manifold, the PCV valve assembly was designed by GM's AC Rochester Division, which supplies the component. The supplier was given system-level specifications and complete responsibility for the performance of this subsystem.

Step 3: Reduce the Costs of Assembly

Design for assembly (DFA) is a fairly well established subset of DFM which involves minimizing the cost of assembly. For most products, assembly contributes a relatively small fraction of the total cost. Nevertheless, focusing attention on assembly costs yields strong indirect benefits. Often as a result of emphasis on DFA, the overall parts count, manufacturing complexity, and support costs are all reduced along with the assembly cost. In this section, we present a few principles useful to guide DFA decisions.

Keeping Score

Boothroyd and Dewhurst (1989) advocate maintaining an ongoing estimate of the cost of assembly. In addition to this absolute score, they propose the concept of *assembly efficiency*. This is measured as an index which is the ratio of the *theoretical minimum assembly time* to an estimate of the actual assembly time for the product. This concept is useful in developing an intuition for what drives the cost of assembly. The expression for the *DFA index* is

$$\text{DFA index} = \frac{(\text{Theoretical minimum number of parts}) \times (3 \text{ seconds})}{\text{Estimated total assembly time}}$$

To determine the theoretical minimum number of parts, ask the following three questions of each part in the proposed assembly. Only parts satisfying one or more of these conditions must "theoretically" be separate.

1. Does the part need to move relative to the rest of the assembly? Small motions that can be accomplished using compliance (e.g., elastic hinges or springs) do not count.
2. Must the part be made of a different material from the rest of the assembly for fundamental physical reasons?
3. Does the part have to be separated from the assembly for assembly access, replacement, or repair?

The "3 seconds" in the numerator reflects the theoretical minimum time required to handle and insert a part that is perfectly suited for assembly. One can think of this as the average time (sustainable over a whole work shift) required to assemble a small part that is easy to grasp, requires no particular orientation, and demands no special insertion effort; such an operation is as fast as placing a ball into a circular hole with adequate clearance.

Integrate Parts

If a part does not qualify as one of those theoretically necessary, then it is a candidate for physical integration with one or more other parts. The resulting multifunctional component is often very complex as a result of the integration of several different geometric features which would otherwise be separate parts. Nevertheless, molded or stamped parts can often incorporate additional features at little or no added cost. Exhibit 11-12 shows the throttle-body end of the redesigned intake manifold. Integrated into this component are the attachments for the EGR return and the vacuum source block. These attachments use a molded "push in and turn" geometry, eliminating the need for several threaded fasteners.

Part integration provides several benefits:

• Integrated parts do not have to be assembled. In effect, the "assembly" of the geometric features of the part is accomplished by the part fabrication process.
• Integrated parts are often less expensive to fabricate than are the separate parts they replace. For molded, stamped, and cast parts, this cost savings occurs because a single complex mold or die is usually less expensive than two or more less complex molds or

Photo by Stuart Cohen

dies and because there is usually less processing time and scrap for the single, integrated part.

- Integrated parts allow the relationships among critical geometric features to be controlled by the part fabrication process (e.g., molding) rather than by an assembly process. This usually means that these dimensions can be more precisely controlled.

Note, however, that part integration is not always a wise strategy and may be in conflict with other sound approaches to minimizing costs. For example, the main intake manifold assembly on the old design was a single cast piece, requiring extensive machining. The team replaced this part with two less-expensive, injection-molded pieces. This is an example of dis-integrating parts in order to achieve benefits in the piece-part production costs.

Maximize Ease of Assembly

Two products with an identical number of parts may nevertheless differ in required assembly time by a factor of two or three. This is because the actual time to grasp, orient, and insert a part depends on the part geometry and the required trajectory of the part insertion. The ideal characteristics of a part for an assembly are (adapted from Boothroyd and Dewhurst, 1989):

- *Part is inserted from the top of the assembly.* This attribute of a part and assembly is called *z-axis assembly.* By using z-axis assembly for all parts, the assembly never has to be inverted, gravity helps to stabilize the partial assembly, and the assembly worker can generally see the assembly location.
- *Part is self-aligning.* Parts that require fine positioning in order to be assembled require slow, precise movements on the part of the assembly worker. Parts and assembly sites can be designed to be self-aligning so that fine motor control is not required of

the worker. The most common self-alignment feature is the *chamfer*. A chamfer can be implemented as a tapered lead on the end of a peg, or a conical widening at the opening of a hole.

• *Part does not need to be oriented.* Parts requiring correct orientation, such as a screw, require more assembly time than parts requiring no orientation, such as a sphere. In the worst case, a part must be oriented correctly in three dimensions. For example, the following parts are listed in order of increasing requirements for orientation: sphere, cylinder, capped cylinder, capped and keyed cylinder.

• *Part requires only one hand for assembly.* This characteristic relates primarily to the size of the part and the effort required to manipulate the part. All other things being equal, parts requiring one hand to assemble require less time than parts requiring two hands, which in turn require less effort than parts requiring a crane or lift to assemble.

• *Part requires no tools.* Assembly operations requiring tools, such as attaching snap rings, springs, or cotter pins, generally require more time than those that do not.

• *Part is assembled in a single, linear motion.* Pushing in a pin requires less time than driving a screw. For this reason, numerous fasteners are commercially available that require only a single, linear motion for insertion.

• *Part is secured immediately upon insertion.* Some parts require a subsequent securing operation, such as tightening, curing, or the addition of another part. Until the part is secured, the assembly may be unstable, requiring extra care, fixtures, or slower assembly.

Consider Customer Assembly

Customers may tolerate completing some of the product assembly themselves, especially if doing so provides other benefits, such as making the purchase and handling of the packaged product easier. However, designing a product such that it can be easily and properly assembled by the most inept customers, many of whom will ignore directions, is a substantial challenge in itself.

Step 4: Reduce the Costs of Supporting Production

In working to minimize the costs of components and the costs of assembly, the team may also achieve reductions in the demands placed on the production support functions. For example, a reduction in the number of parts reduces the demands on inventory management. A reduction in assembly content reduces the number of workers required for production and therefore reduces the cost of supervision and human resource management. Standardized components reduce the demands on engineering support and quality control. There are, in addition, some direct actions the team can take to reduce the costs of supporting production.

It is important to remember that manufacturing cost estimates are often insensitive to many of the factors that actually drive overhead charges. (Recall the discussion of overhead cost estimation above.) Nevertheless, the goal of the design team in this respect should be to reduce the actual costs of production support even if overhead cost estimates do not change.

Drivers of Complexity	Rev. 1	Rev. 2
Number of new parts introduced to the manufacturing system	6	5
Number of new vendors introduced to the manufacturing system	3	2
Number of custom parts introduced to the manufacturing system	2	3
Number of new "major tools" (e.g., molds and dies) introduced to the manufacturing system	2	2
Number of new production processes introduced to the manufacturing system	0	0
Total	**13**	**12**

EXHIBIT 11-13   Scorecard of manufacturing complexity.

Minimize Systemic Complexity

An extremely simple manufacturing system would utilize a single process to transform a single raw material into a single part—perhaps a system extruding a single diameter of plastic rod from plastic pellets. Unfortunately, few such systems exist. Complexity arises from variety in the inputs, outputs, and transforming processes. Many real manufacturing systems involve hundreds of suppliers, thousands of different parts, hundreds of people, dozens of types of products, and dozens of types of production processes. Each variant of suppliers, parts, people, products, and processes introduces complexity to the system. These variants must usually be tracked, monitored, managed, inspected, handled, and inventoried at tremendous cost to the enterprise. Much of this complexity is driven by the design of the product and can therefore be minimized through smart design decisions.

Exhibit 11-13 shows a simple "scorecard" of manufacturing complexity useful for reminding designers of how the product design drives the complexity of the manufacturing system. The team establishes a score for the initial design and then uses changes in the score as a measure of success in reducing complexity. Note that the drivers given in the scorecard shown are generic categories. In practice, the team develops this list (and may prioritize it with weightings) based on the realities and constraints of the firm's production environment. Firms that use activity-based costing usually know quite well their primary drivers of complexity, as these are the cost drivers they use in allocating overhead. As a simple substitute for an accurate support cost model, such a scorecard allows the team to make informed decisions without formally estimating the indirect costs of production.

Error Proofing

An important aspect of DFM is to anticipate the possible failure modes of the production system and to take appropriate corrective actions early in the development process. This strategy is known as *error proofing*. One type of failure mode arises from having slightly different parts that can be easily confused. Examples of slightly different parts are screws differing only in the pitch of the threads (e.g., $4 \times .70$ mm and $4 \times .75$ mm screws) or in the direction of turning (left- and right-handed threads), parts which are mirror images of each other, and parts differing only in material composition.

We recommend either that these subtle differences be eliminated or that slight differences be exaggerated. Exhibit 11-14 shows an example of exaggerating subtle differences between parts: the left and right versions of the reel lock on a videocassette, which are mirror images of each other, are molded in two different colors. Color coding allows the parts to be identified easily and differentiated in materials handling and assembly.

**EXHIBIT
11-14**
Left and right
reel locks inside
a videocassette
(top center).
The two nearly
identical parts
are color coded
to avoid
confusion.

Photo by Stuart Cohen

Step 5: Consider the Impact of DFM Decisions on Other Factors

Minimizing manufacturing cost is not the only objective of the product development process. The economic success of a product also depends on the quality of the product, the timeliness of product introduction, and the cost of developing the product. There may also be situations in which the economic success of a project is compromised in order to maximize the economic success of the entire enterprise. In contemplating a DFM decision, these issues should be considered explicitly.

The Impact of DFM on Development Time

Development time can be precious. For an automobile development project, time may be worth as much as several hundred thousand dollars per day. For this reason, DFM decisions must be evaluated for their impact on development time as well as for their impact on manufacturing cost. While saving $1 in cost on each manifold would be worth perhaps $1 million in annual cost savings, it would almost certainly not be worth causing a six-month delay in an automobile program.

The relationship between DFM and development time is complex. Here, we note a few aspects of the relationship. The application of some of the DFA guidelines may result in very complex parts. These parts may be so complex that their design or the procurement of their tooling becomes the activity that determines the duration of the overall development effort (Ulrich et al., 1993). The cost benefits of the DFM decision may not be worth the delay in project duration. This is particularly true for products competing in dynamic markets.

The Impact of DFM on Development Cost

Development cost closely mirrors development time. Therefore, the same caution about the relationship between part complexity and development time applies to development cost. In general, however, teams that aggressively pursue low manufacturing costs as an integral part of the development process seem to be able to develop products in about the same time and with about the same budget as teams that do not. Part of this phenomenon certainly arises from the correlation between good project management practices and the application of sound DFM methods.

The Impact of DFM on Product Quality

Before proceeding with a DFM decision, the team should evaluate the impact of the decision on product quality. Under ideal circumstances, actions to decrease manufacturing cost would also improve product quality. For example, the new GM manifold resulted in cost reduction, weight reduction, and improved engine performance. It is not uncommon for DFM efforts focused primarily on manufacturing cost reduction to also result in improved serviceability, ease of disassembly, and recycling. However, in some cases actions to decrease manufacturing cost can have adverse effects on product quality (such as reliability or robustness), so it is advisable for the team to keep in mind the many dimensions of quality that are important for the product.

The Impact of DFM on External Factors

Design decisions may have implications beyond the responsibilities of a single development team. In economic terms, these implications may be viewed as externalities. Two such externalities are component reuse and life cycle costs.

- *Component reuse:* Taking time and money to create a low-cost component may be of value to other teams designing similar products. In general, this value is not explicitly accounted for in manufacturing cost estimates. The team may choose to take an action that is actually more costly for their product because of the positive cost implications for other projects.

- *Life cycle costs:* Throughout their life cycles, certain products may incur some company or societal costs which are not (or are rarely) accounted for in the manufacturing cost. For example, products may contain toxic materials requiring special handling in disposal. Products may incur service and warranty costs. Although these costs may not appear in the manufacturing cost analysis, they should be considered before adopting a DFM decision.

Results

During the 1980s, design-for-manufacturing practices were put into place in thousands of firms. Today DFM is an essential part of almost every product development effort. No longer can designers "throw the design over the wall" to production engineers. As a result of this emphasis on improved design quality, some manufacturers claim to have reduced production costs of products by up to 50 percent. In fact, comparing current new product designs with earlier generations, one can usually identify fewer parts in the new product, as well as new materials, more integrated and custom parts, higher-volume standard parts and subassemblies, and simpler assembly procedures.

A sketch of the redesigned intake manifold is shown in Exhibit 11-15. This DFM effort achieved impressive results. Exhibit 11-16 shows the cost estimate for the redesigned

**EXHIBIT
11-15**
The redesigned
intake manifold.

Courtesy of General
Motors Corp.

1993 3800 V-6 COMPOSITE UPPER INTAKE

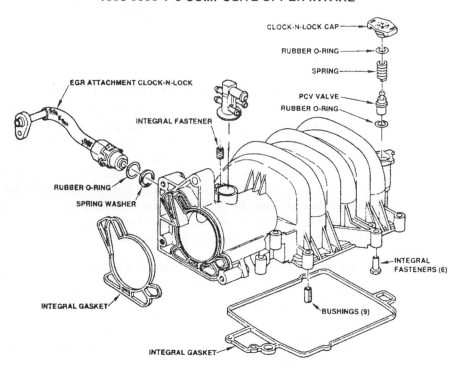

Component	Purchased Materials	Processing (Machine + Labor)	Assembly (Labor)	Total Unit Variable Cost	Tooling and Other NRE, K$	Tooling Lifetime, K units	Total Unit Fixed Cost	Total Cost
Manifold housing	3.85	1.56		5.41	350	1500	0.23	5.65
Intake runner insert	0.83	1.10	0.13	2.05	150	1500	0.10	2.15
Steel inserts (16)	0.32		1.00	1.32				1.32
EGR adapter	1.70		0.13	1.83				1.83
PCV valve								
Valve	0.85		0.04	0.89				0.89
O-rings(2)	0.02		0.16	0.18				0.18
Spring	0.08		0.10	0.18				0.18
Cover	0.02		0.10	0.12				0.12
Vacuum source block	0.04		0.06	0.10				0.10
Total Direct Costs	7.71	2.66	1.71	12.08	500		0.33	12.41
Overhead Charges	1.16	4.79	3.08				0.50	9.52
Total Cost								**21.93**

EXHIBIT 11-16 Cost estimate for the redesigned intake manifold.

intake manifold. (Compare with Exhibit 11-6.) The improvements over the previous design include:

- Unit cost savings of 45 percent.
- Mass savings of 66 percent (3.3 kilograms).
- Simplified assembly and service procedures.
- Improved emissions performance due to routing of EGR into the manifold.
- Improved engine performance due to reduced air induction temperatures.
- Reduced shipping costs due to lighter components.
- Increased standardization across vehicle programs.

For this product, the manufacturing cost savings alone amount to several million dollars annually. The other benefits listed above are also significant, although somewhat more difficult to quantify.

Summary

Design for manufacturing (DFM) is aimed at reducing manufacturing costs while simultaneously improving (or at least not inappropriately compromising) product quality, development time, and development cost.

- DFM begins with the concept development phase and system-level design phase; in these phases important decisions must be made with the manufacturing cost implications in mind.
- DFM utilizes estimates of manufacturing cost to guide and prioritize cost reduction efforts. Cost estimation requires expertise with the relevant production processes. Suppliers and manufacturing experts must be involved in this process.
- Since accurate cost estimation is very difficult, much of DFM practice involves making informed decisions in the absence of detailed cost data.
- Component costs are reduced by understanding what drives these costs. Solutions may involve novel component design concepts or the incremental improvement of existing designs through simplification and standardization.
- Assembly costs can be reduced by following well-established design-for-assembly (DFA) guidelines. Components can be redesigned to simplify assembly operations, or components can be eliminated entirely by integration of their functions into other components.
- Reduction of manufacturing support costs begins with an understanding of the drivers of complexity in the production process. Design decisions have a large impact on the costs of supporting production. Choices should be made with these effects in mind, even though overhead cost estimates are often insensitive to such changes.
- DFM is an integrative method taking place throughout the development process and requiring inputs from across the development team.
- DFM decisions can affect product development lead time, product development cost, and product quality. Trade-offs will frequently be necessary between manufacturing cost and these equally important broader issues.

References and Bibliography

Many current resources are available on the Internet via
www.ulrich-eppinger.net

Two articles describe the needs, methods, and success of DFM in the 1980s.
> Dean, James W., Jr., and Gerald I. Susman, "Organizing for Manufacturable Design," *Harvard Business Review,* January–February 1989, pp. 28–36.

> Whitney, Daniel E., "Manufacturing by Design," *Harvard Business Review,* July–August 1988, pp. 83–91.

There are numerous documented examples of DFM success. One classic example is the story of the IBM Proprinter, described by Dewhurst and Boothroyd.
> Dewhurst, Peter, and Geoffrey Boothroyd, "Design for Assembly in Action," *Assembly Engineering,* January 1987.

There are many references available to aid in component design, materials choice, manufacturing process selection, and understanding of process capabilities. Here are several sources which offer specific guidelines for hundreds of applications, materials, and processes.
> Bralla, James G. (ed.), *Design for Manufacturability Handbook,* McGraw-Hill, New York, 1999.

> Dixon, John R., and Corrado Poli, *Engineering Design and Design for Manufacturing: A Structured Approach,* Field Stone Publishers, Conway, MA, 1995.

> Farag, Mahmoud M., *Materials Selection for Engineering Design,* Prentice Hall, London, 1997.

> Cubberly, William H., and Ramon Bakerjian, *Tool and Manufacturing Engineers Handbook,* Society of Manufacturing Engineers, Dearborn, MI, 1989.

> Trucks, H. E., *Designing for Economical Production,* second edition, Society of Manufacturing Engineers, Dearborn, MI, 1987.

> Bolz, Roger W., *Production Processes: The Productivity Handbook,* fifth edition, Industrial Press, New York, 1981.

Gupta et al. provide a review of state-of-the-art manufacturability analysis methods and related DFM research.
> Gupta, Satyandra K, et al., "Automated Manufacturability Analysis: A Survey," *Research in Engineering Design,* Vol. 9, No. 3, 1997, pp. 168–190.

The *Thomas Register* is useful for identifying suppliers of components, tooling, machinery, and other industrial products.
> *Thomas Register of American Manufacturers,* Thomas Publishing Company, New York, published annually. Also available on the Internet.

The most popular method for DFA is by Boothroyd and Dewhurst. Software is also available to aid in estimating costs for both manual and automatic assembly, as well as a wide range of component costs.
> Boothroyd, Geoffrey, and Peter Dewhurst, *Product Design for Assembly,* Boothroyd Dewhurst, Inc., Wakefield, RI, 1989.

> Boothroyd, Geoffrey, Peter Dewhurst, and Winston A. Knight, *Product Design for Manufacturing,* Marcel Dekker, New York, 1994.

Detailed research on automated assembly has resulted in guidelines for designing products suited for assembly automation.

Nevins, James L., and Daniel E. Whitney, *Concurrent Design of Products and Processes,* McGraw-Hill, New York, 1989.

Boothroyd, Geoffrey, *Assembly Automation and Product Design,* Marcel Dekker, New York, 1992.

Kaplan and others describe the development of activity-based costing systems, which provide insight into a firm's cost drivers and facilitate more accurate cost estimation.

Kaplan, Robert S. (ed.), *Measures for Manufacturing Excellence,* Harvard Business School Press, Boston, 1990.

Clark and Fujimoto conducted a comprehensive study of product development in the world automobile industry. They provide an interesting analysis and discussion of the concept of black box component design.

Clark, Kim B., and Takahiro Fujimoto, *Product Development Performance: Strategy, Organization, and Management in the World Auto Industry,* Harvard Business School Press, Boston, 1991.

Ulrich et al. describe the trade-off between development time and manufacturing cost. They also describe an effort to model support costs in some detail.

Ulrich, Karl, Scott Pearson, David Sartorius, and Mark Jakiela, "Including the Value of Time in Design-for-Manufacturing Decision Making," *Management Science,* Vol. 39, No. 4, April 1993, pp. 429–447.

Ulrich and Pearson present a method for studying products, their costs, and some of the many detail design decisions resulting in the artifacts we see.

Ulrich, Karl T., and Scott Pearson, "Assessing the Importance of Design through Product Archaeology," *Management Science,* Vol. 44, No. 3, March 1998, pp. 352–369.

Exercises

1. Estimate the production cost for a simple product you may have purchased. Try costing a product with fewer than 10 components, such as a floppy disk, a pen, a jackknife, or a baby's toy. Remember that one reasonable upper bound for your estimate, including overhead, is the wholesale price (between 50 percent and 70 percent of retail).

2. Suggest some potential cost-reducing modifications you could make to improve the product costed above. Compute the DFA index before and after these changes.

3. List 10 reasons why reducing the number of parts in a product might reduce production costs. Also list some reasons why costs might increase.

Thought Questions

1. Consider the following 10 "design rules" for electromechanical products. Do these seem like reasonable guidelines? Under what circumstances could one rule conflict with another one? How should such a trade-off be settled?
   a. Minimize parts count.
   b. Use modular assembly.

    c. Stack assemblies.

    d. Eliminate adjustments.

    e. Eliminate cables.

    f. Use self-fastening parts.

    g. Use self-locating parts.

    h. Eliminate reorientation.

    i. Facilitate parts handling.

    j. Specify standard parts.

2. Is it practical to design a product with 100 percent assembly efficiency (DFA index = 1.0)? What conditions would have to be met? Can you think of any products with very high (greater than 75 percent) assembly efficiency?

3. Is it possible to determine what a product really costs once it is put into production? If so, how might you do this?

4. Can you propose a set of metrics that would be useful for the team to predict changes in the actual costs of supporting production? To be effective, these metrics must be sensitive to changes in the design that affect indirect costs experienced by the firm. What are some of the barriers to the introduction of such techniques in practice?

Appendix A

Materials Costs

EXHIBIT 11-17

Range of costs for common engineering materials. Price ranges shown correspond to various grades and forms of each material, purchased in bulk quantities (2003 prices).

Source: Adapted from David G. Ullman, *The Mechanical Design Process,* third edition, McGraw-Hill, New York, 2003

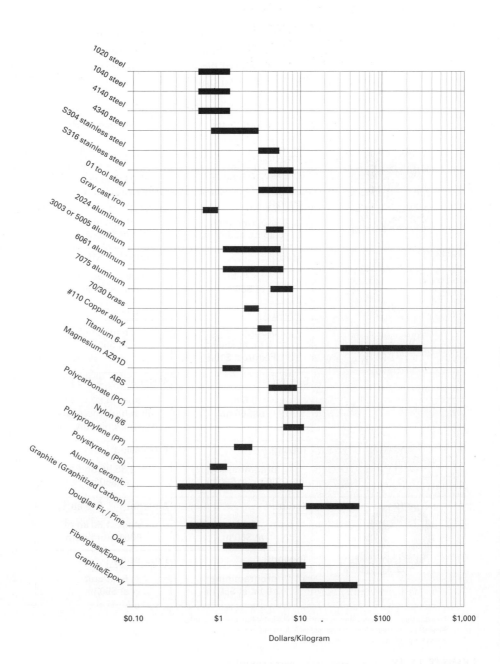

Appendix B

Component Manufacturing Costs

The exhibits in this appendix show example components and their cost data for computer-numerical control (CNC) machining (Exhibit 11-18), injection molding (Exhibit 11–19), progressive die stamping (Exhibit 11-20), and sand casting and investment casting (Exhibit 11-21). The purpose of these examples is to show, in general terms, what typical operations cost and how the cost structure of each process is affected by part complexity.

	Fixed Costs		Variable Costs		Volume	Total Unit Cost
	Setup:	0.75 hr. at $60/hr.	Material:	$9 ea. stock: 1.11 kg of 6061 aluminum	1	$75.00
					10	$21.00
	Tooling:	programming: 0.25 hr. at $60/hr.	Processing:	6 min./unit at $60/hr.	100	$15.50
	Setup:	1.75 hr. at $60/hr.	Material:	$16 ea. stock: 1.96 kg of 6061 aluminum	1	$386.00
					10	$102.50
	Tooling:	programming: 1.0 hr. at $60/hr. Fixtures: $150	Processing:	55 min./unit at $60/hr.	100	$74.15
	Setup:	5.5 hr. at $60/hr.	Material:	$25 ea. stock: 4.60 kg of ultra-high molecular weight polyethylene	1	$646.00
					10	$241.00
	Tooling:	programming: 2.0 hr. at $60/hr.	Processing:	2.85 hr./unit at $60/hr.	100	$200.50
	Setup:	2.0 hr. at $60/hr.	Material:	$12 ea. stock: 1.50 kg of 6061 aluminum	1	$612.00
					10	$396.00
	Tooling:	programming: 2.0 hr. at $60/hr.	Processing:	6 hr./unit at $60/hr.	100	$374.40

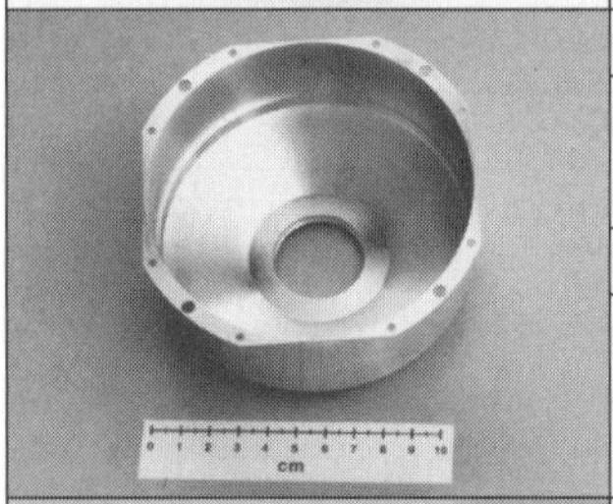

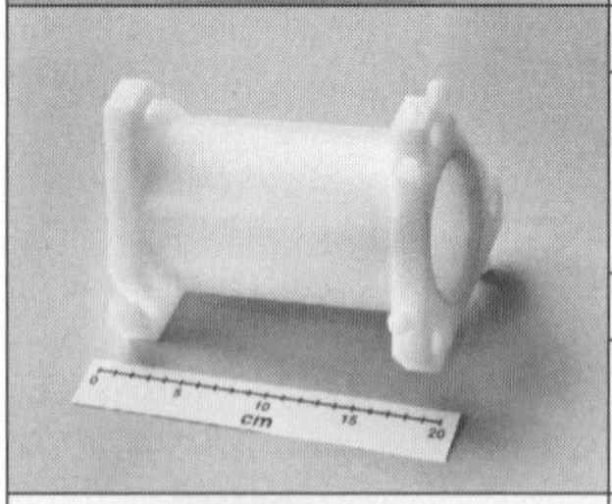

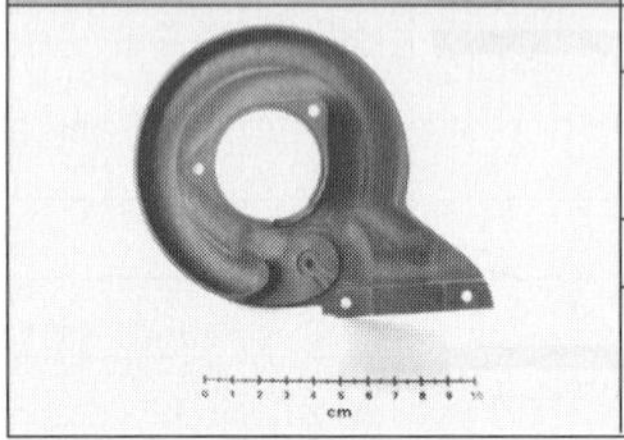

Notes: 1. Programming time is a one-time expense and is included here in tooling costs.
   2. Material prices assume low volumes and include cutting charges.
   3. Processing costs include overhead charges.

EXHIBIT 11-18   **CNC machining cost examples**
CNC machining example components and cost data.

Source: Photos by Stuart Cohen. Examples and data courtesy of Ramco, Inc.

	Fixed Costs		Variable Costs		Volume	Total Unit Cost
	Setup:		Material:	$0.075 ea.		
			45 g of linear low density polyethylene (LLDPF)		10K	$1.915
					100K	$0.295
	Tooling:	$18K 8 cavities/mold no actions	Processing:	1000 pcs/hr. on an 1800 KN press at $40/hr.	1M	$0.133
	Setup:		Material:	$0.244 ea.		
			10 g of steel-filled polycarbonate (PC)		10K	$1.507
					100K	$0.607
	Tooling:	$10K 1 cavity/mold no actions	Processing:	160 pcs/hr. on a 900 KN press at $42/hr.	1M	$0.517
	Setup:		Material:	$0.15 ea.		
			22 g of modified polyphenylene oxide (PPO)		10K	$2.125
					100K	$0.505
	Tooling:	$18K 2 cavities/mold no actions 3 retracting pins	Processing:	240 pcs/hr. on an 800 KN press at $42/hr.	1M	$0.343
	Setup:		Material:	$2.58 ea.		
			227 g of polycarbonate (PC) with 8 brass inserts		10K	$11.085
					100K	$3.885
	Tooling:	$80K 1 cavity/mold 1 action 4 retracting pins	Processing:	95 pcs/hr. on a 2700 KN press at $48/hr.	1M	$3.165

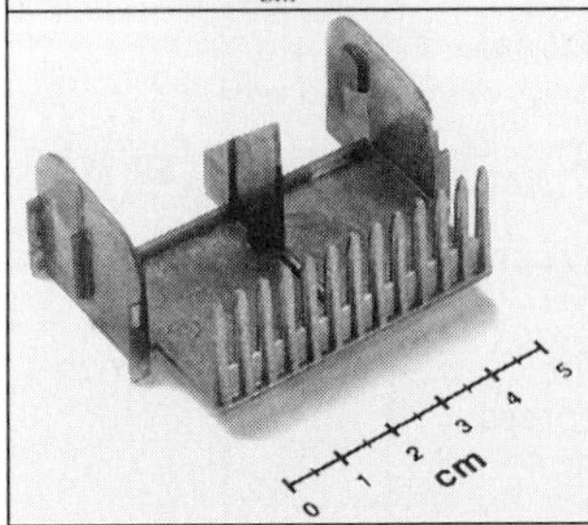

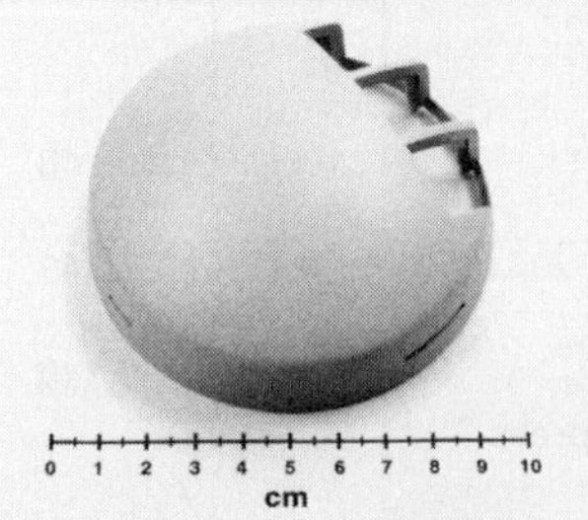

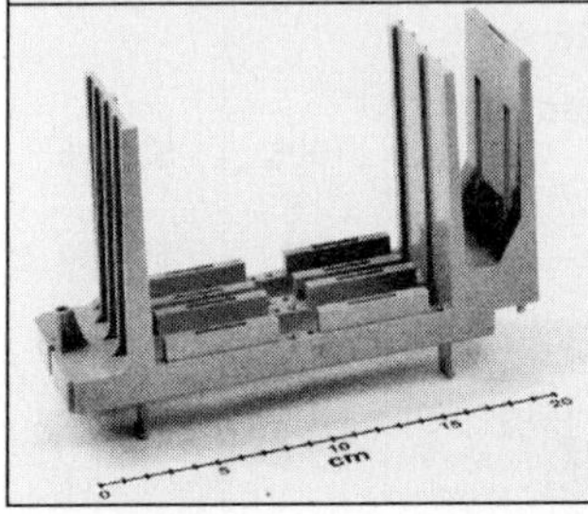

Notes: 1. Setup costs (only a few hours in each case) are negligible for high-volume injection molding.
2. Processing costs include overhead charges.

EXHIBIT 11-19   **Injection molding cost examples**
Injection molding example components and cost data.

Source: Photos by Stuart Cohen. Examples and data courtesy of Lee Plastics, Inc., and Digital Equipment Corporation

	Fixed Costs	Variable Costs	Volume	Total Unit Cost
	Setup:	Material: $0.040 ea. 2.2g 70/30 Brass	100K	$0.281
			1M	$0.083
	Tooling: $22K	Processing: 3000 pcs/hr. on an 550 KN press at $63/hr.	10M	$0.063
	Setup:	Material: $0.032 ea. 3.5 g 304 SST	100K	$0.775
			1M	$0.136
	Tooling: $71K	Processing: 4300 pcs/hr. on a 550 KN press at $140/hr.	10M	$0.072
	Setup:	Material: $0.128 ea. 19.2 g 102 copper	100K	$0.248
			1M	$0.149
	Tooling: $11K	Processing: 4800 pcs/hr. on a 650 KN press at $50/hr.	10M	$0.140
	Setup:	Material: $0.28 ea. 341 g galvanized steel	100K	$2.516
			1M	$0.761
	Tooling: $195K	Processing: 700 pcs/hr. on a 1000 KN press at $200/hr.	10M	$0.585

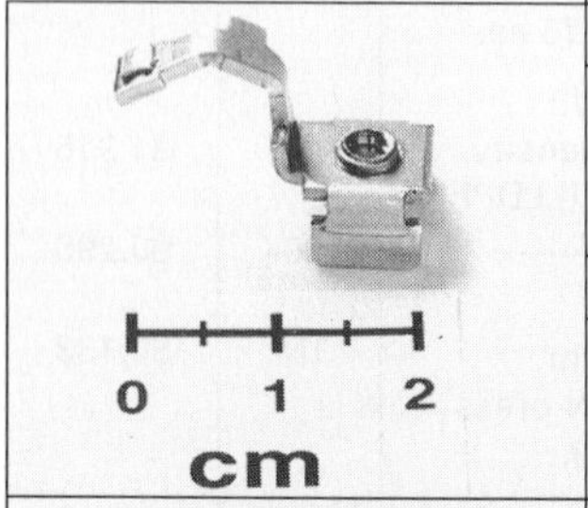

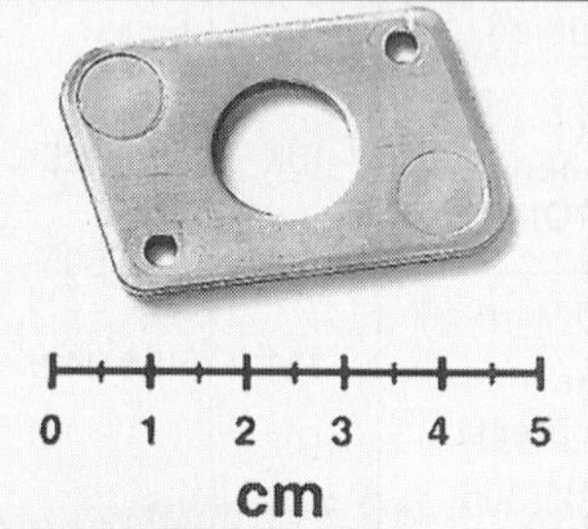

Notes: 1. Setup costs (only a few hours in each case) are negligible for high-volume stamping.
　　　2. Material weights represent the finished stampings. Material costs include scrap.
　　　3. Hourly processing costs are not only driven by press size, but also can include ancillary processing equipment, such as in-die tapping.
　　　4. Processing costs include overhead charges.

EXHIBIT 11-20　**Stamping cost examples**

Volume progressive die stamping example components and cost data.

Source: Photos by Stuart Cohen. Examples and data courtesy of Brainin Advance Industries and other sources

	Fixed Costs	Variable Costs		Volume	Total Unit Cost
Setup:		Material:	$0.53 ea.		
			570 g of gray cast iron	10	$180.91
				100	$18.91
Tooling:	$1.8K 8 impressions/pattern no core	Processing:	120 pcs/hr. at $46/hr.	1000	$2.71
Setup:		Material:	$2.42 ea.		
			2,600 g of gray cast iron	10	$243.95
				100	$27.95
Tooling:	$2.4K 2 impressions/pattern 1 core	Processing:	30 pcs/hr. at $46/hr.	1000	$6.35

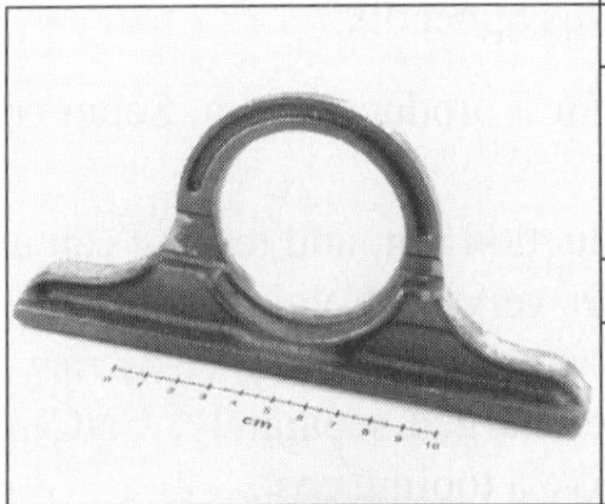

	Fixed Costs	Variable Costs		Volume	Total Unit Cost
Setup:		Material:	$0.713 ea.		
			260 g of yellow brass	10	$163.21
				100	$28.21
Tooling:	$1.5K no cores	Processing:	4 pcs/hr. at $50/hr.	1000	$14.71
Setup:		Material:	$0.395 ea.		
			180 g of 712 aluminum	10	$750.40
				100	$120.40
Tooling:	$7K 3 cores	Processing:	1 pc/hr. at $50/hr.	1000	$57.40

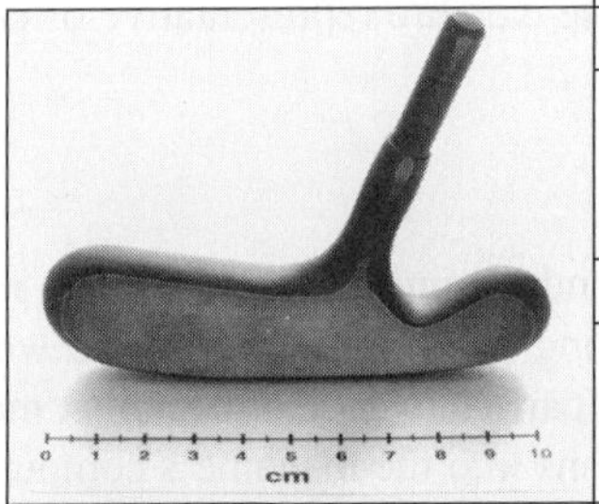

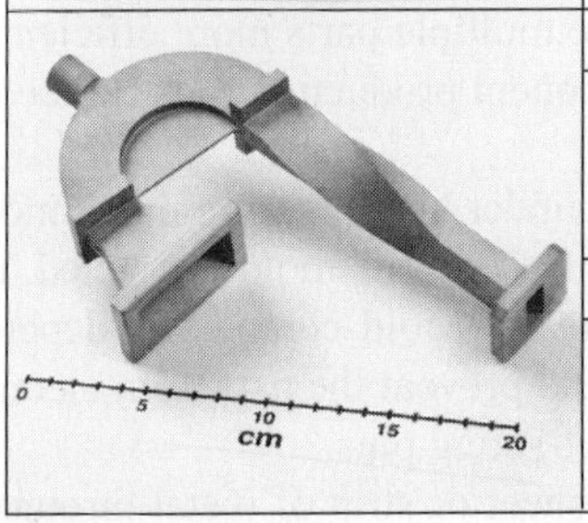

Notes: 1. Setup is not generally charged in costing.
   2. Processing costs include overhead charges.

EXHIBIT 11-21   Casting cost examples

Sand casting (top) and investment casting (bottom) example components and cost data.

Source: Photos by Stuart Cohen. Examples and data courtesy of Cumberland Foundry Co., Inc. (sand casting), and Castronics, Inc. (investment casting)

Terminology

The following terminology applies to all of the tables in this appendix:

- *Setup* is the work required to prepare the equipment for a production run. Setup costs are charged for each run.
- *Tooling costs* are incurred in advance of the first production run, and tooling can usually be reused for later production runs. However, in very high-volume production runs, tooling wears out and therefore is a recurring expense. Tooling costs may be spread over the entire production volume or may be charged separately. CNC programming time is generally also a one-time expense, like a tooling cost.
- *Material types* are listed for each part. Material weights and costs include processing scrap and waste.
- *Processing costs* vary with the type of manufacturing equipment used and include charges for both machine time and labor.

While fixed costs (setup and tooling) are sometimes billed separately from material and processing costs, for these examples, fixed costs are spread over the production volume shown. Unit costs are calculated as

$$\text{Total unit cost} = \frac{\text{Setup costs} + \text{Tooling costs}}{\text{Volume}} + \text{Variable costs}$$

The cost rates given include overhead charges, so these data are representative of custom components purchased from suppliers.

Description of Processes

CNC machining includes computer-controlled milling and turning processes. CNC machines are highly flexible due to automatic tool-changing mechanisms, multiple work axes, and programmable computer control. To produce a particular part, a machinist must first program the cutting tool trajectories and tool selections into the machine's computer. Also, fixtures or other tooling may be utilized to produce multiple parts more efficiently. Once the program is written and fixtures are made, subsequent production runs can be set up much more quickly.

Injection molding is the process of forcing hot plastic under high pressure into a mold, where it cools and solidifies. When the part is sufficiently cool, the mold is opened, the part is ejected, the mold closes, and the cycle begins again. Mold complexity depends highly on the part geometry; undercuts (features that would prevent the part from ejecting out of the mold) are achieved using mold "actions" or "retracting pins."

Progressive die stamping is the process of passing a sheet or strip of metal through a set of dies to cut and/or form it to a desired size and shape. While some stampings require only cutting, formed stampings are made by bending and stretching the metal beyond its yield point, thereby causing permanent deformation.

Sand castings are created by forming a sand mold from master patterns (tooling in the shape of the final part). Special binders are mixed with the sand to allow the sand to retain shape when packed around the pattern to create a single-use mold. Internal cavities in a casting can be created using additional sand cores inside the outer mold. Molten metal is then poured into the mold where the metal cools and solidifies. Once cool, the sand is broken off to reveal the metal casting. Sand castings generally require subsequent machining operations to create finished components.

Investment castings are made by first creating a temporary wax pattern, using master tooling. The wax pattern is then dipped or immersed in plaster or ceramic slurry which is allowed to solidify. The form is then heated, melting out the wax, and leaving behind only the thin shell as a mold. Molten metal is then poured into the mold, where it cools and solidifies. When the metal is cool, the mold is broken off to reveal the metal part.

Detailed process descriptions for the above and numerous other processes, as well as more detailed cost estimating techniques, can be found in the reference books listed for this chapter.

Appendix C

Assembly Costs

Product	Part Data		Assembly Times (Seconds)	
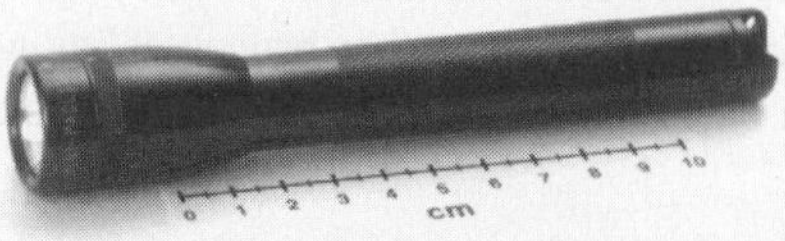	No. of Parts	16	Total	125.7
	No. of Unique Parts	12	Slowest Part	9.7
	No. of Fasteners	0	Fastest Part	2.9
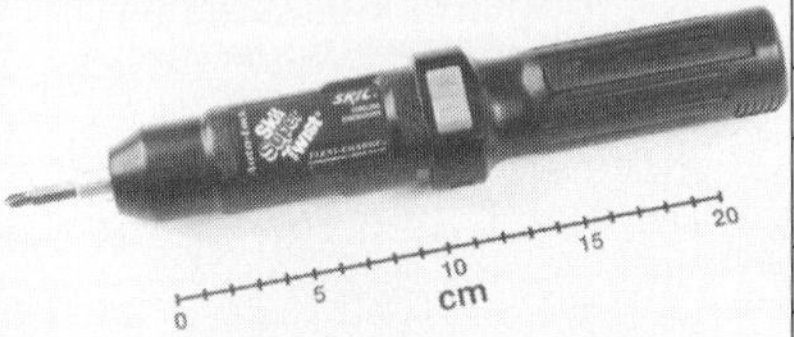	No. of Parts	34	Total	186.5
	No. of Unique Parts	25	Slowest Part	10.7
	No. of Fasteners	5	Fastest Part	2.6
	No. of Parts	49	Total	266.0
	No. of Unique Parts	43	Slowest Part	14.0
	No. of Fasteners	5	Fastest Part	3.5
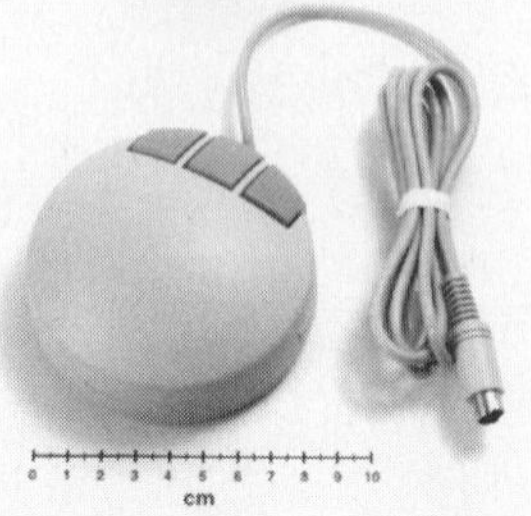	No. of Parts	56/17*	Total	277.0/138.0*
	No. of Unique Parts	44/12*	Slowest Part	8.0/8.0*
	No. of Fasteners	0/0*	Fastest Part	0.75/3.0*

*Data for the mouse are given as: total components (including electronic)/mechanical components only.

Notes: 1. This table gives manual assembly times, which can be converted to assembly costs using applicable labor rates.

2. Assembly times shown include times for individual part handling and insertion, as well as other operations such as subassembly handling and insertion, reorientations, and heat riveting.

EXHIBIT 11-22   **Assembly costs**

Assembly data for common products. Obtained using Boothroyd Dewhurst Inc. DFA Software.

Source: Photos by Stuart Cohen. Data obtained by using Boothroyd Dewhurst Inc. DFA software

Component	Time (Seconds)			Component	Time (Seconds)		
	Min	*Max*	*Avg*		*Min*	*Max*	*Avg*
Screw 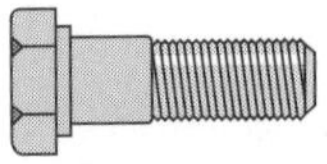	7.5	13.1	10.3	Pin	3.1	10.1	6.6
Snap-fit 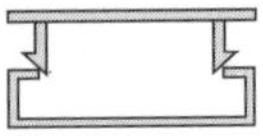	3.5	8.0	5.9	Spring 	2.6	14.0	8.3

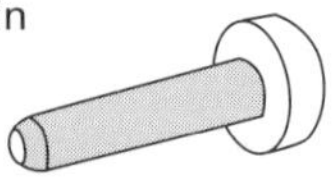

EXHIBIT 11-23 Typical handling and insertion times for common components.

Source: Manual assembly tables in Boothroyd and Dewhurst, 1989

Appendix D

Cost Structures

Type of Firm	Cost Calculation
Electromechanical products manufacturer (Traditional cost structure)	Cost = (113%) × (Materials cost) + (360%) × (Direct labor cost)
Precision valve manufacturer (Activity-based cost structure)	Cost = (108%) × [(Direct labor cost) + (Setup labor cost) + (160%) × (Materials cost) + ($27.80) × (Machine hours) + ($2,000.00) × (Number of shipments)]
Heavy equipment component manufacturer (Activity-based cost structure)	Cost = (110%) × (Materials cost) + (109%) × [(211%) × (Direct labor cost) + ($16.71) × (Machine hours) + ($33.76) × (Setup hours) + ($114.27) × (Number of production orders) + ($19.42) × (Number of material handling loads) + ($487.00) × (Number of new parts added to the system)]

Notes: 1. This table shows total costs per customer order.
   2. Materials costs include costs of raw materials and purchased components.

EXHIBIT 11-24 Typical cost structures for manufacturing firms.

Sources, top to bottom: Unpublished company source; Harvard Business School cases: Destin Brass Products Co., 9-190-089, and John Deere Component Works, 9-187-107

CHAPTER TWELVE

Prototyping

Courtesy of Apple Computer

EXHIBIT 12-1
Powerbook Duo by Apple Computer, Inc.

This chapter was developed in collaboration with Amy Greenlief.

EXHIBIT 12-2
The Duo
trackball.

Photo by Stuart Cohen

Apple Computer sold over 400,000 of the original Powerbook notebook computers in the first year after product launch, making the Powerbook one of the most successful new products in Apple's history. However, the subsequent model of Powerbook, the Duo, had to be even smaller, lighter, and more powerful in order to be competitive. One major obstacle to reducing the thickness of the original Powerbook was the thickness of the trackball mechanism, and so a new trackball became one of the key elements of the Duo development effort. A trackball was commonly used in a portable computer before trackpads became reliable. The ball, recessed in a housing, was manipulated by the user in order to control the position of the cursor on the computer screen. Exhibit 12-1 shows a photograph of the Duo, and Exhibit 12-2 shows a close-up view of the trackball. In addition to the overarching goal of reducing the size of the trackball, the development team at Apple hoped to decrease its cost and increase its quality. Throughout the development of the trackball, the team used a variety of prototypes. This chapter defines *prototype,* explains why prototypes are built, and then presents several principles of prototyping practice. The chapter also describes a method for planning prototypes before they are built. The Duo trackball is used as an illustrative example throughout.

Prototype Basics

What Is a Prototype?

Although dictionaries define *prototype* as a noun only, in product development practice the word is used as a noun, a verb, and an adjective. For example:

- Industrial designers produce *prototypes* of their concepts.
- Engineers *prototype* a design.
- Software developers write *prototype* programs.

We define *prototype* as "an approximation of the product along one or more dimensions of interest." Under this definition, any entity exhibiting at least one aspect of the product that is of interest to the development team can be viewed as a prototype. This definition deviates from standard usage in that it includes such diverse forms of prototypes as concept sketches, mathematical models, and fully functional preproduction versions of the product. *Prototyping* is the process of developing such an approximation of the product.

Types of Prototypes

Prototypes can be usefully classified along two dimensions. The first dimension is the degree to which a prototype is *physical* as opposed to *analytical.* Physical prototypes are tangible artifacts created to approximate the product. Aspects of the product of interest to the development team are actually built into an artifact for testing and experimentation. Examples of physical prototypes include models which look and feel like the product, proof-of-concept prototypes used to test an idea quickly, and experimental hardware used to validate the functionality of a product. Exhibit 12-3 shows three forms of physical prototypes used for diverse purposes. Analytical prototypes represent the product in a nontangible, usually mathematical, manner. Interesting aspects of the product are analyzed, rather than built. Examples of analytical prototypes include computer simulations, systems of equations encoded within a spreadsheet, and computer models of three-dimensional geometry.

The second dimension is the degree to which a prototype is *comprehensive* as opposed to *focused.* Comprehensive prototypes implement most, if not all, of the attributes of a product. A comprehensive prototype corresponds closely to the everyday use of the word *prototype,* in that it is a full-scale, fully operational version of the product. An example of a comprehensive prototype is one given to customers in order to identify any remaining design flaws before committing to production. In contrast to comprehensive prototypes, focused prototypes implement one, or a few, of the attributes of a product. Examples of focused prototypes include foam models to explore the form of a product and wire-wrapped circuit boards to investigate the electronic performance of a product design. A common practice is to use two or more focused prototypes together to investigate the overall performance of a product. One of these prototypes is often a "looks-like" prototype, and the other is a "works-like" prototype. By building two separate focused prototypes, the team may be able to answer its questions much earlier than if it had to create one comprehensive prototype.

Exhibit 12-4 displays a plot with axes corresponding to these two dimensions. Several different prototypes from the trackball example are shown on this plot. Note that focused prototypes can be either physical or analytical, but that for physical products, fully comprehensive prototypes must generally be physical. Prototypes sometimes contain a combination of analytical and physical elements. For example, a mechanical trackball could be linked to a software simulation of the circuitry that would eventually be used to monitor the trackball motion. Some analytical prototypes can be viewed as being more "physical" than others. For example, a video animation of a mechanism generated from a detailed simulation of the physical interactions of the elements in the mechanism is, in one sense, more physical than a set of equations approximating the overall motion of the same mechanism.

EXHIBIT 12-3

Examples of physical prototypes. From top to bottom: a prototype to test the strength of the wings of the Boeing 777, a set of prototypes to evaluate the comfort of screwdriver handles, and an appearance model to communicate the form and style of a product.

Courtesy of Boeing Company

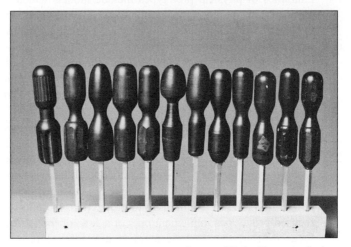

Courtesy of Design Management Institute

Courtesy of Lunar Design

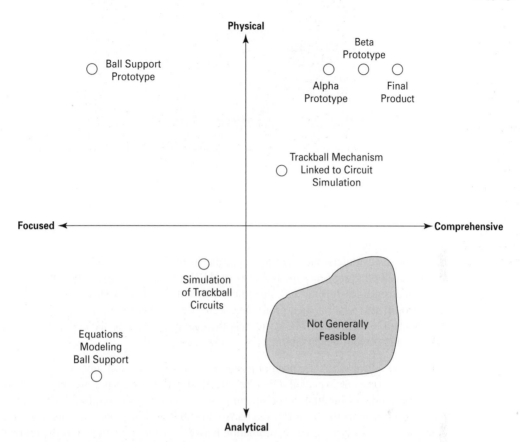

EXHIBIT 12-4 Types of prototypes. Prototypes can be classified according to the degree to which they are physical and the degree to which they implement all of the attributes of the product.

What Are Prototypes Used For?

Within a product development project, prototypes are used for four purposes: learning, communication, integration, and milestones.

Learning Prototypes are often used to answer two types of questions: "Will it work?" and "How well does it meet the customer needs?" When used to answer such questions, prototypes serve as learning tools. In developing the trackball, the ball support system was investigated using focused prototypes. The three support pegs were placed at different radii from the center of the ball socket. In Exhibit 12-5, these support pegs can be seen in the prototype with the 13.00-millimeter radius placement. Other prototypes with spaces for the pegs at different radii are also shown. This is an example of a focused physical prototype used as a learning tool. During the development of the trackball, geometric modeling on a computer was used to check for fit and interference of the trackball components. This is an example of a focused analytical prototype used as a learning tool.

EXHIBIT 12-5
Focused physical prototypes of the trackball support. These prototypes were used for learning about trackball "feel."

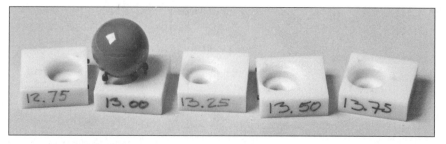

Photo by Stuart Cohen

Communication Prototypes enrich communication with top management, vendors, partners, extended team members, customers, and investors. This is particularly true of physical prototypes: a visual, tactile, three-dimensional representation of a product is much easier to understand than a verbal description or even a sketch of the product. When developing concepts for the Duo trackball, communication between design engineers, industrial designers, and customers was enhanced through the use of "look and feel" prototypes. Several such prototypes were used to communicate possible features of the trackball to customers. The customers then gave feedback to the engineers and industrial designers about which features they preferred.

Integration Prototypes are used to ensure that components and subsystems of the product work together as expected. Comprehensive physical prototypes are most effective as integration tools in product development projects because they require the assembly and physical interconnection of all of the parts and subassemblies that make up a product. In doing so, the prototype forces coordination between different members of the product development team. If the combination of any of the components of the product interferes with the overall function of the product, the problem may be detected only through physical integration in a comprehensive prototype. Common names for these comprehensive physical prototypes are *experimental, alpha, beta,* or *preproduction* prototypes. Two such prototypes (an alpha and a beta) for the Duo trackball are shown in Exhibit 12-6. In the alpha prototype, actuator arms are visible on the switches (lower-left and upper-right corners). In the beta prototype, the actuator arms have been removed. This change was necessary because a tolerance "stack-up" resulted in poor actuation of some switches. This problem was detected as a result of the physical integration forced by the assembly and test of the alpha prototype.

Prototypes also help to integrate the perspectives of the different functions represented on a product development team (Leonard-Barton, 1991). A simple physical model of the form of a product can be used as the medium through which the marketing, design, and manufacturing functions agree on a basic design decision.

Microsoft has refined the use of prototypes to integrate the activities of dozens of software developers. It employs a "daily build," in which a new version of the product is compiled at the end of every day. Software developers "check in" their code by a fixed time of day (e.g., 5:00 P.M.) and a team compiles the code to create a new version of the software. The most recent version of the software is then used by everyone on the team,

EXHIBIT 12-6
Comprehensive
trackball
prototypes. The
alpha prototype,
with the light
gray retaining
ring, is on the
left; the beta
prototype, with
the dark gray
retaining ring, is
on the right.

Photo by Stuart Cohen

EXHIBIT 12-7
Appropriateness
of different types
of prototypes
for different
purposes
(● = more
appropriate, ○ =
less appropriate).
Note that
comprehensive-
analytical
prototypes are
rarely possible.

	Learning	Communication	Integration	Milestones
Focused Analytical	●	○	○	○
Focused Physical	●	●	○	○
Comprehensive Physical	●	●	●	●

in a practice Microsoft calls "eating your own dog food." This practice of creating daily comprehensive prototypes ensures that the efforts of the developers are always synchronized and integrated. Any conflicts are detected immediately and the team can never diverge more than one day from a working version of the product.

Milestones   Particularly in the later stages of product development, prototypes are used to demonstrate that the product has achieved a desired level of functionality. Milestone prototypes provide tangible goals, demonstrate progress, and serve to enforce the schedule. Senior management (and sometimes the customer) often requires a prototype that demonstrates certain functions before allowing the project to proceed. For example, in many government procurements, a prototype must pass a "first-article test" before a contractor can proceed with production. A major milestone for the Duo trackball development was the ability to pass a "50-mile test." During this test the ball is moved randomly by automatic test equipment for a total of 50 miles. There can be no failures of the trackball system for the test to be successful.

While all types of prototypes are used for all four of these purposes, some types of prototypes are more appropriate than others for some purposes. A summary of the relative appropriateness of different types of prototypes for different purposes is shown in Exhibit 12-7.

Principles of Prototyping

Several principles are useful in guiding decisions about prototypes during product development. These principles inform decisions about what type of prototype to build and about how to incorporate prototypes into the development plan.

Analytical Prototypes Are Generally More Flexible than Physical Prototypes

Because an analytical prototype is a mathematical approximation of the product, it will generally contain parameters that can be varied in order to represent design alternatives. In most cases, changing a parameter in an analytical prototype is easier than changing an attribute of a physical prototype. For example, consider an analytical prototype of the trackball that includes a set of equations representing the sliding of the ball on its supports. One of the parameters in the equations is the coefficient of friction between the ball material and the support material. Changing this parameter and then solving the equations is much easier than changing the actual materials in the physical prototype. In most cases, the analytical prototype not only is easier to change than a physical prototype but also allows larger changes than could be made in a physical prototype. For this reason, an analytical prototype frequently precedes a physical prototype. The analytical prototype is used to narrow the range of feasible parameters, and then the physical prototype is used to fine-tune or confirm the design. See Chapter 13, Robust Design, for a detailed example of the use of an analytical prototype to explore several design parameters.

Physical Prototypes Are Required to Detect Unanticipated Phenomena

A physical prototype often exhibits unanticipated phenomena completely unrelated to the original objective of the prototype. One reason for these surprises is that all of the laws of physics are always operating when the team experiments with physical prototypes. Physical prototypes intended to investigate purely geometric issues will also have thermal and optical properties. Some of the incidental properties of physical prototypes are irrelevant to the final product and act as annoyances during testing. However, some of these incidental properties of physical prototypes will also manifest themselves in the final product. In these cases, a physical prototype can serve as a tool for detecting unanticipated detrimental phenomena that may arise in the final product. Analytical prototypes, in contrast, can never reveal phenomena that are not part of the underlying analytical model on which the prototype is based. For this reason at least one physical prototype is almost always built in a product development effort.

A Prototype May Reduce the Risk of Costly Iterations

Exhibit 12-8 illustrates the role of risk and iteration in product development. In many situations, the outcome of a test may dictate whether a development task will have to be repeated. For example, if a molded part fits poorly with its mating parts, the mold may have to be rebuilt. In Exhibit 12-8, a 30 percent risk of returning to the mold-building activity after testing part fit is represented with an arrow labeled with a probability of 0.30. If building and testing a prototype substantially increases the likelihood that the subsequent

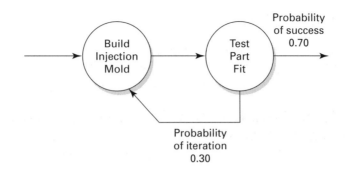

Conventional Process

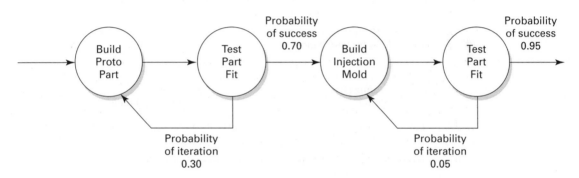

Process with Prototyping

EXHIBIT 12-8   A prototype may reduce the risk of costly iteration. Taking time to build and test a prototype may allow the development team to detect a problem that would otherwise not have been detected until after a costly development activity, such as building an injection mold.

activities will proceed without iteration (e.g., from 70 percent to 95 percent, as indicated in Exhibit 12-8), the prototype phase may be justified.

The anticipated benefits of a prototype in reducing risk must be weighed against the time and money required to build and evaluate the prototype. Products that are high in risk or uncertainty, due to high costs of failure, new technology, or the revolutionary nature of the product, will benefit from such prototypes. On the other hand, products for which failure costs are low and the technology is well known do not derive as much risk-reduction benefit from prototyping. Most products fall between these extremes. Exhibit 12-9 represents the range of situations that can be encountered in different types of development projects.

A Prototype May Expedite Other Development Steps

Sometimes the addition of a short prototyping phase may allow a subsequent activity to be completed more quickly than if the prototype were not built. If the required time for the prototype phase is less than the savings in duration of the subsequent activity, then this strategy is appropriate. One of the most common occurrences of this situation is in mold design, as illustrated in Exhibit 12-10. The existence of a physical model of a geometrically complex part allows the mold designer to more quickly visualize and design the mold.

EXHIBIT 12-9

The use of comprehensive prototypes depends on the relative level of technical or market risk and the cost of building a comprehensive prototype.

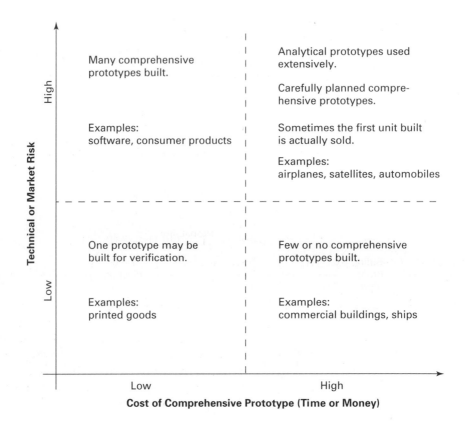

Many comprehensive prototypes built.

Examples: software, consumer products

Analytical prototypes used extensively.

Carefully planned compre-hensive prototypes.

Sometimes the first unit built is actually sold.

Examples: airplanes, satellites, automobiles

One prototype may be built for verification.

Examples: printed goods

Few or no comprehensive prototypes built.

Examples: commercial buildings, ships

High

Low

Technical or Market Risk

Low   High

Cost of Comprehensive Prototype (Time or Money)

EXHIBIT 12-10

Role of a prototype in expediting another step. Taking time to build a prototype may enable more rapid completion of a subsequent step.

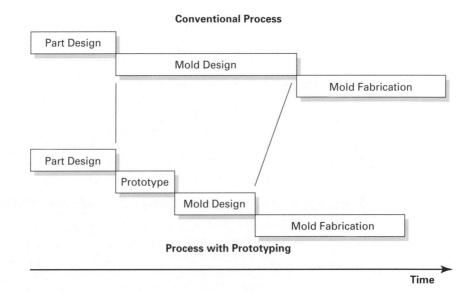

Conventional Process

Part Design

Mold Design

Mold Fabrication

Part Design

Prototype

Mold Design

Mold Fabrication

Process with Prototyping

Time

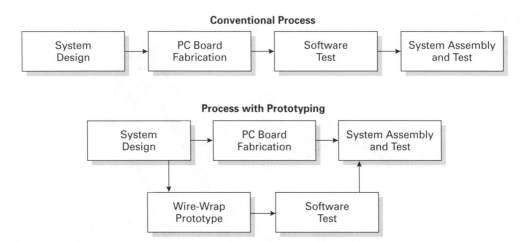

EXHIBIT 12-11   Use of a prototype to remove a task from the critical path.

A Prototype May Restructure Task Dependencies

The top part of Exhibit 12-11 illustrates a set of tasks that are completed sequentially. It may be possible to complete some of the tasks concurrently by building a prototype. For example, a software test may depend on the existence of a physical circuit. Rather than waiting for the production version of the printed circuit board to use in the test, the team may be able to rapidly fabricate a prototype (e.g., a wire-wrapped board) and use it for the test while the production of the printed circuit board proceeds.

Prototyping Technologies

Hundreds of different production technologies are used to create prototypes, particularly physical prototypes. Two technologies have emerged as particularly important in the past 10 years: three-dimensional (3D) computer modeling and free-form fabrication.

3D Computer Modeling

Since the 1990s, the dominant mode of representing designs has shifted dramatically from drawings, often created using a computer, to 3D computer models. These models represent designs as collections of 3D entities, each usually constructed from geometric primitives, such as cylinders, blocks, and holes. Exhibit 12-12 is an image of a 3D computer model.

The advantages of 3D computer modeling include the ability to easily visualize the three-dimensional form of the design; the ability to automatically compute physical properties such as mass and volume; and the efficiency arising from the creation of one and only one canonical description of the design, from which other, more focused descriptions, such as cross-sectional views, can be created. 3D computer models can also be used to detect geometric interference among parts and are the underlying representation for more focused analyses of, for example, kinematics or stress. These 3D computer models have

**EXHIBIT
12-12**
3D computer
model of a fully
suspended
mountain
bicycle.

Courtesy of SolidWorks

begun to serve as prototypes. In some settings the use of 3D computer modeling has eliminated one or more physical prototypes. For example, in the development of the Boeing
777, the development team was able to avoid building a full-scale wooden prototype of the
plane, which had historically been used to detect geometric interferences among structural
elements and the components of various other systems, such as hydraulic lines. A 3D computer model of an entire product is known, depending on the industry setting, as a "digital
mock up," "digital prototype," or "virtual prototype."

Free-Form Fabrication

In 1984, the first commercial free-form fabrication system was introduced by 3D Systems. This technology, called *stereolithography,* and dozens of competing technologies
which followed it, create physical objects directly from 3D computer models, and can be
thought of as "three-dimensional printers." This collection of technologies is often called
rapid prototyping. Most of the technologies work by constructing an object, one cross-
sectional layer at a time, by depositing a material or by selectively solidifying a liquid.
The resulting parts are most often made from plastics, but other materials are available,
including wax, paper, ceramics, and metals. In some cases the parts are used directly for
visualization or in working prototypes. However, the parts are often used as patterns to
make molds or dies from which parts with particular material properties can then be
molded or cast. Exhibit 12-13 shows enclosure parts for a consumer product that were
made by casting polyurethane into a silicon rubber mold which had been created from a
stereolithography part.

Free-form fabrication technologies enable realistic three-dimensional prototypes to be
created earlier and less expensively than was possible before. When used appropriately,

Courtesy of 3D Systems, Inc.

these prototypes can reduce product development time and/or improve the resulting product. In addition to enabling the rapid construction of working prototypes, these technologies can be used to embody product concepts quickly and inexpensively, increasing the ease with which concepts can be communicated to other team members, senior managers, development partners, or potential customers.

Planning for Prototypes

A potential pitfall in product development is what Clausing calls the "hardware swamp" (Clausing, 1994). The swamp is caused by misguided prototyping efforts, that is, the building and debugging of prototypes (physical or analytical) that do not substantially contribute to the goals of the overall product development project. One way to avoid the swamp is to carefully define each prototype before embarking on an effort to build and test it. This section presents a four-step method for planning a prototype during a product development effort. The method applies to all types of prototypes: focused, comprehensive, physical, and analytical. A template for recording the information generated from the method is given in Exhibit 12-14. We use the ball-support prototype shown in Exhibit 12-5 as an example to illustrate the method.

Step 1: Define the Purpose of the Prototype

Recall the four purposes of prototypes: learning, communication, integration, and milestones. In defining the purpose of a prototype, the team lists its specific learning and communication needs. Team members also list any integration needs and whether or not

Name of Prototype	Ball Support
Purpose (Communication, Learning, Integration, Milestones)	What combination of spacing between ball supports and ball inertia yields the best trackball feel? (Learning) How much variation is there in users' preferences for feel? (Learning)
Level of Approximation	Ball surface material as planned for production design. Support material as planned for production design. Support contact geometry as planned for production design.
Outline of Test Plan	Build two sets of five different spacings for ball support. Test two balls with different inertias. Test spacings of 12.75, 13.00, 13.25, 13.50, 13.75 mm for each of two ball inertias. Verify that all the spacings provide at least minimally acceptable performance. Have at least 20 users rank order the prototypes according to feel.
Schedule	12 August Parts available. 12 August Parts assembled. 20 August Tests completed. 22 August Analysis of results completed.

EXHIBIT 12-14 Prototype planning template.

the prototype is intended to be one of the major milestones of the overall product development project.

For the ball-support prototype, the purpose of the prototype was to determine the spacing of the support bearings and the inertia of the ball that would result in the best ball feel. This "learning" prototype was very focused and there were no other major purposes of the prototype.

Step 2: Establish the Level of Approximation of the Prototype

Planning a prototype requires that the degree to which the final product is approximated be defined. The team should consider whether a physical prototype is necessary or whether an analytical prototype would best meet its needs. In most cases, the best prototype is the simplest prototype that will serve the purposes established in Step 1. In some cases, an existing prototype or a prototype being built for another purpose can be borrowed.

For the ball support, the team decided that the only attributes of the product that needed to be approximated were the ball diameter, the ball inertia, the ball material, the bearing support geometry and stiffness, and the bearing support materials. All of the other aspects of the trackball could be ignored, including the color, the surrounding enclosure, the ball retention device, and the position-sensing mechanism of the trackball.

A member of the team had previously explored an analytical prototype of the trackball support and felt that the physical prototype was necessary to verify his analysis. He had discovered that there was a basic trade-off between minimizing friction, which would tend to drive the supports closer together, and maximizing the stability of the ball, which would tend to drive the supports farther apart. The team used the analytical prototype to determine the range of ball support dimensions that would be investigated with the physical prototype.

Step 3: Outline an Experimental Plan

In most cases, the use of a prototype in product development can be thought of as an experiment. Good experimental practice helps to ensure the extraction of maximum value from the prototyping activity. The experimental plan includes the identification of the variables of the experiment (if any), the test protocol, an indication of what measurements will be performed, and a plan for analyzing the resulting data. When many variables must be explored, efficient experimental design greatly facilitates this process. Chapter 13, Robust Design, discusses experimental design in detail.

For the ball-support prototype, the team decided to vary only the support spacing and the inertia of the ball. The members of the team agreed that the spacing would be tested at values of 12.75, 13.00, 13.25, 13.50, and 13.75 millimeters. The team also prepared two different balls with different inertias, one with a steel core and one of solid plastic. With only two parameters to explore, the team could test all combinations of diameters and inertias. The test protocol was to have a person subjectively evaluate the feel of the trackball at each of the 10 combinations of spacings and inertias and rank them in order of preference. The team agreed to select the setting preferred by the most people. The team members also decided to analyze the variation in the responses as an indication of the degree to which people agree on what constitutes good feel. In an attempt to allow the use of analytical prototypes in the future, the team also planned to measure the rolling friction for each of the test conditions and to attempt to model the relationship between the ball and support characteristics and the resulting friction.

Step 4: Create a Schedule for Procurement, Construction, and Testing

Because the building and testing of a prototype can be considered a subproject within the overall development project, the team benefits from a schedule for the prototyping activity. Three dates are particularly important in defining a prototyping effort. First, the team defines when the parts will be ready to assemble. (This is sometimes called the "bucket of parts" date.) Second, the team defines the date when the prototype will first be tested. (This is sometimes called the "smoke test" date, because it is the date the team will first apply power and "look for smoke" in products with electrical systems.) Third, the team defines the date when it expects to have completed testing and produced the final results.

For the ball-support prototype, there was little assembly involved, so when parts were available the prototypes could immediately be assembled and tested. The team planned for eight days of testing and two days of analysis.

Planning Milestone Prototypes

The above method for planning a prototype applies to all prototypes, including those as simple as the ball support and those as complex as the beta prototype of the entire trackball. Nevertheless, the comprehensive prototypes the team uses as development milestones benefit from additional planning. This planning activity typically occurs in conjunction with the overall product development planning activity at the end of the concept development phase. In fact, planning the milestone dates is an integral part of establishing an overall product development project plan. (See Chapter 16, Managing Projects.)

All other things being equal, the team would prefer to build as few milestone prototypes as possible because designing, building, and testing prototypes consumes a great deal of time and money. However, in reality, few highly engineered products are developed with fewer than two milestone prototypes, and many efforts require four or more. As a base case, the team should consider using alpha, beta, and preproduction prototypes as milestones. The team should then consider whether any of these milestones can be eliminated or whether in fact additional prototypes are necessary.

Alpha prototypes are typically used to assess whether the product works as intended. The parts in alpha prototypes are usually similar in material and geometry to the parts that will be used in the production version of the product, but they are usually made with prototype production processes. For example, plastic parts in an alpha prototype may be machined or rubber molded instead of injection molded as they would be in production.

Beta prototypes are typically used to assess reliability and to identify remaining bugs in the product. These prototypes are often given to customers for testing in the intended use environment. The parts in beta prototypes are usually made with actual production processes or supplied by the intended parts suppliers, but the product is usually not assembled with the intended final assembly facility. For example, the plastic parts in a beta prototype might be molded with the production injection molds but would probably be assembled by a technician in a prototype shop rather than by production workers or automated equipment.

Preproduction prototypes are the first products produced by the entire production process. At this point the production process is not yet operating at full capacity but is making limited quantities of the product. These prototypes are used to verify production process capability, are subjected to further testing, and are often supplied to preferred customers. Preproduction prototypes are sometimes called *pilot-production prototypes.*

The most common deviations from the standard prototyping plan are to eliminate one of the standard prototypes or to add additional early prototypes. Eliminating a prototype (usually the alpha) may be possible if the product is very similar to other products the firm has already developed and produced, or if the product is extremely simple. Additional early prototypes are common in situations where the product embodies a new concept or technology. These early prototypes are sometimes called *experimental* or *engineering prototypes.* They usually do not look like the final product, and many of the parts of the prototype are not designed with the intention of eventually being produced in quantity.

Once preliminary decisions have been made about the number of prototypes, their characteristics, and the time required to assemble and test them, the team can place these milestones on the overall time line of the project. When the team attempts to schedule these milestones, the feasibility of the overall product development schedule can be assessed. Frequently a team will discover, when working backward from the target date for the product launch, that the assembly and test of one milestone prototype overlaps or is perilously close to the design and fabrication of the next milestone prototype. If this overlapping happens in practice, it is the worst manifestation of the "hardware swamp." When prototyping phases overlap, there is little transfer of learning from one prototype to the next, and the team should consider omitting one or more of the prototypes to allow the remaining prototypes to be spread out more in time. During project planning, overlapping prototyping phases can be avoided by beginning the project sooner, delaying product launch, eliminating a milestone prototype, or devising a way to accelerate the development activities preceding each prototype. (See Chapter 16, Managing Projects, for some techniques for achieving this acceleration.)

Summary

Product development almost always requires the building and testing of prototypes. A prototype is an approximation of the product on one or more dimensions of interest.

- Prototypes can be usefully classified along two dimensions: (1) the degree to which they are physical as opposed to analytical and (2) the degree to which they are comprehensive as opposed to focused.
- Prototypes are used for learning, communication, integration, and milestones. While all types of prototypes can be used for all of these purposes, physical prototypes are usually best for communication, and comprehensive prototypes are best for integration and milestones.
- Several principles are useful in guiding decisions about prototypes during product development: Analytical prototypes are generally more flexible than physical prototypes. Physical prototypes are required to detect unanticipated phenomena. A prototype may reduce the risk of costly iterations. A prototype may expedite other development steps. A prototype may restructure task dependencies.
- 3D computer modeling and free-form fabrication technologies have reduced the relative cost and time required to create prototypes.
- A four-step method for planning a prototype is:

  1. Define the purpose of the prototype.
  2. Establish the level of approximation of the prototype.
  3. Outline an experimental plan.
  4. Create a schedule for procurement, construction, and testing.

- Milestone prototypes are defined in the product development project plan. The number of such prototypes and their timing is one of the key elements of the overall development plan.

References and Bibliography

Many current resources are available on the Internet via
www.ulrich-eppinger.net

Clausing describes some of the pitfalls in prototyping, including the "hardware swamp."
Clausing, Don, *Total Quality Development,* ASME Press, New York, 1994.

Leonard-Barton describes how prototypes are used for the integration of different product development functions.
Leonard-Barton, Dorothy, "Inanimate Integrators: A Block of Wood Speaks,"
Design Management Journal, Vol. 2, No. 3, Summer 1991, pp. 61–67.

Cusumano describes Microsoft's use of the "daily build" in its software development process. The daily build is an extreme example of using comprehensive prototypes to force integration.
Cusumano, Michael A., "How Microsoft Makes Large Teams Work Like Small Teams," *Sloan Management Review,* Fall 1997, pp. 9–20.

Schrage presents a view of product development centered around the role of prototyping and simulation in the innovation process.

Schrage, Michael, *Serious Play: How the World's Best Companies Simulate to Innovate,* Boston, Harvard Business School Press, 2000.

Two books written for general audiences contain very interesting accounts of prototyping. Walton's book on the development of the 1996 Ford Taurus contains fascinating descriptions of prototyping and testing in the automobile industry. Particularly engaging is the description of testing heaters in northern Minnesota in midwinter, using development engineers as subjects. Sabbagh's book on the development of the Boeing 777 contains riveting accounts of brake system tests and wing strength tests, among others.

Walton, Mary, *Car: A Drama of the American Workplace,* Norton, New York, 1997.

Sabbagh, Karl, *Twenty-First-Century Jet: The Making and Marketing of the Boeing 777,* Scribner, New York, 1996.

Wall, Ulrich, and Flowers provide a formal definition of the quality of a prototype in terms of its fidelity to the production version of a product. They use this definition to evaluate the prototyping technologies available for plastic parts.

Wall, Matthew B., Karl T. Ulrich, and Woodie C. Flowers, "Evaluating Prototyping Technologies for Product Design," *Research in Engineering Design,* Vol. 3, 1992, pp. 163–177.

Wheelwright and Clark describe the use of prototypes as a managerial tool for major product development programs. Their discussion of periodic prototyping cycles is particularly interesting.

Wheelwright, Stephen C., and Kim B. Clark, *Revolutionizing Product Development: Quantum Leaps in Speed, Efficiency, and Quality,* The Free Press, New York, 1992.

Exercises

1. A furniture manufacturer was considering a line of seating products to be fabricated by cutting and bending a recycled plastic product available in large sheets. Create a prototype of at least one possible chair design by cutting and bending a sheet of paper or cardboard. (You may wish to design the chair with a sketch first, or just start working with the sheet directly.) What can you learn about the chair design from your prototype? What can't you learn about the chair design from such a prototype?

2. Position the chair prototype described in Exercise 1 on the plot in Exhibit 12-4. For which of the four major purposes would a product development team use such a prototype?

3. Devise a prototyping plan (similar to that in Exhibit 12-14) for investigating the comfort of different types of handles for kitchen knives.

4. Position the prototypes shown in Exhibits 12-3, 12-12, and 12-13 on the plot in Exhibit 12-4. Briefly explain your reasoning for each placement.

Thought Questions

1. Many product development teams separate the "looks like" prototype from the "works like" prototype. They do this because integrating both function and form is difficult in the early phases of development. What are the strengths and weaknesses of this approach? For what types of products might this approach be dangerous?

2. Over the past 10 years, several technologies have been developed to create physical parts directly from computer-aided design files (e.g., stereolithography and selective laser sintering). How might a team use such rapid prototyping technologies during the concept development phase of the product development process? Might the technologies facilitate identifying customer needs, establishing specifications, generating product concepts, selecting product concepts, and/or testing product concepts?

3. Some companies have reportedly abandoned the practice of doing a customer test with the early prototypes of their products, preferring instead to go directly and quickly to market in order to observe the actual customer response. For what types of products and markets might this practice make sense?

4. Is a drawing a physical or analytical prototype?

5. Microsoft uses frequent comprehensive prototypes in its development of software. In fact, in some projects there is a "daily build," in which a new version of the product is integrated and compiled *every day*. Is this approach only viable for software products, or could it be used for physical products as well? What might be the costs and benefits of such an approach for physical products?

Robust Design

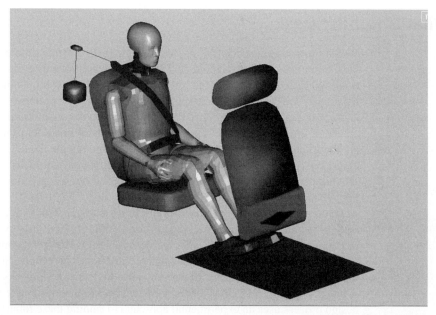

EXHIBIT 13-1

Rear seat belt experiment. This experiment was run on a simulation model to explore many design parameters and noise conditions.

Ford Motor Company safety engineers were working with a supplier to better understand the performance of rear seat belts. In any conventional seat belt system with lap and shoulder belts, if the lap portion of the belt rides upward, the passenger may slide beneath it, potentially resulting in abdominal injury. This phenomenon, called "submarining," is related to a large number of factors, including the nature of the collision, the design of the vehicle, the properties of the seats and seat belts, and other conditions. Based on experimentation, simulation, and analysis, Ford engineers hoped to determine which of the many factors were most critical to passenger safety and to avoiding submarining. The image shown in Exhibit 13-1 depicts the model used in Ford's simulation analysis.

This chapter presents a method for designing and conducting experiments to improve the performance of products even in the presence of uncontrollable variations. This method is known as robust design.

What Is Robust Design?

We define a *robust* product (or process) as one that performs as intended even under non-ideal conditions such as manufacturing process variations or a range of operating situations. We use the term *noise* to describe uncontrolled variations that may affect performance, and we say that a quality product should be robust to noise factors.

Robust design is the product development activity of improving the desired performance of the product while minimizing the effects of noise. In robust design we use experiments and data analysis to identify robust setpoints for the design parameters we can control. A *robust setpoint* is a combination of design parameter values for which the product performance is as desired under a range of operating conditions and manufacturing variations.

Conceptually, robust design is simple to understand. For a given performance target (safely restraining rear-seat passengers, for example), there may be many combinations of parameter values that will yield the desired result. However, some of these combinations are more sensitive to uncontrollable variation than others. Since the product will likely operate in the presence of various noise factors, we would like to choose the combination of parameter values that is least sensitive to uncontrollable variation. The robust design process uses an experimental approach to finding these robust setpoints.

To understand the concept of robust setpoints, consider two hypothetical factors affecting some measure of seat belt performance, as shown in Exhibit 13-2. Assume that factor A has a linear effect, f_A, on performance and factor B has a nonlinear effect, f_B. Further consider that we can choose setpoints for each factor: A1 or A2 for factor A, and B1 or B2 for factor B. Assuming that the effects of f_A and f_B are additive, a combination of A1 and B2 will provide approximately the same level of overall performance as a combination of A2 and B1. Manufacturing variations will be present at any chosen setpoint, so that the actual value may not be exactly as specified. By choosing the value of B1 for factor B, where the sensitivity of the response to factor B is relatively small, unintended variation in factor B has a relatively small influence on overall product performance. Therefore, the choice of B1 and A2 is a more robust combination of setpoints than the combination of B2 and A1.

The robust design process can be used at several stages of the product development process. As with most product development issues, the earlier that robustness can be considered in the product development process, the better the robustness results can be. Ro-

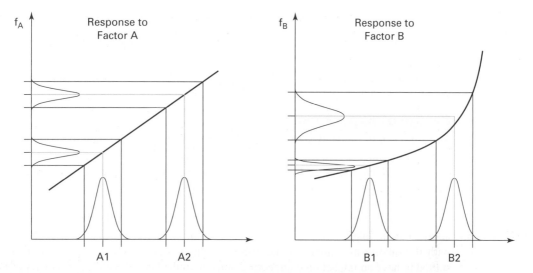

EXHIBIT 13-2   Robust design exploits nonlinear relationships to identify setpoints where the product performance is less sensitive to variations. In this example, the chosen value for the factor A setpoint does not affect robustness, whereas that of factor B does. Choosing B1 minimizes the effect of variation in factor B on overall performance.

bust design experiments can be used within the concept development phase as a way to refine the specifications and set realistic performance targets. While it is beneficial to consider product robustness as early as the concept stage, experiments for robust design are used most frequently during the detail design phase as a way to ensure the desired product performance under a variety of conditions. In detail design, the robust design activity is also known as *parameter design,* as this is a matter of choosing the right setpoints for the design parameters under our control. These include the product's materials, dimensions, tolerances, manufacturing processes, and operating instructions.

For many engineering design problems, equations based on fundamental physical principles can be solved for robust parameter choices. However, engineers generally cannot fully model the kinds of uncertainties, variations, and noise factors that arise under real conditions. Furthermore, the ability to develop accurate mathematical models is limited for many engineering problems. For example, consider the difficulty of accurately modeling the seat belt submarining problem under a wide variety of conditions. In such situations, empirical investigation through designed experiments is necessary. Such experiments can be used to directly support decision making and can also be used to improve the accuracy of mathematical models.

In the case of the seat belt design problem, Ford's engineers wished to test a range of seat belt design parameters and collision conditions. However, crash testing is very expensive, so Ford worked with its seat belt supplier to develop a simulation model which was calibrated using experimental crash data. Considering the hundreds of possible design parameter combinations, collision conditions, and other factors of interest, the engineers chose to explore the simulation model using a carefully planned experiment. Although simulation requires a great deal of computational effort, the simulation model still allowed Ford engineers to run dozens of experiments under a wide variety of conditions, which would not have been possible using physical crash testing.

For the Ford seat belt design team, the goals of this designed experiment were to learn:

- What combination of seat, seat belt, and attachment parameters minimizes rear-seat passenger submarining during a crash.
- How submarining is affected by uncontrollable conditions. What combination of design parameters is most robust to such noise factors?

Design of Experiments

The approach to robust design presented in this chapter is based on a method called *design of experiments* (DOE). In this method, the team identifies the parameters that can be controlled and the noise factors it wishes to investigate. The team then designs, conducts, and analyzes experiments to help determine the parameter setpoints to achieve robust performance.

In Japan during the 1950s and 1960s, Dr. Genichi Taguchi developed techniques to apply DOE to improve the quality of products and manufacturing processes. Beginning with the quality movement of the 1980s, Taguchi's approach to experimental design started to have an impact on engineering practice in the United States, particularly at Ford Motor Company, Xerox Corporation, AT&T Bell Laboratories, and through the American Supplier Institute (which was created by Ford).

Taguchi receives credit for promoting several key ideas of experimental design for the development of robust products and processes. These contributions include introducing noise factors into experiments to observe these effects and the use of a *signal-to-noise ratio* metric including both the desired performance (signal) and the undesired effects (noise). While statisticians had been showing engineers how to run experiments for decades, it was not until Taguchi's methods were widely explained to the U.S. manufacturing industry during the 1990s that experiments became commonly utilized to achieve robust design.

DOE is not a substitute for technical knowledge of the system under investigation. In fact, the team should use its understanding of the product and how it operates to choose the right parameters to investigate by experiment. The experimental results can be used in conjunction with technical knowledge of the system in order to make the best choices of parameter setpoints. Furthermore, the experimental results can be used to build better mathematical models of the product's function. In this way, experimentation complements technical knowledge. For example, Ford engineers have basic mathematical models of seat belt performance as a function of passenger sizes and collision types. These models allow Ford to size the mechanical elements and to determine the belt attachment geometry. Based on empirical and simulation data, Ford's analytical models and seat belt design guidelines gain precision over time, reducing the need for time-consuming empirical and simulation studies. Eventually, this technical knowledge may improve to the point where only confirming tests of new seat belt configurations are required.

Basic experimental design and analysis for product development can be successfully planned and executed by the development team. However, the field of DOE has many advanced methods to address a number of complicating factors and yield more useful experimental results. Development teams thus can benefit from consulting with a statistician or DOE expert who can assist in designing the experiment and choosing the best analytical approach.

The Robust Design Process

To develop a robust product through DOE, we suggest this seven-step process:

1. Identify control factors, noise factors, and performance metrics.
2. Formulate an objective function.
3. Develop the experimental plan.
4. Run the experiment.
5. Conduct the analysis.
6. Select and confirm factor setpoints.
7. Reflect and repeat.

Step 1: Identify Control Factors, Noise Factors, and Performance Metrics

The robust design procedure begins with identification of three lists: control factors, noise factors, and performance metrics for the experiment:

- *Control factors:* These are the design variables to be varied in a controlled manner during the experiment, in order to explore the product's performance under the many combinations of parameter setpoints. Experiments are generally run at two or three discrete levels (setpoint values) of each factor. These parameters are called control factors because they are among the variables that can be specified for production and/or operation of the product. For example, the webbing stiffness and coefficient of friction are control factors of interest for the experiment.

- *Noise factors:* Noise factors are variables that cannot be explicitly controlled during the manufacturing and operation of the product. Noise factors may include manufacturing variances, changes in materials properties, multiple user scenarios or operating conditions, and even deterioration or misuse of the product. If through special techniques the team can control the noise factors during the experiment (but not in production or operation), then variance can deliberately be induced during the experiment to assess its impact. Otherwise, the team simply lets the noise take place during the experiment, analyzes the results in the presence of typical variation, and seeks to minimize the effects of this variation. For seat belts to be used with a range of seats, the shape of the seat and the seat fabric must be considered noise factors. The goal is to design a seat belt system that works well regardless of the values of these factors.

- *Performance metrics:* These are the product specifications of interest in the experiment. Usually the experiment is analyzed with one or two key product specifications as the performance metrics in order to find control factor setpoints to optimize this performance. These metrics may be derived directly from key specifications where robustness is of critical concern. (See Chapter 5, Product Specifications.) For example, how far the passenger's back or buttocks move forward during the collision would be possible performance metrics for the seat belt experiment.

For the seat belt design problem, the team held a meeting to list the control factors, noise factors, and performance metrics. As Taguchi teaches, they placed these lists into a single graphic, called a *parameter diagram* (or *p-diagram*), as shown in Exhibit 13-3.

EXHIBIT 13-3
Parameter
diagram used to
design the seat
belt experiment.
Bold text
indicates the
performance
metric used and
the control
factors and noise
factors chosen
for exploration.

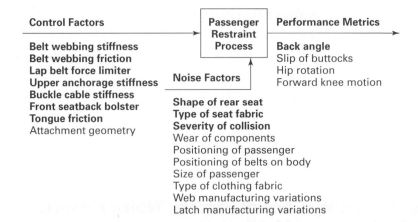

Control Factors | Passenger Restraint Process | Performance Metrics

Belt webbing stiffness
Belt webbing friction
Lap belt force limiter
Upper anchorage stiffness
Buckle cable stiffness
Front seatback bolster
Tongue friction
Attachment geometry

Noise Factors

Shape of rear seat
Type of seat fabric
Severity of collision
Wear of components
Positioning of passenger
Positioning of belts on body
Size of passenger
Type of clothing fabric
Web manufacturing variations
Latch manufacturing variations

Back angle
Slip of buttocks
Hip rotation
Forward knee motion

After listing the various factors, the team must decide which ones will be explored by experiment. When a large number of parameters are suspected of potentially affecting performance, the selection of critical variables can be substantially narrowed by using analytical models and/or by running a *screening experiment* with two levels for each of many factors. Then a finer experiment is run with two or more levels of the few parameters believed to affect performance.

Ford engineers considered the lists shown in Exhibit 13-3. They chose to focus the experiment on exploration of seven seat belt parameters, holding constant the geometric locations of the three attachment points. They decided to use "back angle at peak" as the output metric, the angle that the passenger's back makes with respect to vertical at the moment of maximum restraint. Back angle is a smaller-is-better performance metric, measured in radians.

A primary concern in this experiment was the effect of three particular noise factors: seat shape, fabric type, and severity of collision. Through preliminary analysis, the team found the best and worst combinations of these noise conditions with respect to the submarining effect. These three noise factors were thereby combined into two extreme noise conditions for the purposes of the experiment. This approach, known as *compounded noise,* can be helpful when many noise factors must be considered. (See Testing Noise Factors in Step 3.)

Step 2: Formulate an Objective Function

The experiment's performance metric(s) must be transformed into an *objective function* that relates to the desired robust performance. Several objective functions are useful in robust design for different types of performance concerns. They can be formulated either as functions to be maximized or minimized, and they include:

- *Maximizing:* This type of function is used for performance dimensions where larger values are better, such as maximum deceleration before belt slippage. Common forms of this objective function η are $\eta = \mu$ or $\eta = \mu^2$, where μ is the mean of the experimental observations under a given test condition.

- *Minimizing:* This type of function is used for performance dimensions where smaller values are better, such as back angle at peak deceleration. Common forms of this objective function are $\eta = \mu$ or $\eta = \sigma^2$, where σ^2 is the variance of the experimental observations under a given test condition. Alternatively, such minimization objectives can be formulated as functions to be maximized, such as $\eta = 1/\mu$ or $\eta = 1/\sigma^2$.
- *Target value:* This type of function is used for performance dimensions where values closest to a desired setpoint or target are best, such as amount of belt slackening before restraint. A common maximizing form of this objective function is $\eta = 1(\mu - t)^2$, where t is the target value.
- *Signal-to-noise ratio:* This type of function is used particularly to measure robustness. Taguchi formulates this metric as a ratio with the desired response in the numerator and the variance in the response as the denominator. Generally the mean value of the desired response, such as the mean back angle at peak, is not difficult to adjust by changing control factors. In the denominator, we place the variance of this response (the noise response), which is to be minimized, such as the variance in back angle resulting from noise conditions. In practice, reducing variance is more difficult than changing the mean. By computing this ratio, we can highlight robust factor settings for which the noise response is relatively low as compared to the signal response. A common maximizing form of this objective function is $\eta = 10 \log (\mu^2/\sigma^2)$.

The Ford statistician consulting with the team suggested two objective functions: the average back angle at peak and the range of the back angle at peak (the difference between the maximum and minimum back angle at peak at the two noise conditions to be tested). Both of these are objectives to be minimized. Together these two metrics would provide deeper insight into the behavior of the system than either one alone.

Step 3: Develop the Experimental Plan

Statisticians have developed many types of efficient experimental plans. These plans lay out how to vary the *factor levels* (values of the control factors and possibly also some of the noise factors) in a series of experiments in order to explore the system's behavior. Some DOE plans are more efficient for characterizing certain types of systems, while others provide more complete analysis.

Experimental Designs

A critical concern in designing experiments is the cost of setting up and running the experimental trials. In situations where this cost is low, running a large number of trials and using an experimental design with resolution high enough to explore more factors, factor combinations, and interactions may be feasible. On the other hand, when the cost of experimentation is high, efficient DOE plans can be used that simultaneously change several factors at once. Some of the most popular experimental designs are listed below and depicted in Exhibit 13-4. Each one has important uses.

- *Full factorial:* This design involves the systematic exploration of every combination of levels of each factor. This allows the team to identify all of the multifactor interaction

Full-Factorial Matrix

			A1								A2						
			B1				B2				B1				B2		
			C1		C2		C1		C2		C1		C2		C1		C2	
			D1	D2	D1	D2	D1	D2	D1	D2	D1	D2	D1	D2	D1	D2	D1	D2
E1	F1	G1	x	x	x	x	x	x	x	x	x	x	x	x	x	x	x	x
		G2	x	x	x	x	x	x	x	x	x	x	x	x	x	x	x	x
	F2	G1	x	x	x	x	x	x	x	x	x	x	x	x	x	x	x	x
		G2	x	x	x	x	x	x	x	x	x	x	x	x	x	x	x	x
E2	F1	G1	x	x	x	x	x	x	x	x	x	x	x	x	x	x	x	x
		G2	x	x	x	x	x	x	x	x	x	x	x	x	x	x	x	x
	F2	G1	x	x	x	x	x	x	x	x	x	x	x	x	x	x	x	x
		G2	x	x	x	x	x	x	x	x	x	x	x	x	x	x	x	x

1/2 Fractional Factorial Matrix

			A1								A2						
			B1				B2				B1				B2		
			C1		C2		C1		C2		C1		C2		C1		C2	
			D1	D2	D1	D2	D1	D2	D1	D2	D1	D2	D1	D2	D1	D2	D1	D2
E1	F1	G1	x			x		x	x				x	x		x			x
		G2		x	x		x				x				x		x	x
	F2	G1		x	x		x				x				x		x	x
		G2	x			x		x	x				x	x		x			x
E2	F1	G1		x	x		x				x	x			x		x	x
		G2	x			x		x	x				x	x		x			x
	F2	G1	x			x		x	x				x	x		x			x
		G2		x	x		x				x	x			x		x	x

1/4 Fractional Factorial Matrix

			A1								A2						
			B1				B2				B1				B2		
			C1		C2		C1		C2		C1		C2		C1		C2	
			D1	D2	D1	D2	D1	D2	D1	D2	D1	D2	D1	D2	D1	D2	D1	D2
E1	F1	G1	x			x		x	x			x	x		x			x
		G2																
	F2	G1																
		G2	x			x		x	x			x	x		x			x
E2	F1	G1																
		G2	x	x		x			x	x		x		x	x		
	F2	G1	x	x		x			x	x		x		x	x		
		G2																

1/8 Fractional Factorial Matrix

			A1								A2						
			B1				B2				B1				B2		
			C1		C2		C1		C2		C1		C2		C1		C2	
			D1	D2	D1	D2	D1	D2	D1	D2	D1	D2	D1	D2	D1	D2	D1	D2
E1	F1	G1	x										x					
		G2				x										x		
	F2	G1						x					x					
		G2								x								
E2	F1	G1						x					x					
		G2				x										x		
	F2	G1					x											x
		G2	x														x	

L8 Orthogonal Array
(1/16 Fractional Factorial Matrix)

			A1								A2						
			B1				B2				B1				B2		
			C1		C2		C1		C2		C1		C2		C1		C2	
			D1	D2	D1	D2	D1	D2	D1	D2	D1	D2	D1	D2	D1	D2	D1	D2
E1	F1	G1	x															
		G2												x				
	F2	G1											x					
		G2						x										
E2	F1	G1							x									
		G2													x			
	F2	G1												x				
		G2		x														

One Factor at a Time

			A1								A2						
			B1				B2				B1				B2		
			C1		C2		C1		C2		C1		C2		C1		C2	
			D1	D2	D1	D2	D1	D2	D1	D2	D1	D2	D1	D2	D1	D2	D1	D2
E1	F1	G1	x	x	x		x				x							
		G2	x															
	F2	G1	x															
		G2																
E2	F1	G1	x															
		G2																
	F2	G1																
		G2																

EXHIBIT 13-4   Several alternative experimental plans for seven factors (A, B, C, D, E, F, and G) at two levels each. The full-factorial experiment contains $2^7 = 128$ trials, while the L8 orthogonal array design contains only 8 trials, denoted by the × marks in the matrices. The L8 orthogonal array plan is the one used for the seat belt experiment and is shown in conventional row/column format in Exhibit 13-5.

Source: Fractional factorial layouts adapted from Ross (1996) effects, in addition to the primary (main) effect of each factor on performance. This type of experiment is generally practical only for a small number of factors and levels and when experiments are inexpensive (as with fast software-based simulations or very flexible hardware). For an investigation of k factors at n levels each, the number of trials in the full-factorial experiment is n^k. Full factorial experimentation is typically infeasible for an experiment with greater than four to five factors.

- *Fractional factorial:* This design uses only a small fraction of the combinations used above. In exchange for this efficiency, the ability to compute the magnitudes of all the interaction effects is sacrificed. Instead, the interactions are confounded with other interactions or with some of the main factor effects. Note that the fractional factorial layout still maintains *balance* within the experimental plan. This means that for the several trials at any given factor level, each of the other factors is tested at every level the same number of times.

- *Orthogonal array:* This design is the smallest fractional factorial plan that still allows the team to identify the main effects of each factor. However, these main effects are confounded with many interaction effects. Nevertheless, orthogonal array layouts are widely utilized in technical investigations because they are extremely efficient. Taguchi popularized the orthogonal array DOE approach, even though statisticians had developed such plans several decades earlier and the roots of these designs can be traced back many centuries. Orthogonal array plans are named according to the number of rows (experiments) in the array: L4, L8, L9, L27, and so on. The appendix to this chapter shows several orthogonal array experimental plans.

- *One factor at a time:* This is an unbalanced experimental plan because each trial is conducted with all but one of the factors at nominal levels (and the first trial having all the factors at the nominal level). This is generally considered to be an ineffective way to explore the factor space, even though the number of trials is small, $1 + k\,(n - 1)$. However, for parameter optimization in systems with significant interactions, an adaptive version of the one-at-a-time experimental plan has been shown to be generally more efficient than orthogonal array plans (Frey et al., 2003).

The Ford team chose to use the L8 orthogonal array experiment design because this plan would be an efficient way to explore seven factors at two levels each. Subsequent rounds of experimentation could later be used to explore additional levels of key parameters as well as interaction effects if necessary. The orthogonal array experimental plan is shown in Exhibit 13-5.

Testing Noise Factors

Several methods are used to explore the effects of noise factors in experiments. If some noise factors can be controlled for the purpose of the experiment, then it may be possible to directly assess the effect of these noise factors. If the noise factors cannot be controlled during the experiment, we allow the noise to vary naturally and simply assess the product's performance in the presence of noise. Some common ways to test noise factors are:

- Assign additional columns in the orthogonal array or fractional factorial layout to the noise factors, essentially treating the noise as another variable. This allows the effects of the noise factors to be determined along with the control factors.

EXHIBIT 13-5

Factor assignments and the L8 orthogonal array experiment design used for the seat belt experiment. This DOE plan tests seven factors at two levels each. Each row was replicated twice, under the two compounded noise conditions, yielding 16 test data points for analysis.

Factor	Description
A	**Belt webbing stiffness:** Compliance characteristic of the webbing measured in a tensile load machine
B	**Belt webbing friction:** Coefficient of friction, which is a function of the belt weave and surface coating
C	**Lap belt force limiter:** Allows controlled release of the seat belt at a certain force level
D	**Upper anchorage stiffness:** Compliance characteristic of the structure to which the upper anchorage (D-loop) is mounted
E	**Buckle cable stiffness:** Compliance characteristic of the cables by which the buckle is attached to the vehicle body
F	**Front seatback bolster:** Profile and stiffness of seat back where the knees may contact
G	**Tongue friction:** Coefficient of friction for the bearing area of the tongue which slides along the webbing

	A	B	C	D	E	F	G	N–	N+
1	1	1	1	1	1	1	1		
2	1	1	1	2	2	2	2		
3	1	2	2	1	1	2	2		
4	1	2	2	2	2	1	1		
5	2	1	2	1	2	1	2		
6	2	1	2	2	1	2	1		
7	2	2	1	1	2	2	1		
8	2	2	1	2	1	1	2		

- Use an *outer array* for the noise factors. This method tests several combinations of the noise factors for each row in the main (inner) array. An example of this approach is shown in the appendix, where the outer array consists of an L4 design, testing combinations of three noise factors by replicating each row four times.

- Run replicates of each row, allowing the noise to vary in a natural, uncontrolled manner throughout the experiment, resulting in measurable variance in performance for each row. With this approach, it is particularly important to randomize the order of the trials so that any trends in the noise are unlikely to be correlated with the systematic changes in the control factors. (See Step 4.)

- Run replicates of each row with *compounded noise.* In this method, selected noise factors are combined to create several representative noise conditions or extreme noise conditions. This approach also yields measurable variance for each row, which can be attributed to the effect of noise.

The Ford team chose to utilize the compounded noise approach in the seat belt experiment. The team tested each row using the two combinations of the three noise factors representing the best- and worst-case conditions. This resulted in 16 experimental runs for the L8 DOE plan, as shown in Exhibit 13-5.

	A	B	C	D	E	F	G	N–	N+	Avg	Range
1	1	1	1	1	1	1	1	0.3403	0.2915	0.3159	0.0488
2	1	1	1	2	2	2	2	0.4608	0.3984	0.4296	0.0624
3	1	2	2	1	1	2	2	0.3682	0.3627	0.3655	0.0055
4	1	2	2	2	2	1	1	0.2961	0.2647	0.2804	0.0314
5	2	1	2	1	2	1	2	0.4450	0.4398	0.4424	0.0052
6	2	1	2	2	1	2	1	0.3517	0.3538	0.3528	0.0021
7	2	2	1	1	2	2	1	0.3758	0.3580	0.3669	0.0178
8	2	2	1	2	1	1	2	0.4504	0.4076	0.4290	0.0428

EXHIBIT 13-6   Data obtained from the seat belt experiment.

Step 4: Run the Experiment

To execute the experiment, the product is tested under the various treatment conditions described by each row in the experimental plan. Randomizing the sequence of the experimental runs ensures that any systematic trend over the duration of the experiment is not correlated with the systematic changes to the levels of the factors. For example, if the experiments of the L8 plan are not randomized, and the test conditions drift over time, this effect may be incorrectly attributed to factor A since this column changes halfway through the experiment. For some experiments, changing certain factors may be so difficult that all trials at each level of that factor are run together and only partial randomization may be achieved. In practice, randomize the trials whenever practical, and when not possible, validate the results with a confirmation run. (See Step 6.)

In the seat belt experiment, each of the eight factor combinations in the L8 design was tested under the two compounded noise conditions. The 16 data points containing the back angle data are shown in Exhibit 13-6 in the columns titled N– and N+.

Step 5: Conduct the Analysis

There are many ways to analyze the experimental data. For all but the most basic analysis, the team benefits from consulting with a DOE expert or from referring to a good book on statistical analysis and experimental design. The basic analytical method is summarized here.

Computing the Objective Function

The team will have already devised the objective functions for the experiment and will generally have an objective related to the mean performance and the variance in performance. Sometimes the mean and variance will be combined and expressed as a single objective in the form of a signal-to-noise ratio. The values of the objective function can be computed for each row of the experiment. For the seat belt experiment, the columns on the right side of the table in Exhibit 13-6 show the computed objective function values (average back angle and range of back angle) for each row. Recall that these are both objectives to be minimized.

Computing Factor Effects by Analysis of Means

The most straightforward analysis to conduct will simply yield the main effect of each factor assigned to a column in the experiment. These main effects are called the *factor effects*. The *analysis of means* involves simply averaging all the computed objective functions for each factor level. In the L8 DOE example, the effect of factor level A1 (factor A at level 1) is the average of trials 1, 2, 3, and 4. Similarly, the effect of factor level E2 is the average of trials 2, 4, 5, and 7. The results of an analysis of means are conventionally shown on factor effects charts.

Exhibit 13-7 presents the factor effects charts for the seat belt example. These effects are plotted for each of the objective functions. Exhibit 13-7(a) plots the *average performance* at each factor level (the first objective function). This chart shows which factor levels can be used to raise or lower the mean performance. Recall that back angle at peak is to be minimized, and note that the chart suggests that factor levels [A1 B2 C2 E1 F1 G1] will

EXHIBIT 13-7

Factor effects charts for the seat belt experiment.

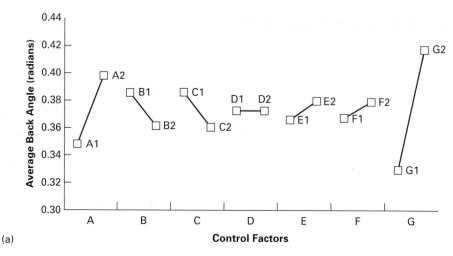

(a)

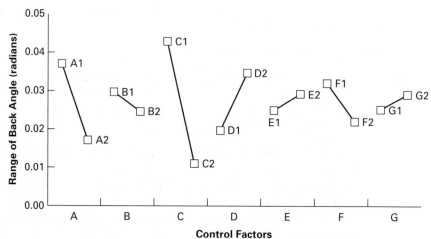

(b)

minimize the average back angle metric. (Factor D appears to have no effect upon mean performance.) However, these levels will not necessarily achieve robust performance. Exhibit 13-7(b) is based on the *range of performance* at each factor level (the second objective function). This chart suggests that levels [A2 B2 C2 D1 E1 F2 G1] will minimize the range of back angle at peak.

Taguchi recommends that the signal-to-noise ratio for each factor level be plotted in order to identify robust setpoints. Since the signal-to-noise ratio includes the mean performance in the numerator and the variance in the denominator, it represents a combination of these two objectives or a trade-off between them. Rather than specifically plotting the signal-to-noise ratio, many engineers and statisticians prefer to simply interpret the two objectives together, giving more control over the trade-off. To do so, the factor effects charts shown in Exhibit 13-7 can be compared in order to choose a robust setpoint in the next step.

Step 6: Select and Confirm Factor Setpoints

Analysis of means and the factor effects charts help the team determine which factors have a strong effect on mean performance and variance, and therefore how to achieve robust performance. These charts help to identify which factors are best able to reduce the product's variance (robustness factors) and which factors can be used to improve the performance (scaling factors). By choosing setpoints based on these insights, the team should be able to improve the overall robustness of the product.

For example, consider the effects of factor A on both average and range of back angle in the experiment. The charts in Exhibit 13-7 show that level A1 would minimize back angle, but level A2 would minimize the range of back angle, representing a trade-off between performance and robustness. A similar trade-off is evident in factor F. However, for factors B, C, D, E, and G, there is no such trade-off, and levels B2, C2, D1, E1, and G1 minimize both objectives.

Using factors B, C, D, E, and G to achieve the desired robustness and factors A and F to increase performance, Ford engineers selected the setpoint [A1 B2 C2 D1 E1 F1 G1]. As is usually the case, the chosen setpoint is not one of the eight orthogonal array rows tested in the experiment. Given that this setpoint has never been tested, a confirmation run should be used to ensure that the expected robust performance has been achieved.

Step 7: Reflect and Repeat

One round of experiments may be sufficient to identify appropriately robust setpoints. Sometimes, however, further optimization of the product's performance is worthwhile, and this may require several additional rounds of experimentation.

In subsequent experimentation and testing, the team may choose to:

- Reconsider the setpoints chosen for factors displaying a trade-off of performance versus robustness.
- Explore interactions among some of the factors in order to further improve the performance.
- Fine-tune the parameter setpoints using values between the levels tested or outside this range.

- Investigate other noise and/or control factors that were not included in the initial experiment.

As with all product development activities, the team should take some time to reflect on the DOE process and the robust design result. Did we run the right experiments? Did we achieve an acceptable result? Could it be better? Should we repeat the process and seek further performance/robustness improvement?

Caveats

Design of experiments is a well-established field of expertise. This chapter summarizes only one very basic approach in order to encourage the use of experimentation in product design to achieve more robust product performance. Most product development teams should include team members with DOE training or have access to engineers and/or statisticians with specialized expertise in design and analysis of experiments.

Obviously many assumptions underly the type of analysis used in DOE. One basic assumption made in interpreting analysis of means is that the factor effects are independent, without interactions across the factors. In fact, most actual systems exhibit many interactions, but these interactions are often smaller than the main effects. Verification of this assumption is another motive for running confirming experiments at the chosen setpoints.

If necessary, experiments can be designed to specifically test interaction effects. This type of experiment is outside the scope of this chapter. DOE texts generally provide a number of ways to explore interactions across the factors, including the following:

- Assign specific interactions to be explored in certain columns of the orthogonal array (instead of using the column for a control factor).
- Execute a larger fractional factorial design.
- Use an adaptive one-at-a-time experimental plan (Frey et al., 2003).

Many advanced graphical and analytical techniques are available to assist in interpretation of the experimental data. Analysis of variance (ANOVA) provides a way to assess the significance of the factor effects results in light of the experimental error observed in the data. ANOVA takes into account the number of observations made of each degree of freedom in the experiment and the scale of the results to determine whether each effect is statistically significant. This helps determine to what extent detailed design decisions should be based on the experimental results. However, ANOVA makes many more assumptions and can be difficult to set up properly, so it is also beyond the scope of this chapter. Refer to a DOE text (Ross, 1988; Montgomery, 2001) or consult with a DOE expert to assist with ANOVA.

Summary

Robust design is a set of engineering design methods used to create robust products and processes.

- A robust product (or process) is one that performs properly even in the presence of noise effects. Noises are due to many kinds of uncontrolled variation that may affect performance, such as manufacturing variations, operating conditions, and product deterioration.

- We suggest an approach to the development of robust products based on design of experiments (DOE). This seven-step process for robust design is:

  1. Identify control factors, noise factors, and performance metrics.
  2. Formulate an objective function.
  3. Develop the experimental plan.
  4. Run the experiment.
  5. Conduct the analysis.
  6. Select and confirm factor setpoints.
  7. Reflect and repeat.

- Orthogonal array experimental plans provide a very efficient method for exploring the main effects of each factor chosen for the experiment.
- To achieve robust performance, use of objective functions helps in capturing both mean performance due to each control factor and variance of performance due to noise factors.
- Analysis of means and factor effects charts facilitate the choice of robust parameter setpoints.
- Because many nuances are involved in successful DOE, most teams applying these methods will benefit from assistance by a DOE expert.

References and Bibliography

Many current resources are available on the Internet via
www.ulrich-eppinger.net

Taguchi's methods for experimental design and details about orthogonal array experimentation plans are explained in several texts, including Taguchi's classic two-volume text translated into English. Phadke provides numerous examples and practical advice on application of DOE. Ross emphasizes insights gained through ANOVA analysis.

Taguchi, Genichi, *System of Experimental Design: Engineering Methods to Optimize Quality and Minimize Costs,* two volumes, Louise Watanabe Tung (trans.), White Plains, NY, 1987.

Taguchi, Genichi, *Introduction to Quality Engineering: Designing Quality into Products and Processes,* Asian Productivity Organization (trans. and pub.), Tokyo, 1986.

Phadke, Madhav S., *Quality Engineering Using Robust Design,* Prentice Hall, Englewood Cliffs, NJ, 1989.

Ross, Phillip J., *Taguchi Techniques for Quality Engineering,* McGraw-Hill, New York, 1996.

Grove and Davis present a thorough explanation of experimental design techniques in engineering, including planning, running, and analyzing the experiments. A different analysis of Ford's seat belt experiment is included in this text, as well as many more automotive applications of robust design.

Grove, Daniel M., and Timothy P. Davis, *Engineering, Quality and Experimental Design,* Addison Wesley Longman, Edinburgh Gate, UK, 1992.

Several excellent texts provide detailed explanations of the use of statistical methods, fractional factorial experimental plans, analytical and graphical interpretations, and response surface methods.

Box, George E. P., J. Stuart Hunter, and William G. Hunter, *Statistics for Experimenters: An Introduction to Design, Data Analysis, and Model Building,* John Wiley and Sons, New York, 1978.

Box, George E. P., and Norman R. Draper, *Empirical Model Building and Response Surfaces,* John Wiley and Sons, New York, 1987.

Montgomery, Douglas C., *Design and Analysis of Experiments,* fifth edition, John Wiley and Sons, New York, 2001.

Recent research has renewed interest in one-at-a-time DOE plans. An adaptive one-factor-at-a-time approach has been shown to yield better performance optimization than the corresponding orthogonal array design for systems where the interaction effects are more significant than the noise and error effects.

Frey, Daniel D., Fredrik Engelhardt, and Edward M. Greitzer, "A Role for One-Factor-at-a-Time Experimentation in Parameter Design," *Research in Engineering Design,* 2003.

DOE can be used in many aspects of product development. Almquist and Wyner explain how carefully planned experiments are effective in evaluating and tuning parameters of sales campaigns.

Almquist, Eric, and Gordon Wyner, "Boost Your Marketing ROI with Experimental Design," *Harvard Business Review,* Vol. 79, No. 9, October 2001, pp. 135–141.

Exercises

1. Design an experiment to determine a robust process for making coffee.
2. Explain why the 1/4-fractional-factorial and orthogonal array plans shown in Exhibit 13-4 are balanced.
3. Formulate an appropriate signal-to-noise ratio for the seat belt experiment. Analyze the experimental data using this metric. Is signal-to-noise ratio a useful objective function in this case? Why or why not?

Thought Questions

1. If you are able to afford a larger experiment (with more runs), how might you best utilize the additional runs?
2. When would you choose not to randomize the order of the experiments? How would you guard against bias?
3. Explain the importance of balance in an experimental plan.

Appendix

Orthogonal Arrays

DOE texts provide several orthogonal array plans for experiments. The simplest arrays are for two-level and three-level factor experiments. Using advanced techniques, DOE plans can also be created for mixed two-, three-, and/or four-level factor experiments and many other special situations. This appendix shows some of the basic orthogonal arrays from Taguchi's text *Introduction to Quality Engineering* (1986). These plans are shown in row/column format, with the factor level assignments in the columns and the experimental runs in the rows. The numbers 1, 2, and 3 in each cell indicate the factor levels. (Alternatively, factor levels can be labeled as – and + for two-level factors or –, 0, and + for three levels.) Recall that the orthogonal arrays are named according to the number of rows in the design. Included here are the two-level arrays L4, L8, and L16 and the three-level arrays L9 and L27. Also shown is a DOE plan using the L8 inner array for seven control factors and the L4 outer array for three noise factors. This plan allows analysis of the effects of the three noise factors.

Two-Level Orthogonal Arrays

L4: 3 Factors at 2 Levels Each

	A	B	C
1	1	1	1
2	1	2	2
3	2	1	2
4	2	2	1

L8: 7 Factors at 2 Levels Each

	A	B	C	D	E	F	G
1	1	1	1	1	1	1	1
2	1	1	1	2	2	2	2
3	1	2	2	1	1	2	2
4	1	2	2	2	2	1	1
5	2	1	2	1	2	1	2
6	2	1	2	2	1	2	1
7	2	2	1	1	2	2	1
8	2	2	1	2	1	1	2

L16: 15 Factors at 2 Levels Each

	A	B	C	D	E	F	G	H	I	J	K	L	M	N	O
1	1	1	1	1	1	1	1	1	1	1	1	1	1	1	1
2	1	1	1	1	1	1	1	2	2	2	2	2	2	2	2
3	1	1	1	2	2	2	2	1	1	1	1	2	2	2	2
4	1	1	1	2	2	2	2	2	2	2	2	1	1	1	1
5	1	2	2	1	1	2	2	1	1	2	2	1	1	2	2
6	1	2	2	1	1	2	2	2	2	1	1	2	2	1	1
7	1	2	2	2	2	1	1	1	1	2	2	2	2	1	1
8	1	2	2	2	2	1	1	2	2	1	1	1	1	2	2
9	2	1	2	1	2	1	2	1	2	1	2	1	2	1	2
10	2	1	2	1	2	1	2	2	1	2	1	2	1	2	1
11	2	1	2	2	1	2	1	1	2	1	2	2	1	2	1
12	2	1	2	2	1	2	1	2	1	2	1	1	2	1	2
13	2	2	1	1	2	2	1	1	2	2	1	1	2	2	1
14	2	2	1	1	2	2	1	2	1	1	2	2	1	1	2
15	2	2	1	2	1	1	2	1	2	2	1	2	1	1	2
16	2	2	1	2	1	1	2	2	1	1	2	1	2	2	1

Three-Level Orthogonal Arrays

L9: 4 Factors at 3 Levels Each

	A	B	C	D
1	1	1	1	1
2	1	2	2	2
3	1	3	3	3
4	2	1	2	3
5	2	2	3	1
6	2	3	1	2
7	3	1	3	2
8	3	2	1	3
9	3	3	2	1

L27: 13 Factors at 3 Levels Each

	A	B	C	D	E	F	G	H	I	J	K	L	M
1	1	1	1	1	1	1	1	1	1	1	1	1	1
2	1	1	1	1	2	2	2	2	2	2	2	2	2
3	1	1	1	1	3	3	3	3	3	3	3	3	3
4	1	2	2	2	1	1	1	2	2	2	3	3	3
5	1	2	2	2	2	2	2	3	3	3	1	1	1
6	1	2	2	2	3	3	3	1	1	1	2	2	2
7	1	3	3	3	1	1	1	3	3	3	2	2	2
8	1	3	3	3	2	2	2	1	1	1	3	3	3
9	1	3	3	3	3	3	3	2	2	2	1	1	1
10	2	1	2	3	1	2	3	1	2	3	1	2	3
11	2	1	2	3	2	3	1	2	3	1	2	3	1
12	2	1	2	3	3	1	2	3	1	2	3	1	2
13	2	2	3	1	1	2	3	2	3	1	3	1	2
14	2	2	3	1	2	3	1	3	1	2	1	2	3
15	2	2	3	1	3	1	2	1	2	3	2	3	1
16	2	3	1	2	1	2	3	3	1	2	2	3	1
17	2	3	1	2	2	3	1	1	2	3	3	1	2
18	2	3	1	2	3	1	2	2	3	1	1	2	3
19	3	1	3	2	1	3	2	1	3	2	1	3	2
20	3	1	3	2	2	1	3	2	1	3	2	1	3
21	3	1	3	2	3	2	1	3	2	1	3	2	1
22	3	2	1	3	1	3	2	2	1	3	3	2	1
23	3	2	1	3	2	1	3	3	2	1	1	3	2
24	3	2	1	3	3	2	1	1	3	2	2	1	3
25	3	3	2	1	1	3	2	3	2	1	2	1	3
26	3	3	2	1	2	1	3	1	3	2	3	2	1
27	3	3	2	1	3	2	1	2	1	3	1	3	2

Combined Inner and Outer Arrays

L8 × L4: 7 Control Factors and 3 Noise Factors at 2 Levels Each

								1	1	2	2	Na
								1	2	1	2	Nb
	A	B	C	D	E	F	G	1	2	2	1	Nc
1	1	1	1	1	1	1	1					
2	1	1	1	2	2	2	2					
3	1	2	2	1	1	2	2					
4	1	2	2	2	2	1	1					
5	2	1	2	1	2	1	2					
6	2	1	2	2	1	2	1					
7	2	2	1	1	2	2	1					
8	2	2	1	2	1	1	2					

CHAPTER FOURTEEN

Patents and Intellectual Property

EXHIBIT 14-1
Hot beverage insulating sleeve by David W. Coffin Sr. (U.S. Patent 5,205,473).

David Coffin, an individual inventor, developed a product concept and prototype for an insulating sleeve that would make a hot beverage cup more comfortable to hold (Exhibit 14-1). The product opportunity arose in the 1980s after many food vendors had abandoned polystyrene foam hot beverage cups in favor of paper cups. The inventor was interested in commercialization and/or licensing his invention and sought protection of the intellectual property that he had created. This chapter provides an overview of intellectual property in the context of product development and provides specific guidance for preparing an invention disclosure or provisional patent application.

Within the context of product development, the term *intellectual property* refers to the legally protectable ideas, concepts, names, designs, and processes associated with a new product. Intellectual property can be one of the most valuable assets of firms. Unlike physical property, intellectual property cannot be secured with lock and key to prevent its unwanted transfer. Therefore, legal mechanisms have been developed to protect the rights of intellectual property owners. These mechanisms are intended to provide an incentive and reward to those who create new useful inventions, while at the same time encouraging the dissemination of information for the long-run benefit of society.

What Is Intellectual Property?

Four types of intellectual property are relevant to product design and development. Exhibit 14-2 presents a taxonomy of types of intellectual property. Although some areas overlap, and all four types of intellectual property may be present in a single product, a particular invention usually falls into one of these categories.

- *Patent:* A patent is a temporary monopoly granted by a government to an inventor to exclude others from using an invention. In the United States, a patent expires 20 years from the filing date. Most of the balance of this chapter focuses on patents.
- *Trademark:* A trademark is an exclusive right granted by a government to a trademark owner to use a specific name or symbol in association with a class of products or services. In the context of product development, trademarks are typically brands or product names. For example, *JavaJacket* is a trademark for an insulated cup holder, and companies other than Java Jacket, Inc., may not make unauthorized use of the word JavaJacket to refer to their own cup-holder products. In the United States, registration of a trademark is possible, but not strictly necessary to preserve the trademark rights. In most other countries, the rights of a trademark are gained through registration.
- *Trade secret:* A trade secret is information used in a trade or business that offers its owner a competitive advantage and that can be kept secret. A trade secret is not a right conferred by a government but is the result of vigilance on the part of an organization in preventing the dissemination of its proprietary information. Perhaps the most famous trade secret is the formula for the beverage Coca-Cola.
- *Copyright:* A copyright is an exclusive right granted by a government to copy and distribute an original work of expression, whether literature, graphics, music, art, entertainment, or software. Registration of a copyright is possible but not necessary. A copyright comes into being upon the first tangible expression of the work and lasts for up to 95 years.

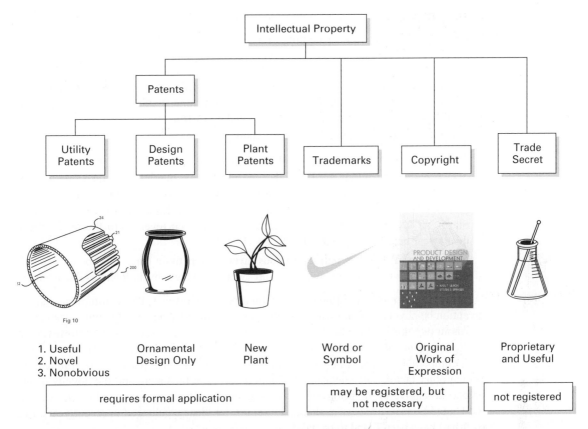

EXHIBIT 14-2   Taxonomy of types of intellectual property relevant to product design and development.

This chapter focuses on patents. Appendix A to this chapter briefly discusses trademarks. We do not devote substantial attention here to copyrights and trade secrets, but several references to other resources appear at the end of the chapter.

Overview of Patents

For most engineered goods, two basic types of patents are relevant: *design patents* and *utility patents*. (A third type of patent covers plants.) Design patents provide the legal right to exclude someone from producing and selling a product with the identical ornamental design described by the design patent. A design patent can be thought of as a "copyright" for the ornamental design of a product. Because design patents must be limited to ornamental design, for most engineered goods, design patents are of very limited value. For this reason, the chapter focuses further on utility patents.

Patent law in most of the world evolved from English law and so patent laws in different countries are somewhat similar. This chapter uses U.S. law as a reference point, and so readers with intentions to obtain patents in other countries should carefully investigate the laws in those countries.

Utility Patents

United States law allows for patenting of an invention that relates to a new process, machine, article of manufacture, composition of matter, or a new and useful improvement of one of these things. Fortunately, these categories include almost all inventions embodied by new products. Note that inventions embodied in software are sometimes patented, but usually the invention is described as a process or machine. Exhibit 14-3 shows the first page of a patent for the insulating sleeve invented by Coffin.

In addition, the law requires that patented inventions be:

- *Useful:* The patented invention must be useful to someone in some context.
- *Novel:* Novel inventions are those that are not known publicly and therefore are not evident in existing products, publications, or prior patents. The definition of novelty relates to disclosures of the actual invention to be patented as well. In the United States, an invention to be patented must not have been revealed to the public more than a year before the patent is filed.
- *Nonobvious:* Patent law defines obvious inventions as those that would be clearly evident to those with "ordinary skill in the art" who faced the same problem as the inventor.

Usefulness is rarely a hurdle to obtaining a patent. However, the requirements that an invention be novel and nonobvious are the most common barriers to obtaining a patent.

About two thirds of applications filed for patents result in issued patents. However, an issued patent is not necessarily *valid.* A patent may be challenged in a government court by a competitor at some point in the future. The validity of a patent is determined by, among other factors, the adequacy of the description in the patent and the novelty of the invention relative to the prior art. A tiny fraction of patents—a few hundred per year in the United States—are ever challenged in court. Of those challenged in recent years, just over half have been found to be valid.

An inventor associated with a patent is a person who actually created the invention individually or in collaboration with other inventors. In some cases the inventor is also the owner of the intellectual property. However, in most cases, the patent is *assigned* to some other entity, usually the inventor's employer. The actual intellectual property rights associated with a patent belong to the owner of the patent and not necessarily to the inventor. (Appendix B to this chapter provides some advice to individual inventors interested in commercializing their inventions.)

A patent owner has the right to exclude others from using, making, selling, or importing an infringing product. This is an *offensive right,* which requires that the patent owner sue the infringer. There are also *defensive rights* associated with patents. Any invention described in a patent, whether part of the claimed invention or not, is considered by the legal system to be known publicly and forms part of the *prior art.* This disclosure is a defensive act blocking a competitor from patenting the disclosed invention.

Preparing a Disclosure

This chapter is focused on a process for preparing an *invention disclosure*—in essence a detailed description of an invention. This disclosure will be in the form of a patent application, which can serve as a provisional patent application and with relatively little additional work could be a regular patent application. It is possible, even typical, for a patent

US005205473A

United States Patent [19]
Coffin, Sr.

[11]	Patent Number:	5,205,473
[45]	Date of Patent:	Apr. 27, 19993

[54] RECYCLABLE CORRUGATED BEVERAGE CONTAINER AND HOLDER

[75] Inventor: David W. Coffin, Sr., Fayetteville, N.Y.

[73] Assignee: Design By Us Company, Philadelphia, Pa.

[21] Appl. No.: 854,425

[22] Filed: Mar. 19, 1992

[51] Int. Cl. B65D 3/28
[52] U.S. Cl. .... 229/1.5 B; 206/813; 220/441; 220/DIG. 30; 229/1.5 H; 229/DIG. 2; 493/296; 493/907
[38] Field of Search .... 229/1.5 B, 1.3 H, 4.5, 229/DIG. 2; 220/441, 671, 737–739, DIG. 30; 493/287, 296, 907, 908; 209/8, 47, 215; 206/813

[56] References Cited

U.S. PATENT DOCUMENTS

1,732,322	10/1929	Wilson et al. 220/DIG. 30
1,771,765	7/1930	Benson 229/4.5
2,266,828	12/1941	Sykes .... 229/1.5 B
2,300,473	11/1942	Winkle 229/4.5
2,503,815	3/1950	Harman
2,617,549	11/1952	Egger
2,641,402	6/1953	Bruun 229/4.5
2,661,889	12/1953	Phinney 229/4.5
2,969,901	1/1961	Behrens .... 229/1.3 B
3,237,834	3/1966	Davis et al. .... 229/1.3 B
3,779,157	12/1973	Ross, Jr. et al. 53/527
3,785,254	1/1974	Mann
3,890,762	6/1975	Ernst et al.
3,908,523	9/1975	Shikays .... 229/1.5 B
4,080,880	3/1978	Shikay 493/296
4,146,660	3/1979	Hall et al.
4,176,034	11/1979	Kelley 209/8
5,009,326	4/1991	Reaves et al.
5,092,485	3/1992	Lee .... 229/1.3 B

OTHER PUBLICATIONS

"The Wiley Encyclopedia of Packaging Technology", John Wiley & Sons, pp. 66-69, 1986.

Primary Examiner—Gary E. Elkins
Attorney, Agent, or Firm—Synnestvedt & Lochner

[57] ABSTRACT

Corrugated beverage containers and holders are which employ recyclable materials, but provide fluting structures for containing insulating air. These products are easy to hold and have a lesser impact on the environment than polystyrene containers.

18 Claims, 8 Drawing Sheets

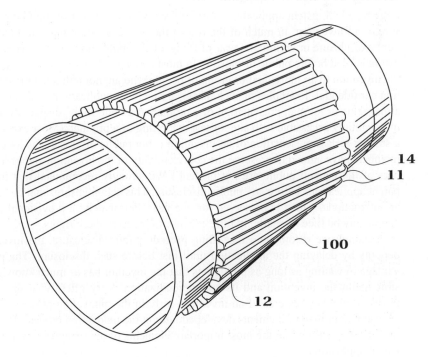

EXHIBIT 14-3 The first page of U.S. Patent 5,205,473.

attorney to do much of the work described in the chapter. However, our belief is that having the inventor draft a detailed disclosure is the best way to communicate the inventor's knowledge, even though in most cases a patent attorney will revise the disclosure to prepare the formal patent application. Although many readers will be able to complete a provisional patent application from the guidance provided here, this chapter is not a substitute for competent legal advice. Inventors pursuing serious commercial opportunities should consult with a patent attorney after preparing their disclosure.

The steps in the process are:

1. Formulate a strategy and plan.
2. Study prior inventions.
3. Outline claims.
4. Write the description of the invention.
5. Refine claims.
6. Pursue application.
7. Reflect on the results and the process.

Step 1: Formulate a Strategy and Plan

In formulating a patent strategy and plan, a product development team must decide on the timing of the filing of a patent application, the type of application to be filed, and the scope of the application.

Timing of Patent Applications

Legally, a U.S. patent application must be filed within one year of the first public disclosure of an invention. In much of the rest of the world, a patent must be filed before any public disclosure or within one year of filing a U.S. application, so long as the U.S. application is filed before public disclosure. In most cases, public disclosure is a description of the invention to an individual or group of people who are not obligated to keep the invention confidential. Examples of such disclosure include: publication of invention details in a magazine or journal, presentation of a product at a trade show, display of the invention on a publicly accessible web site, or test marketing of a product. (Most experts agree that a student's class presentation of an invention is not public disclosure, as long as the class members have agreed to preserve the confidentiality of the invention and as long as members of the general public are not present.) We strongly recommend that inventors file patent applications before any public disclosure. This action ensures that the option to file an international patent is preserved for one year. Fortunately, a provisional patent application may be filed at relatively little expense to preserve these rights.

Although we recommend that filing precede public disclosure, the inventor usually benefits by delaying the application until just before such disclosure. The principal advantage to waiting as long as possible is that the inventor has as much knowledge as possible about the invention and its commercialization. Very often what the inventor believes are the key features of an invention early in the innovation process turn out to be less important than refinements developed later in the innovation process. By waiting, the inventor can ensure that the most important elements of the invention are captured in the patent application.

The U.S. patent system grants priority among competing patent applications based on the date of invention, whereas the patent systems in much of the rest of the world grant priority based on the date of the patent application. Therefore, in the U.S., inventors should carefully document the date of their inventions. This is best done by describing an invention in a bound notebook and then having a witness sign and date each page.

Type of Application

A team faces two basic choices about the type of patent application to be pursued. First, the team must decide whether to file a *regular patent application* or a *provisional patent application*. Second, the team must decide whether to pursue domestic and/or foreign patents.

A regular patent application was the only option available to an inventor in the United States until substantial changes were made to patent law in 1995. Under current U.S. patent law, an inventor may file a provisional patent application. A provisional patent application needs only to fully describe the invention. It does not need to contain claims or comply with the formal structure and language of a regular patent application. The principal advantage of a provisional patent application is that it requires less cost and effort to prepare and file than a regular patent application, but it preserves all options to pursue further patent filings for a period of one year. Once a provisional patent application has been filed, a company may label its products "patent pending," and it retains the right to file a foreign patent application and/or a regular patent application, as long as it does so within one year. The only fundamental disadvantage of a provisional patent application is that it delays the eventual issuance of a patent by up to one year, as the process of examining a patent application does not begin until a regular patent application is filed. Another possible disadvantage is that the preliminary nature of a provisional patent application may lead to the use of less care in preparing the description of the invention than might be the case with a regular application. The description of the invention must be complete in a provisional patent application, and the regular patent application that follows cannot contain features that were not described in the provisional application.

Filing patents internationally is expensive and somewhat complex. The team should therefore consult with a patent professional about international patent strategy, as patent law varies somewhat from country to country. To obtain foreign patent rights, an application must eventually be filed in each country in which a patent is sought. (The European Community, however, acts as a single entity with respect to patent filing.) Foreign applications can be expensive, costing up to $15,000 per country for filing fees, translation fees, and patent agent fees.

The expense of filing for foreign patents can be delayed, generally by 30 months, by filing a *Patent Cooperation Treaty* (PCT) application. A PCT application is filed in one country (e.g., the United States) but is designated as a PCT application, which is the beginning of a process by which foreign patents can be pursued. A PCT application costs only slightly more than a regular patent application in filing fees, but it allows for a substantial delay before application fees must be paid in the countries in which foreign patents are sought.

The provisional patent application and the PCT application together provide a vehicle for a small company or individual inventor to preserve most patent rights with relatively little cost. A typical strategy is to file a provisional patent application before any public disclosure of the invention; then, within one year, to file a PCT application with the United States patent office; then, when forced to act or abandon the application at some point in the future (usually a year or more away), to pursue actual foreign applications. This strategy allows a delay of two or more years before substantial legal and application fees must be paid. During this period, the team can assess the true commercial potential of the products embodying the invention and can estimate the value of more extensive patent protection.

Scope of Application

The team should evaluate the overall product design and decide which elements embody inventions that are likely to be patentable. Typically the process of reviewing the product design will result in a list of elements that the team considers to be novel and nonobvious. The team should focus on those elements that present substantial barriers to competition, which are typically the elements that in the opinion of the team represent a substantial improvement over the publicly known methods of addressing similar problems.

Complex products often embody several inventions. For example, a printer may embody novel signal processing methods and novel paper handling techniques. Sometimes these inventions fall into very different *classes* within the patent system. As a result, a product development team may need to file multiple applications corresponding to the distinct classes of invention. For simple products or for products that embody a single type of invention, a single patent application usually suffices. The decision about whether to divide an application into multiple parts is complex and is best made in consultation with a patent attorney. However, all intellectual property rights are preserved even if a patent application is filed that contains multiple classes of inventions. In such cases, the patent office will inform the inventor that the application must be divided.

While defining the scope of the patent, the team should also consider who the inventors are. An inventor is a person who contributed substantially to the creation of the invention. The definition of an inventor for the purposes of patent law is subjective. For example, a technician who only ran experiments would not typically be an inventor, but a technician who ran experiments and then devised a solution to an observed problem with the device could be considered an inventor. There is no limit to the number of inventors named in a patent application. We believe that product development and invention are most often team efforts and that many members of the team who participated in concept generation and the subsequent design activities could be considered inventors. Failing to name a person who is an inventor can result in a patent being declared not valid.

Step 2: Study Prior Inventions

There are three key reasons for studying prior inventions, the so-called *prior art*. First, by studying the prior patent literature, design teams can learn whether an invention may infringe on existing unexpired patents. Although there is no legal barrier to patenting an invention that infringes on an existing patent, if anyone manufactures, sells, or uses a product that infringes upon an existing patent without a license, the patent owner may sue for damages. Second, by studying the prior art, the inventors get a sense of how similar their invention is to prior inventions and therefore how likely they are to be granted a broad patent. Third, the team will develop background knowledge enabling the members to craft novel claims.

In the course of product development efforts, most teams accumulate a variety of references to prior inventions. Some of the sources of information on prior inventions include:

- Existing and historical product literature.
- Patent searches.
- Technical and trade publications.

Several good online reference sources can be used for searching patents. Simple keyword searches are often sufficient to find most relevant patents. It is important for the team to keep a file containing the prior art they are aware of. This information must be provided to the patent office shortly after filing the patent application.

In the Coffin patent for the cup holder shown in Exhibit 14-1, references to 19 other U.S. patents are cited along with a reference to a book. (The references cited by the inventor and by the patent examiner are listed on the first page of a patent. The first page of the Coffin patent is reproduced as Exhibit 14-3.) Among the prior art for the Coffin patent, for example, is a 1930 patent by Benson (1,771,765; "Waterproof Paper Receptacle") in which a corrugated holder insulates a paper liner cup. The Benson patent describes a cup holder that fits underneath and into the bottom of the liner cup. This is one reason that the invention in the Coffin patent is described as a tube with an opening at the top and bottom.

Step 3: Outline Claims

Issuance of a patent gives the owner a legal right to exclude others from infringing on the invention specifically described in the patent's claims. Claims describe certain characteristics of the invention; they are written in formal legal language and must adhere to some rules of composition. In step 5 we describe how the formal legal language works. However, at this point in the process of preparing the disclosure, the team benefits from thinking carefully about what it believes is unique about the invention. We therefore recommend that the team outline the claims. Don't worry about legal precision at this point. Instead, make a list of the features and characteristics of the invention that the team believes are unique and valuable. For example, an outline of the claims for the Coffin invention might be:

- Use of corrugations as insulation, in many possible forms
  - Corrugations on the inside surface of the tube
  - Corrugations on the outside surface of the tube
  - Corrugations sandwiched between two flat layers of sheet material
  - Vertical orientation of flutes
    - Flutes open at top and bottom of holder
  - Corrugations with "triangle wave" cross section
  - Corrugations with "sine wave" cross section
- Tubular form with openings at both ends
  - In shape of truncated cone
- Recyclable materials
  - Recyclable adhesive
  - Recyclable sheeting
    - Cellulose material

- Biodegradable adhesive
- Surface to print on
- Holder folds flat along two fold lines

The outline of the claims provides guidance about what must be described in detail in the description.

Step 4: Write the Description of the Invention

The bulk of a patent application is formally known as the *specification.* To avoid confusion with our use of the word "specifications" in this book, we call the body of the patent application the *description,* because this is the part of the application that actually describes the invention. The description must present the invention in enough detail that someone with "ordinary skill in the art" (i.e., someone with the skills and capabilities of a typical practitioner working in the same basic field as the invention) could implement the invention. The description should also be a marketing document promoting the value of the invention and the weaknesses in existing solutions. The patent application will be read by a patent examiner, who will search and study prior patents. The description must convince the examiner that the inventors developed something useful that is different from existing inventions and that is nonobvious. In these respects one can think of the description as essentially a technical report on the invention. There are some formatting conventions for patent applications, although these are not strictly necessary for an invention disclosure or a provisional patent application.

Patent law requires that the application "teach" with sufficient detail that someone "skilled in the art" could practice the invention. For example, in the Coffin patent, the inventor discloses that the adhesive to bond the flutes is "a recyclable, and preferably a biodegradable adhesive, for example, R130 adhesive by Fasson Inc., Grand Rapids, MI." The requirement to completely teach the invention may be somewhat counterintuitive for someone accustomed to treating inventions confidentially. Patent law requires that inventors disclose what they know about the invention, but in exchange they are granted the right to exclude others from practicing the invention for a limited time period. This requirement reflects the basic tension in the patent system between granting a temporary monopoly to inventors in exchange for publication of information that will eventually be available for use by anyone.

A typical description includes the following elements:

- *Title:* Provide a short descriptive label for the invention, for example, "Recyclable Corrugated Beverage Container and Holder."
- *List of inventors:* All inventors must be listed. A person should be listed as an inventor if he or she originated any of the inventions claimed in the application. There are no legal limits to the number of inventors and no requirements about the order in which inventors are listed. A failure to list an inventor could result in a patent eventually being declared not valid.
- *Field of the invention:* Explain what type of device, product, machine, or method this invention relates to. For example, the Coffin patent reads, "This invention relates to insulating containers, and especially to those which are recyclable and made of cellulosic materials."

- ***Background of the invention:*** State the problem that the invention solves. Explain the context for the problem, what is wrong with existing solutions, why a new solution is needed, and what advantages are offered by the invention.

- ***Summary of the invention:*** This section should present the substance of the invention in summarized form. The summary may point out the advantages of the invention and how it solves the problems described in the background.

- ***Brief description of the drawings:*** List the figures in the description along with a brief description of each drawing. For example, "Figure 10 is a perspective view of a preferred embodiment illustrating internal flute portions in breakaway views."

- ***Detailed description of the invention:*** This section of the description is usually the most detailed and contains detailed descriptions of embodiments of the invention along with an explanation of how these embodiments work. Further discussion of the detailed description is provided below.

Figures

Formal figures for patents must comply with a variety of rules about labeling, line weight, and types of graphical elements. However, for an invention disclosure or provisional patent application, informal figures are sufficient and hand sketches or CAD drawings are perfectly appropriate. At some point after filing a regular patent application, the patent office will request formal figures, at which time a professional drafter may be hired to prepare formal versions of the necessary figures. Prepare enough figures to clearly show the key elements of the invention in the preferred embodiments that have been considered. A simple invention like the cup holder would probably require 5 to 15 figures.

The features shown in the figures may be labeled with words (e.g., "outer layer"), although to facilitate preparation of a regular patent application, the team may wish to use "reference numerals" on the figures right from the beginning. No rule stipulates that reference numerals must be uninterrupted and consecutive, so a convenient numbering scheme uses reference numerals 10, 11, 12, and so on, for features that first appear in Figure 1; reference numerals 20, 21, 22, and so on, for features that first appear in Figure 2; and so on. This way adding numerals to one figure does not influence the use of numerals in another figure. The same feature shown in more than one figure must be labeled with the same reference numeral, so some numerals will be carried over from one figure to another.

Writing the Detailed Description

The detailed description describes *embodiments* of the invention. An embodiment is a physical realization of the claimed invention. Patent law requires that the application describe the *preferred embodiment*—that is, the best way of practicing the invention. Typically, a detailed description is organized as a collection of paragraphs, each describing an embodiment of the invention in terms of its physical structure along with an explanation of how that embodiment works.

A good strategy for writing the detailed description is to first create the figures that show embodiments of the invention. Next describe the embodiment by labeling each feature of the embodiment in the figure and explaining the arrangement of these features. Finally, explain how the embodiment works and why the features are important to this function. This process is repeated for each of the embodiments described in the detailed description.

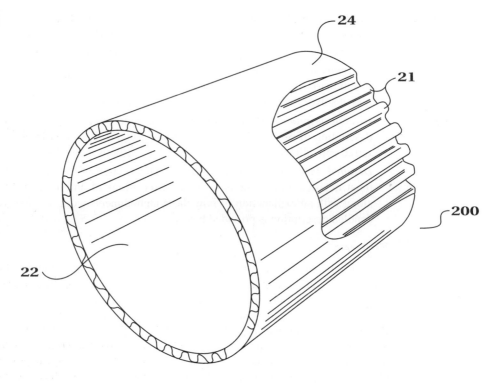

FIG. 10

Consider Figure 10 from the Coffin patent, shown here as Exhibit 14-4. A detailed description might include language like the following:

> A preferred embodiment of the invention is shown in Figure 10. A liner surface 22 and an outer surface 24 sandwich a corrugation 21. The assembly 200 forms a tubular shape whose diameter changes linearly with length so as to form a section of a truncated cone. The smooth outer surface 22 provides a smooth surface onto which graphics may be printed. Corrugation 21 is bonded to outer surface 24 and liner surface 22 with a recyclable adhesive.

The detailed description should show alternative embodiments of the invention. For example, in the Coffin patent, the invention features "flutes" that create an insulating air gap. In a preferred embodiment these flutes are formed by smooth wavy corrugations, with the smooth surface on the outside to allow for graphics to be easily printed on the sleeve. Alternative embodiments include triangle waves and/or sheet materials on either or both sides of the tube. These alternative embodiments are described in the detailed description and shown in the figures. (See Exhibit 14-5.)

Defensive Disclosure

The primary benefit of a patent is that it grants the owner *offensive rights.* That is, the owner has the right to prevent others from practicing the invention. However, patents also offer a subtle mechanism for taking *defensive* actions. A patent is considered prior art and so an invention that appears in a patent may not be patented in the future. For this reason, inventors

EXHIBIT 14-5
Figures 6a, 6b, 7a, and 7b from the Coffin patent showing alternative embodiments of the invention.

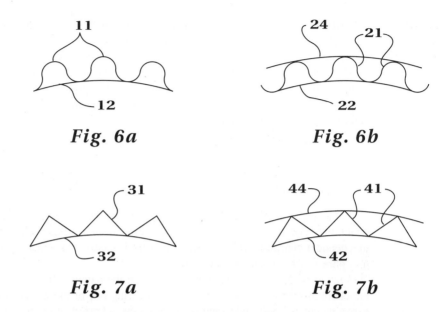

Fig. 6a

Fig. 6b

Fig. 7a

Fig. 7b

may benefit from disclosing essentially every invention they considered that relates to the claimed invention no matter how wide-ranging. This may be done in the detailed description. Even though these inventions may not actually be reflected in the claims of the patent, their disclosure becomes part of the prior art and therefore prevents others from patenting them. This defensive strategy may offer competitive advantages in fields of emerging technologies.

Step 5: Refine Claims

The claims are a set of numbered phrases that precisely define the essential elements of the invention. The claims are the basis for all offensive patent rights. A patent owner can prevent others from practicing the invention described by the claims only. The rest of the patent application is essentially background and context for the claims.

Writing the Claims

Although claims must be expressed verbally, they adhere to a strict mathematical logic. Almost all claims are formulated as a recursive expression of the form

$$X = A + B + C. \ldots , \qquad \text{where } A = u + v + w \ldots , \quad B = \ldots$$

This is expressed verbally as:

An X comprising an A, a B, and a C, wherein said A is comprised of a u, a v, and a w and wherein said B is . . .

Note that claims comply with some verbal conventions. The word *comprising* means "including but not limited to" and is almost always used as the equal sign in the expression. The first time an element, say a *liner sheet,* is named in a claim the inventor uses the indefinite article *a* as in "comprises *a* liner sheet." Once this element has been named it is never referred to as *the* liner sheet, but always as *said* liner sheet. This is true for every subsequent instance in which *liner sheet* is used in the claims. Although these conventions are not difficult to remember once learned, inventors preparing a disclosure for subsequent editing by a patent attorney should not worry too much about formal correctness of the language. The language is easily corrected when the formal patent application is prepared.

Multiple claims are arranged hierarchically into *independent* claims and *dependent* claims. Independent claims stand alone and form the root nodes of a hierarchy of claims. Dependent claims always add further restrictions to an independent claim. Dependent claims are typically written in this form:

The invention of Claim N, further comprising *Q, R,* and *S* . . .

or

The invention of Claim N, wherein said *A* . . .

Dependent claims essentially inherit all of the properties of the independent claim on which they depend. In fact, a dependent claim can be read as if all of the language of the independent claim on which it depends were inserted as a replacement for the introductory phrase "The invention of Claim N."

The dependent claims are important in that the patent office may reject the independent claim as obvious or not novel while allowing one or more dependent claims. In such cases, patentable material remains; the original independent claim can be deleted and the original dependent claim can be rewritten as an independent claim.

The elements of a claim form a *logical and* relationship. To infringe on a claim, a device must include each and every element named in the claim. If, for example, a competitive product were to use only three of four elements named in a claim, it would not infringe on the claim.

Consider this example from the Coffin patent (edited slightly for clarity).

Claim 1

A beverage container holder, comprising a corrugated tubular member comprising cellulosic material and at least a first opening therein for receiving and retaining a beverage container, said corrugated tubular member comprising fluting means for containing insulating air; said fluting means comprising fluting adhesively attached to a liner with a recyclable adhesive.

Claim 1 is an independent claim. Consider Claim 2, which is dependent on Claim 1.

Claim 2

The holder of claim 1, wherein said tubular member further comprises a second opening wherein said first opening and said second opening are of unequal cross-sectional dimensions.

This claim adheres to the logical structure shown in Exhibit 14-6.

Let us reinforce the idea that a claim is formed of a logical "and" relation among its elements. Claim 1 is for a holder that includes all of these elements:

- Corrugated tube
  - Made of cellulosic material
  - With a first opening
  - With fluting means
    - Made of flutes adhesively attached to a liner
      - Using recyclable adhesive

EXHIBIT 14-6
The logical structure of Claims 1 and 2 from the Coffin patent. Note that Claim 2 is dependent on Claim 1 and simply adds further restrictions, a second opening and a relationship between the first and second openings.

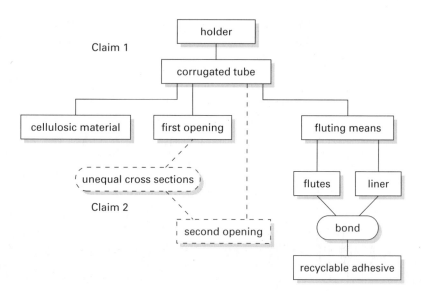

If a competing cup holder does not have every one of these elements, it does not infringe upon this claim. So, for example, if it were made of polystyrene, it would not infringe on this claim (no cellulosic material). Consider a patent by Jay Sorensen filed shortly after the Coffin patent. Sorensen's patent is for a cup holder with a dimpled surface. (See Exhibit 14-7.) Because this invention does not have "fluting means" it does not infringe upon the Coffin patent. Sorensen's claims include the following (edited slightly for clarity):

Claim 4

A cup holder comprising a band of material formed with an open top and an open bottom through which a cup can extend and an inner surface immediately adjacent to said cup, said band comprising a plurality of discrete, spaced-apart, approximately semi-spherically shaped depressions distributed on substantially the entire inner surface of said band so that each depression defines a non-contacting region of said band creating an air gap between said band and said cup, thereby reducing the rate of heat transfer through said holder.

At least two lessons can be derived from comparing the Coffin and Sorenson inventions. First, patents often provide relatively limited commercial advantages. In this case, by creating a cup holder with dimples instead of flutes, Sorensen was able to avoid infringing the Coffin patent. In fact, both inventions are embodied by successful commercial products, but neither patent provides ironclad protection from competition. The second lesson is that the inventor should invest in considering as many ways as possible to achieve the desired function of the invention, in this case an insulating layer. Had Coffin thought of a dimpled surface, then this feature could have been described in his patent application. In the best case, the dimpled invention may have formed the basis for additional claims in the patent. In the worst case, the description of the dimpled embodiment in the patent application would have been prior art and therefore prevented Sorensen from obtaining his patent. (It would not prevent Sorenson and others from practicing the dimpled invention unless it were claimed.)

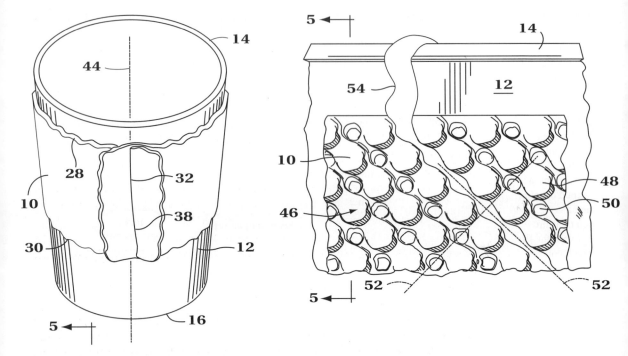

EXHIBIT 14-7 Figure from the Sorensen patent (U.S. Patent 5,425,497, "dimpled cup holder").

Guidelines for Crafting Claims

Several guidelines are helpful in crafting claims. Writing great claims is tricky, so we advise inventors to seek the help of an experienced patent attorney in refining patent application.

- Always try to make a claim as general as possible. When a specific descriptor is used, try making it general. For example, Coffin's patent speaks of a "tubular member" and not a "tube."
- Avoid absolute definitions by using modifiers like "substantially," "essentially," and "approximately."
- Attempt to create an invention that does not infringe on the draft claim, and then try to rewrite the claim or add an additional claim such that the hypothetical invention would infringe.

Step 6: Pursue Application

In most cases, the inventor will deliver the draft application to a patent attorney or other intellectual property professional for refinement and formal application. It is possible to file a patent application as an individual if severely budget constrained. Pressman provides detailed guidelines for doing this (Pressman, 2002). Note that the statutory requirements are administratively complex, and so we highly recommend that commercial product development teams retain a competent specialist to pursue any application to the patent office.

Once an invention disclosure is prepared, the team can proceed in four different ways, with the specific course of action dictated by the business context.

- *The team can file a provisional patent application.* An individual or small company can file a provisional patent application for less than $100 in filing fees. The application need contain only a description of the invention and need not comply with the formalities of a regular patent application. Once a provisional application is filed, a product may be labeled "patent pending." If the team wishes to pursue a regular patent application, this application must be filed within one year of the filing of the provisional patent application. A provisional patent therefore acts as an option to pursue a regular patent application and allows the team time to pursue licensing or further investigation before incurring the expense of a regular patent application.
- *The team may file a regular patent application in the United States.* This process costs about $500 in filing fees for a small company or individual, in addition to the legal fees for a patent attorney.
- *The team may file a patent cooperation treaty or PCT application.* A PCT application allows a single patent application in a single country, say the United States, to initiate the process of pursuing international patent protection. Eventually, the inventor must pursue patent protection in individual countries or collections of countries (e.g., European Union). However, the PCT process allows the first steps of the process to be carried out relatively efficiently and with a single point of contact. The entire process of pursuing foreign patent rights is beyond the scope of this chapter. Consult a patent attorney for details.
- *The team can defer application indefinitely.* The team may delay in hope that future information will make a course of action obvious. In some cases, the team may decide not to pursue the invention and therefore may decide to abandon the patent application process. The consequences of delay may be substantial. If the invention is disclosed publicly, then all international patent rights are forgone. If a year passes after public disclosure without filing a regular patent application, then U.S. patent rights are also forgone. Nevertheless, the team may be able to defer any action for several months before these eventualities are realized.

At some point after the team files a regular patent application or a PTO application, the patent office will issue an *office action* responding to the application. In almost all cases, a patent examiner will reject many or all of the claims as either obvious or not novel. This is the norm and it is part of a back-and-forth exchange between the patent office and inventor that should eventually result in claims that are patentable. Next, the inventor and patent attorney sharpen arguments, edit claims to reflect comments from the examiner, and respond to the office action with an amended application. Most applications eventually result in an issued patent, although the claims rarely remain exactly as originally written.

The patent office does not review or act on provisional patent applications. It merely records their filing and stores the application for review when and if a regular application is filed. Note that most foreign patents must be filed within one year of the provisional patent application filing date.

Step 7: Reflect on the Results and the Process

In reflecting on the patent application or invention disclosure, the team should consider at least the following questions:

- What are the essential and distinctive features of the product concept, and therefore the invention? Are these features reflected in the description of the invention and in the claims? Does the description communicate the best way of practicing the invention?
- What is the timing of future required actions? The team's patent attorney will typically maintain a *docket*—essentially a calendar indicating when further actions must be taken to preserve patent rights. However, the inventor or someone within the team's company should also be responsible for thinking about the actions that must be taken in the coming months.
- Which aspects of the process of preparing the patent application or invention disclosure went smoothly and which aspects require further efforts in the future?
- What did the team learn about the prior art that may inform future product development efforts? For example, are there valuable technologies that might be licensed from existing patent holders? Are competitors' patents expiring, possibly allowing the team to use a convenient solution to a long-standing problem?
- How strong an intellectual property position does the team have? Are the features of the invention in the patent application so novel and valuable that they really prevent competitors from direct competition, or is the patent likely to be merely a deterrent to the most direct copies of the products that embody the invention?
- Did the team begin the process too early or too late? Was the effort rushed? What is the ideal timing for the next effort to prepare a patent application?

Summary

- A patent is a temporary monopoly granted by a government to exclude others from using, making, or selling an invention. Patent law is intended to balance an incentive for invention with the free dissemination of information.
- Utility patents are the central element of the intellectual property for most technology-based product development efforts.
- An invention can be patented if it is useful, novel, and nonobvious.
- The final invention that is patented is defined by the patent claims. The rest of the patent application essentially serves as background and explanation in support of the claims.
- We recommend a seven-step process for pursuing a patent:
  1. Formulate a strategy and plan.
  2. Study prior inventions.
  3. Outline claims.
  4. Write the description of the invention.
  5. Refine claims.

6. Pursue application.

7. Reflect on the results and the process.

- Provisional patent applications and patent cooperation treaty (PCT) applications can be used to minimize the costs of pursuing patent protection while preserving all future options.

References and Bibliography

Many current resources are available on the Internet via
www.ulrich-eppinger.net

The examples in the chapter are derived from the Coffin and Sorenson patents.

Coffin, David W., *Recyclable Corrugated Beverage Container and Holder,* United States Patent 5,205,473, April 27, 1993.

Sorenson, Jay, *Cup Holder,* United States Patent 5,425,497, June 20, 1995.

Pressman's book is a comprehensive guide to the details of patent law and provides a step-by-step process for writing a patent application and pursuing the application with the patent office. The book also contains valuable related information on licensing inventions.

Pressman, David, *Patent It Yourself,* ninth edition, Nolo Press, Berkeley, CA, 2002.

Stim provides an in-depth discussion of most aspects of intellectual property, including trademarks and copyrights.

Stim, Richard, *Intellectual Property: Patents, Trademarks, and Copyrights,* second edition, Delmar Learning, Clifton Park, NY, 2000.

Exercises

1. Find a patent number on a product that interests you. Look up the patent using an on-line reference tool.

2. Draft a claim for the self-stick note pad invention marketed by 3M Corporation as the Post-it note.

3. Draw a logic diagram of two claims for the patent in Exercise 1.

4. Generate one or more product concepts that are very different from the Coffin and Sorensen inventions to address the problem of handling a hot coffee cup and that don't infringe the Coffin and Sorensen patents.

Thought Questions

1. Controversy erupted in 1999 when the J. M. Smucker Company sued a Michigan bakery, Albie's, for infringing on its patent by marketing a crustless peanut butter and jelly sandwich crimped on the edges. (See U.S. patent 6,004,596.) Albie's argued that the patent was issued in error because the invention was obvious. Look up the Smucker patent. Do you believe that the Smucker's invention is nonobvious? Why or why not?

2. Why might an inventor describe but not claim an invention in a patent?

Appendix A

Trademarks

A trademark is a word or symbol associated with the products of a particular manufacturer. Trademarks can form an important element of the portfolio of intellectual property owned by a company. Trademarks can be words, "word marks" (stylized graphics spelling out words), and/or symbols. Trademarks usually correspond to brands, product names, and sometimes to company names.

Trademark law is intended to prevent unfair competition, which could result if one manufacturer named its products like those of another in an attempt to confuse the public. In fact, to avoid confusion, when a manufacturer uses a competitor's trademark in advertising, say for the purposes of comparison, it must by law indicate that the name is a trademark of the competitor.

Trademarks must not be purely descriptive. For example, although a company could not obtain a trademark on "Insulating Sleeve," it could for names that are suggestive but not purely descriptive such as "Insleev," "ThermaJo," or "CupPup."

In the United States, a federal trademark may be established merely by using the mark in interstate commerce. This is done by attaching "TM" to the word or symbol when used in advertising or labeling the product (e.g., JavaJacket™). Trademarks may also be registered at modest cost through the United States Patent and Trademark Office using a simple process. When registered, a trademark is denoted with the symbol ® (e.g., Coke®).

Given the importance of the Internet for communicating with customers, when creating new product names, the team should strive to create trademarks that correspond exactly to domain names on the Internet.

Appendix B

Advice to Individual Inventors

Most students of product development and product development professionals have at some point had an idea for a novel product. Often further thought results in a product concept, which sometimes embodies a patentable invention. A common misconception among inventors is that a raw idea or even a product concept is highly valuable. Here is some advice based on observations of many inventors and product commercialization efforts.

- A patent can be a useful element of a plan for developing and commercializing a product. However, it is not really a central element of that activity. Patenting an invention can usually wait until many of the technical and market risks have been addressed.
- A patent by itself rarely has any commercial value. (An idea by itself has even less value.) To extract value from a product opportunity, an inventor must typically complete a product design, resolving the difficult trade-offs associated with addressing customer needs while minimizing production costs. Once this hard work is completed, a product design may have substantial value. In most cases, pursuing a patent is not

worth the effort except as part of a larger effort to take a product concept through to a substantial development milestone such as a working prototype. If the design is proven through prototyping and testing, a patent can be an important mechanism for increasing the value of this intellectual property.

• Licensing a patent to a manufacturer as an individual inventor is very difficult. If you are serious about your product opportunity, be prepared to pursue commercialization of your product on your own or in partnership with a smaller company. Once you have demonstrated a market for the product, licensing to a larger entity becomes much more likely.

• File a provisional patent application. For very little money, an individual using the guidelines in this chapter can file a provisional application. This action provides patent protection for a year, while you evaluate whether your idea is worth pursuing.

CHAPTER FIFTEEN

Product Development Economics

Courtesy of Polaroid Corporation

EXHIBIT 15-1
One of Polaroid's digital color photo printers.

This chapter was developed in collaboration with Thomas P. Foody.

A product development team at Polaroid Corporation was in the midst of developing a new photograph printer, the CI-700. Exhibit 15-1 shows one of Polaroid's color photo printers. The CI-700 would produce instant full-color photographs from digital images stored in a computer. The primary markets for the product are the graphic arts, insurance, and real estate industries. During the CI-700's development, the Polaroid product development team was faced with several decisions which it knew could have a significant impact on the product's profitability:

- Should the team take more time for development in order to make the product available on multiple computer "platforms," or would a delay in bringing the CI-700 to market be too costly?
- Should the product use print media (instant film) from Polaroid's consumer camera business or new and specialized premium-quality print media?
- Should the team increase development spending in order to increase the reliability of the CI-700?

The product development team needed tools to help it make these and other development decisions. This chapter presents an economic analysis method for supporting the decisions of product development teams. The process consists of two types of analysis, *quantitative* and *qualitative.* The emphasis in this chapter is on quick, approximate methods for supporting decision making within the project team.

Elements of Economic Analysis

Quantitative Analysis

There are several basic cash inflows (revenues) and cash outflows (costs) in the life cycle of a successful new product. Cash inflows come from product sales. Cash outflows include spending on product and process development; costs of production ramp-up such as equipment purchases and tooling; costs of marketing and supporting the product; and ongoing production costs such as raw materials, components, and labor. The cumulative cash inflows and outflows over the life cycle of a typical successful product are presented schematically in Exhibit 15-2.

Economically successful products are profitable; that is, they generate more cumulative inflows than cumulative outflows. A measure of the degree to which inflows are greater than outflows is the *net present value* (NPV) of the project, or the value in today's dollars of all of the expected future cash flows. The quantitative part of the economic analysis method described in this chapter estimates the NPV of a project's expected cash flows. The method uses NPV techniques because they are easily understood and used widely in business. (Appendix A provides a brief tutorial on NPV.) The value of quantitative analysis is not only in providing objective evaluations of projects and alternatives but also in bringing a measure of structure and discipline to the assessment of product development projects.

Qualitative Analysis

Quantitative analysis can capture only those factors that are measurable, yet projects often have both positive and negative implications that are difficult to quantify. Also, quantita-

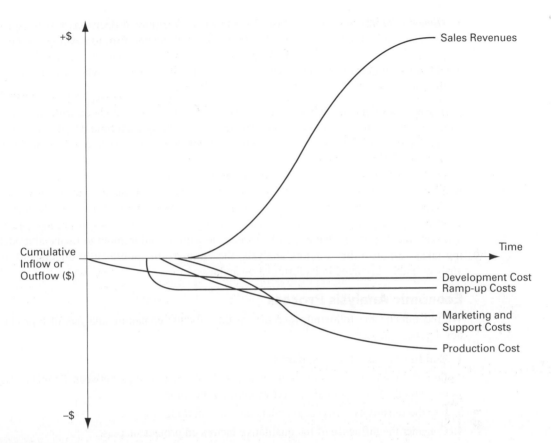

EXHIBIT 15-2   Typical cash flows for a successful new product.

tive analysis rarely captures the characteristics of a dynamic and competitive environment. The chief executive of a major American corporation underscores this point: "I've had MBAs argue with me that a capital expenditure is wrong because it doesn't have a payback within two or three years; they ignore the fact that if we don't make the move, we'll fall behind the rest of our industry in four or five years" (Linder and Smith, 1992). The method in this chapter uses qualitative analysis to capture some of these issues. Our approach to qualitative analysis is to consider specifically the interactions between the project and (1) the firm, (2) the market, and (3) the macroeconomic environment.

When Should Economic Analysis Be Performed?

Economic analysis, which includes both quantitative and qualitative approaches, is useful in at least two different circumstances:

- *Go/no-go milestones:* For example, should we try to develop a product to address this market opportunity? Should we proceed with the implementation of a selected concept? Should we launch the product we have developed? These decisions typically arise at the end of each phase of development.

- ***Operational design and development decisions:*** Operational decisions involve questions such as: Should we spend $100,000 to hire an outside firm to develop this component in order to save two months of development time? Should we launch the product in four months at a unit cost of $450 or wait until six months when we can reduce the cost to $400?

The analysis done at the beginning of a project can usually be updated with current information so that it does not have to be created in its entirety each time. Used in this way, the analysis becomes one of the information systems the team uses to manage the development project.

Economic analysis can be carried out by any member of the development team. In small companies, the project leader or one of the members of the core project team will implement the details of the analysis. In larger companies, a representative from a finance or planning group may be appointed to assist the development team in performing the analysis. We emphasize that even when someone with formal training in financial modeling takes responsibility for this analysis, the entire team should fully understand the analysis and be involved in its formulation and use.

Economic Analysis Process

We recommend the following four-step method for the economic analysis of a product development project:

1. Build a base-case financial model.
2. Perform a sensitivity analysis to understand the relationships between financial success and the key assumptions and variables of the model.
3. Use the sensitivity analysis to understand project trade-offs.
4. Consider the influence of the qualitative factors on project success.

The balance of this chapter is organized around these four steps.

Step 1: Build a Base-Case Financial Model

Constructing the base-case model consists of estimating the timing and magnitude of future cash flows and then computing the NPV of those cash flows.

Estimate the Timing and Magnitude of Future Cash Inflows and Outflows

The timing and magnitude of the cash flows is estimated by merging the project schedule with the project budget, sales volume forecasts, and estimated production costs. The level of detail of cash flows should be coarse enough to be convenient to work with, yet it should contain enough resolution to facilitate effective decision making. The most basic categories of cash flow for a typical new product development project are:

- Development cost (all remaining design, testing, and refinement costs up to production ramp-up).
- Ramp-up cost.
- Marketing and support cost.

- Production cost.
- Sales revenues.

Depending on the types of decisions the model will support, greater levels of detail for one or more areas may be required. More detailed modeling may consider these same five cash flows in greater detail, or it may consider other flows. Typical refinements include:

- Breakdown of production costs into direct costs and indirect costs (i.e., overhead).
- Breakdown of marketing and support costs into launch costs, promotion costs, direct sales costs, and service costs.
- Inclusion of tax effects, including depreciation and investment tax credits. (Tax effects are typically considered in even simple financial modeling. For the sake of clarity, however, we omit the tax effects in our examples.)
- Inclusion of such miscellaneous inflows and outflows as working capital requirements, cannibalization (the impact of the new product on existing product sales), salvage costs, and opportunity costs.

The financial model we use in this chapter is simplified to include only the major cash flows that are typically considered in practice, but conceptually it is identical to more complex models. The numerical values of the cash flows come from budgets and other estimates obtained from the development team, the manufacturing organization, and the marketing organization. Exhibit 15-3 shows the relevant financial estimates for the CI-700. (These data have been disguised to protect Polaroid's proprietary financial information.) For a more detailed discussion of manufacturing costs, see Chapter 11, Design for Manufacturing. Note that all revenues and expenses to date are *sunk costs* and are irrelevant to NPV calculations. (The concept of sunk costs is reviewed in Appendix A.)

In order to complete the model, the financial estimates must be merged with timing information. This can be done by considering the project schedule and sales plan. Exhibit 15-4 shows the project timing information in Gantt chart form for the CI-700. (For most projects, a time increment of months or quarters is most appropriate.) The remaining time to market is estimated to be five quarters, and product sales are anticipated to last 11 quarters.

A common method of representing project cash flow is a table. The rows of the table are the different cash flow categories, while the columns represent successive time periods. Usually this table is encoded in a computer spreadsheet to facilitate analysis. For this example, we assume that the rate of cash flow for any category is constant across any time period (e.g., total development spending of $5 million over one year is allocated equally to each of the four quarters); however, the values can be arranged in any way that best represents the team's forecast of the cash flows. We multiply the unit sales quantity

EXHIBIT 15-3

CI-700 project budgets, sales volume forecasts, and production costs.

1. Development cost	$5 million
2. Ramp-up cost	$2 million
3. Marketing and support cost	$1 million/year
4. Unit production cost	$400/unit
5. Sales and production volume	20,000 units/year
6. Unit price	$800/unit

	Year 1				Year 2				Year 3				Year 4			
	Q1	Q2	Q3	Q4	Q1	Q2	Q3	Q4	Q1	Q2	Q3	Q4	Q1	Q2	Q3	Q4
Development	▓	▓														
Ramp-up					▓											
Marketing and support					▓	▓	▓	▓	▓	▓	▓	▓	▓	▓	▓	▓
Production and sales window						▓	▓	▓	▓	▓	▓	▓	▓	▓	▓	▓

EXHIBIT 15-4   CI-700 project schedule from inception through market withdrawal.

($ values in thousands)	Year 1				Year 2				Year 3				Year 4			
	Q1	Q2	Q3	Q4	Q1	Q2	Q3	Q4	Q1	Q2	Q3	Q4	Q1	Q2	Q3	Q4
Development cost	−1,250	−1,250	−1,250	−1,250												
Ramp-up cost				−1,000	−1,000											
Marketing & support cost					−250	−250	−250	−250	−250	−250	−250	−250	−250	−250	−250	−250
Production cost						−2,000	−2,000	−2,000	−2,000	−2,000	−2,000	−2,000	−2,000	−2,000	−2,000	−2,000
Production volume						5,000	5,000	5,000	5,000	5,000	5,000	5,000	5,000	5,000	5,000	5,000
Unit production cost						−0.4	−0.4	−0.4	−0.4	−0.4	−0.4	−0.4	−0.4	−0.4	−0.4	−0.4
Sales Revenue						4,000	4,000	4,000	4,000	4,000	4,000	4,000	4,000	4,000	4,000	4,000
Sales volume						5,000	5,000	5,000	5,000	5,000	5,000	5,000	5,000	5,000	5,000	5,000
Unit price						0.8	0.8	0.8	0.8	0.8	0.8	0.8	0.8	0.8	0.8	0.8

EXHIBIT 15-5   Merging the project financials and schedule into a cash flow table (all dollar values are in thousands in this and subsequent tables).

by the unit price to find the total product revenues in each period. We also multiply the unit production quantity by the unit production cost to find the total production cost in each period. Exhibit 15-5 illustrates the resulting table.

Compute the Net Present Value of the Cash Flows

Computing the NPV requires that the net cash flow for each period be determined, and then that this cash flow be converted to its present value (its value in today's dollars), as shown in Exhibit 15-6. Consider, for example, the calculations for year 3, first quarter:

1. The period cash flow is the sum of inflows and outflows:

Marketing cost	$ −250,000
Product revenues	4,000,000
Production cost	−2,000,000
Period cash flow	$1,750,000

2. The present value of this period cash flow discounted at 10 percent per year (2.5 percent per quarter) back to the first quarter of year 1 (a total of eight quarters) is

($ values in thousands)	Year 1				Year 2				Year 3				Year 4			
	Q1	Q2	Q3	Q4	Q1	Q2	Q3	Q4	Q1	Q2	Q3	Q4	Q1	Q2	Q3	Q4
Development cost	−1,250	−1,250	−1,250	−1,250												
Ramp-up cost				−1,000	−1,000											
Marketing & support cost					−250	−250	−250	−250	−250	−250	−250	−250	−250	−250	−250	−250
Production cost						−2,000	−2,000	−2,000	−2,000	−2,000	−2,000	−2,000	−2,000	−2,000	−2,000	−2,000
Production volume						5,000	5,000	5,000	5,000	5,000	5,000	5,000	5,000	5,000	5,000	5,000
Unit production cost						−0.4	−0.4	−0.4	−0.4	−0.4	−0.4	−0.4	−0.4	−0.4	−0.4	−0.4
Sales Revenue						4,000	4,000	4,000	4,000	4,000	4,000	4,000	4,000	4,000	4,000	4,000
Sales volume						5,000	5,000	5,000	5,000	5,000	5,000	5,000	5,000	5,000	5,000	5,000
Unit price						0.8	0.8	0.8	0.8	0.8	0.8	0.8	0.8	0.8	0.8	0.8
Period Cash Flow	−1,250	−1,250	−1,250	−2,250	−1,250	1,750	1,750	1,750	1,750	1,750	1,750	1,750	1,750	1,750	1,750	1,750
PV Year 1, r = 10%	−1,250	−1,220	−1,190	−2,089	−1,132	1,547	1,509	1,472	1,436	1,401	1,367	1,334	1,301	1,269	1,239	1,208
Project NPV	**8,203**															

EXHIBIT 15-6 Total cash flows, present values, and net present value.

$1,436,306. (The concepts of *present value, net present value,* and *discount rate* are reviewed in Appendix A.)

$$\frac{\$1,750,000}{1.025^8} = \$1,436,306$$

3. Project NPV is the sum of the discounted cash flows for each of the periods, or $8,203,000. (Here and in the rest of the chapter we round financial figures to the nearest one thousand dollars.)

The Base-Case Financial Model Can Support Go/No-Go Decisions and Major Investment Decisions

The NPV of this project, according to the base-case model, is positive, so the model supports and is consistent with the decision to proceed with development. Such modeling can also be used to support major investment decisions. Say, for example, that Polaroid were deciding between two different production facilities with different ramp-up, production, and support costs. The team could develop a model for each of the two scenarios and then compare the NPVs. The scenario with the higher NPV would better support the investment decision. We now consider sensitivity analysis as a technique for readily understanding multiple scenarios for ongoing product development decisions.

Step 2: Perform Sensitivity Analysis

Sensitivity analysis uses the financial model to answer "what if" questions by calculating the change in NPV corresponding to a change in the factors included in the model. Both internal and external factors influence project value. *Internal factors* are those over which the development team has a large degree of influence, including development program expense, development speed, production cost, and product performance. *External factors* are those which the team cannot arbitrarily change, including the competitive environment

EXHIBIT 15-7

Key factors influencing product development profitability.

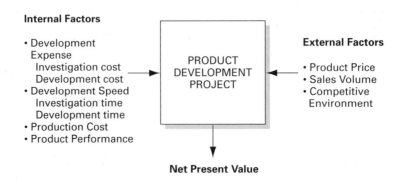

Internal Factors

- Development Expense
  Investigation cost
  Development cost
- Development Speed
  Investigation time
  Development time
- Production Cost
- Product Performance

PRODUCT DEVELOPMENT PROJECT

External Factors

- Product Price
- Sales Volume
- Competitive Environment

Net Present Value

($ values in thousands)	Year 1				Year 2				Year 3				Year 4			
	Q1	Q2	Q3	Q4	Q1	Q2	Q3	Q4	Q1	Q2	Q3	Q4	Q1	Q2	Q3	Q4
Development cost	−1,000	−1,000	−1,000	−1,000												
Ramp-up cost				−1,000	−1,000											
Marketing & support cost					−250	−250	−250	−250	−250	−250	−250	−250	−250	−250	−250	−250
Production cost						−2,000	−2,000	−2,000	−2,000	−2,000	−2,000	−2,000	−2,000	−2,000	−2,000	−2,000
Production volume						5,000	5,000	5,000	5,000	5,000	5,000	5,000	5,000	5,000	5,000	5,000
Unit production cost						−0.4	−0.4	−0.4	−0.4	−0.4	−0.4	−0.4	−0.4	−0.4	−0.4	−0.4
Sales Revenue						4,000	4,000	4,000	4,000	4,000	4,000	4,000	4,000	4,000	4,000	4,000
Sales volume						5,000	5,000	5,000	5,000	5,000	5,000	5,000	5,000	5,000	5,000	5,000
Unit price						0.8	0.8	0.8	0.8	0.8	0.8	0.8	0.8	0.8	0.8	0.8
Period Cash Flow	−1,000	−1,000	−1,000	−2,000	−1,250	1,750	1,750	1,750	1,750	1,750	1,750	1,750	1,750	1,750	1,750	1,750
PV Year 1, r = 10%	−1,000	−976	−952	−1,857	−1,132	1,547	1,509	1,472	1,436	1,401	1,367	1,334	1,301	1,269	1,239	1,208
Project NPV	9,167															

EXHIBIT 15-8   CI-700 financial model with 20 percent decrease in development spending.

(e.g., market response, actions of competitors), sales volume, and product price. (There may be disagreement over whether price is an internal or external factor. In either case, there is little disagreement that price is strongly influenced by the prices of competitive products and that it is coupled to sales volume.) While external factors are not directly controlled by product development teams, they are often influenced by the internal factors. The external and internal factors are shown in Exhibit 15-7.

Development Cost Example

As a first example, let us consider the sensitivity of NPV to changes in development cost. By making incremental changes to development cost while holding other factors constant, we can see the incremental impact on project NPV. For example, what will be the change in NPV if development cost is decreased by 20 percent? A 20 percent decrease would lower the total development spending from $5 million to $4 million. If development time remains one year, then the spending per quarter would decrease from $1.25 million to $1 million. This change is simply entered in the model and the resulting NPV is calculated. This change to the CI-700's base-case model is shown in Exhibit 15-8.

Change in Development Cost, %	Development Cost, $ Thousands	Change in Development Cost, $ Thousands	Change in NVP, %	NPV, $ Thousands	Change in NPV, $ Thousands
50	7,500	2,500	−29.4	5,791	−2,412
20	6,000	1,000	−11.8	7,238	−964
10	5,500	500	−5.9	7,721	−482
base	5,000	base	0.0	8,203	0
−10	4,500	−500	5.9	8,685	482
−20	4,000	−1,000	11.8	9,167	964
−50	2,500	−2,500	29.4	10,615	2,412

EXHIBIT 15-9   CI-700's development cost sensitivities.

A 20 percent decrease in development cost will increase NPV to $9,167,000. This represents a dollar increase of $964,000 and a percentage increase of 11.8 in NPV. This is an extremely simple case: we assume we can achieve the same project goals by spending $1 million less on development, and we therefore have increased the project value by the present value of the $1 million in savings accrued over a time period of one year. The CI-700 development cost sensitivity analysis for a range of changes is shown in Exhibit 15-9. The values in the table are computed by entering the changes corresponding to each scenario into the base-case model and noting the results. It is often useful to know the absolute dollar changes in NPV as well as the relative percentage changes, so we show both in the sensitivity table.

Development Time Example

As a second example, we calculate the development time sensitivities for the CI-700 model. Consider the impact on project NPV of a 25 percent increase in development time. A 25 percent increase in development time would raise the time from four quarters to five quarters. This increase in development time would also delay the start of production ramp-up, marketing efforts, and product sales. To perform the sensitivity analysis, we must make several assumptions about the changes. We assume the same total amount of development cost, even though we will increase the time period over which the spending occurs, thus lowering the rate of spending from $1.25 million to $1.0 million per quarter. We also assume that there is a fixed window for sales which starts as soon as the product enters the market and ends in the fourth quarter of year 4. In effect, we assume we can sell product from the time we are able to introduce it until a fixed date in the future. Note that these assumptions are unique to this development project. Different product development projects would require different assumptions as appropriate. For example, we might have instead assumed that the sales window simply shifts in time by one quarter. The change to the CI-700's financial model is shown in Exhibit 15-10.

Exhibit 15-11 presents the development time sensitivities for a range of changes. We can see that a 25 percent increase in development time will decrease NPV to $6,764,000. This represents a decrease in NPV of $1,439,000, or 17.5 percent.

We recommend that sensitivities be computed for each of the external and internal factors, with the exception of the competitive environment, which is not explicitly contained in the base-case model. These sensitivity analyses inform the team about which factors in

($ values in thousands)	Year 1				Year 2				Year 3				Year 4			
	Q1	Q2	Q3	Q4	Q1	Q2	Q3	Q4	Q1	Q2	Q3	Q4	Q1	Q2	Q3	Q4
Development cost	−1,000	−1,000	−1,000	−1,000	−1,000											
Ramp-up cost					−1,000	−1,000										
Marketing & support cost						−250	−250	−250	−250	−250	−250	−250	−250	−250	−250	−250
Production cost							−2,000	−2,000	−2,000	−2,000	−2,000	−2,000	−2,000	−2,000	−2,000	−2,000
Production volume							5,000	5,000	5,000	5,000	5,000	5,000	5,000	5,000	5,000	5,000
Unit production cost							−0.4	−0.4	−0.4	−0.4	−0.4	−0.4	−0.4	−0.4	−0.4	−0.4
Sales Revenue							4,000	4,000	4,000	4,000	4,000	4,000	4,000	4,000	4,000	4,000
Sales volume							5,000	5,000	5,000	5,000	5,000	5,000	5,000	5,000	5,000	5,000
Unit price							0.8	0.8	0.8	0.8	0.8	0.8	0.8	0.8	0.8	0.8
Period Cash Flow	−1,000	−1,000	−1,000	−1,000	−2,000	−1,250	1,750	1,750	1,750	1,750	1,750	1,750	1,750	1,750	1,750	1,750
PV Year 1, r = 10%	−1,000	−976	−952	−929	−1,812	−1,105	1,509	1,472	1,436	1,401	1,367	1,334	1,301	1,269	1,239	1,208
Project NPV	6,764															

EXHIBIT 15-10   CI-700 financial model with 25 percent increase in development time.

Change in Development Time, %	Development Time, Quarters	Change in Development Time, Quarters	Change in NVP, %	NPV, $ Thousands	Change in NPV, $ Thousands
50	6	2	−34.6	5,363	−2,840
25	5	1	−17.5	6,764	−1,439
base	4	base	−0.0	8,203	0
−25	3	−1	18.0	9,678	1,475
−10	2	−2	36.4	11,190	2,987

EXHIBT 15-11   CI-700's development time sensitivities.

the model have a substantial influence on NPV. This information is useful in helping the team understand which factors should be studied in more detail in order to refine and improve the base-case model. The information is also useful in supporting the operating decisions of the team, as discussed in the next step.

Step 3: Use Sensitivity Analysis to Understand Project Trade-Offs

Why would a product development team want to change the factors under its control? For instance, why should development time be increased if the change lowers the NPV of the project? Typically, the development team will make such a change only if some other offsetting gain is expected, such as a better-quality product with higher sales volumes. We therefore need to understand the relative magnitude of these financial interactions.

Six Potential Interactions

Development teams attempt to manage six potential interactions between the internally driven factors. These potential interactions are shown schematically in Exhibit 15-12. The potential interaction between any two internal factors depends on the characteristics

**EXHIBIT
15-12**
Potential
interactions
between
internally driven
factors.

Adapted from Smith
and Reinertsen, 1991

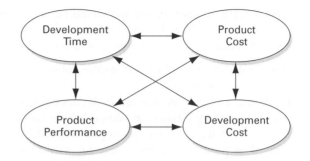

Change in Sales Volume, %	Sales Volume, $ Thousands	Change in Sales Volume	Change in NVP, %	NPV, $ Thousands	Change in NPV, $ Thousands
30	6,500	1,500	63.0	13,375	5,172
20	6,000	1,000	42.0	11,651	3,448
10	5,500	500	21.0	9,927	1,724
base	5,000	base	0.0	8,203	0
−10	4,500	−500	−21.0	6,479	−1,724
−20	4,000	−1,000	−42.0	4,755	−3,448
−30	3,500	−1,500	−63.0	3,031	−5,172

EXHIBIT 15-13   CI-700's sales volume sensitivities.

of the specific product context. In many cases the interactions are trade-offs. For example, decreasing development time may lead to lower product performance. Increased product performance may require additional product cost. However, some of these interactions are more complex than a simple trade-off. For example, decreasing product development time may require an increase in development spending, yet extending development time may also lead to an increase in cost if the extension is caused by a delay in a critical task rather than by a planned extension of the schedule.

In general, these interactions are important because of the linkage between the internal factors and the external factors. For example, increasing development cost or time may enhance product performance and therefore increase sales volume or allow higher prices. Decreasing development time may allow the product to reach the market sooner and thus increase sales volume.

While accurate modeling of externally driven factors (e.g., price, sales volume) is often very difficult, the quantitative model can nevertheless support decision making. Recall from our initial examples that the CI-700 development team was considering increasing development spending in order to develop a higher-quality product, which they hoped would lead to greater sales volume. The quantitative model can support this decision by answering the question of how much the sales volume would have to increase to justify the additional spending on development. We have calculated the sensitivity of NPV to changes in development cost (see Exhibit 15-9). We can also calculate the sensitivity of NPV to changes in sales volume (Exhibit 15-13). Say that the CI-700 development team is considering a 10 percent increase in development cost. From Exhibit 15-9 we found that this increase in spending will decrease NPV by 5.9 percent.

Now, what increase in volume would be necessary to at least compensate for the decrease in NPV? From Exhibit 15-13 we know that a 10 percent increase in sales volume would increase NPV by 21 percent. It follows, under assumptions of linearity, that a 2.8 percent $\left(\dfrac{10 \times 5.9}{21.0} \right)$ increase in sales volume would increase NPV by 5.9 percent. To summarize, a 10 percent *increase* in development cost would *decrease* NPV by 5.9 percent. A 2.8 percent *increase* in sales volume is needed to offset the drop. While the precise impact of the increased development spending on sales volume is not known, the model does provide a helpful guide for what magnitude of sales volume increase is needed to sustain particular increases in development cost.

Trade-Off Rules

The near linearity of many sensitivity analyses allows the team to compute some *trade-off rules* to inform day-to-day decision making. These rules take the form of the cost per unit change in the internal and external factors. For example, what is the cost of a one-month delay in development time? What is the cost of a 10 percent development budget overrun? What is the cost of a $1 per unit increase in manufacturing cost? The trade-off rules are easily computed from the base-case model and can be used to inform the team of the relative magnitude of the sensitivities of the project profitability on factors under its control. Exhibit 15-14 contains the rules for the CI-700.

The trade-off rules inform the original questions posed in the chapter introduction. The team decided that waiting for software that would allow the printer to be used with both Apple Macintosh and Microsoft Windows operating systems would delay product introduction by two months. It was calculated that the delay would have an approximate cost of $960,000. Rather than wait, the team reasoned that it could introduce the product before all printer drivers were available as long as the drivers were offered as soon as they were available. The film development group at Polaroid estimated that the development of a new print medium would cost more than $1 million and would take at least one year. The team decided that these time and budget penalties did not warrant the marginal increase in print size and quality that would be enabled by the new medium. Finally, the team felt that the reliability of the product could be dramatically improved with the addition of only one engineer and one technician to the team. This additional cost was ex-

Factor	Trade-Off Rule	Comments
Development time	$480,000 per month change	Assumes a fixed window of opportunity for sales.
Sales volume	$1,724,000 per 10% change	Increasing sales is a powerful way to increase profits; 10% is 500 units/quarter.
Product cost or sales price	$43,000 per $1 change in cost or price	A $1 increase in price or a $1 decrease in cost; each results in a $1 increase in unit profit margins.
Development cost	$482,000 per 10% change	A dollar spent or saved on development is worth the present value of that dollar; 10% is $500,000.

EXHIBIT 15-14 Trade-off rules for the CI-700 project.

pected to be approximately $100,000 for the remainder of the project. The team noted that they would have to increase sales by only 0.6 percent to justify this investment. Reliability was identified as a key customer need, and so the team chose to aggressively pursue increased reliability.

Limitations of Quantitative Analysis

Financial modeling and sensitivity analysis are powerful tools for supporting product development decisions, but these techniques have important limitations. One school of thought believes that rigorous financial analyses are required to bring discipline and control to the product development process. However, detractors argue that quantitative analysis suffers from some of the following problems:

- *It focuses only on measurable quantities.* Quantitative techniques like NPV emphasize and rely on that which is measurable. However, many critical factors impacting product development projects are difficult to measure accurately. In effect, quantitative techniques encourage investment in measurable assets and discourage investment in intangible assets.

- *It depends on validity of assumptions and data.* Product development teams may be lulled into a sense of security by the seemingly precise result of an NPV calculation. Financial analyses such as the ones we have shown in this chapter may seem to provide precise estimates of product development project value. However, *such precision in no way implies accuracy*. We can develop a highly sophisticated financial model of a product development project which computes project NPV to the fifth decimal place; yet if the assumptions and data of our model are not correct, the value calculated will not be correct. Consider the CI-700 development time sensitivity example's assumption of a fixed product sales window. This assumption was useful, but its integrity can easily be questioned. Indeed, a different assumption could give dramatically different results.

- *Bureaucracy reduces productivity.* Detractors of financial analysis assert that such activities provide a high level of planning and control at the expense of product development productivity. According to detractors, extensive planning and review guarantee that a brilliantly conceived, well-engineered product will reach the market after its market window has already closed. Detractors also argue that overzealously applied "professional" management techniques stifle the product development process. Potentially productive development time is devoted to preparation of analyses and meetings. The cumulative effect of this planning and review can be a ballooning development process.

These concerns are generally quite valid. However, in our opinion, they are largely associated with the blind application of the results of the quantitative analysis or arise from the combination of financial analysis with an already stifling bureaucracy. We reject the notion that quantitative analysis should not be done just because problems can arise from the blind application of the results. Rather, development teams should understand the strengths and limitations of the techniques and should be fully aware of how the models work and on what assumptions they are based. Furthermore, qualitative analysis, as discussed in the next section, can remedy some of the inherent weaknesses in the quantitative techniques.

Step 4: Consider the Influence of the Qualitative Factors on Project Success

Many factors influencing development projects are difficult to quantify because they are complex or uncertain. We refer to such factors as *qualitative* factors. After providing a conceptual framework for qualitative analysis, we use examples from the CI-700 to illustrate how the analysis is carried out.

Consider the following questions about the CI-700 project: Will knowledge gained from the CI-700 development spill over and be of benefit to other Polaroid development projects? How will competitors react to the introduction of the CI-700? Will competitors modify their own development efforts in response to Polaroid's actions? Will there be significant fluctuations in the dollar/yen exchange rate which would change the cost of component parts?

Our quantitative model implicitly accounts for these and many other issues with several broad assumptions. The model assumes that decisions made by the project team do not affect actions of groups external to the project, or alternatively that the external forces do not change the team's actions. This important assumption of our model is common to many other financial models and is called the *ceteris paribus* (other things being equal) assumption.

Projects Interact with the Firm, the Market, and the Macro Environment

Decisions made within a project in general do have important consequences for the firm as a whole, for competitors and customers in the market, and even for the macroeconomic (macro) environment in which the market operates (Exhibit 15-15). Similarly, events and actions outside of a development project often significantly impact its value. Qualitative analysis focuses largely on these interactions. The most basic approach to qualitative analysis is to consider (1) the interactions between the project and the firm as a whole, (2) the interactions between the project and the market in which the product will be sold, and (3) the interactions between the project and the macro environment.

Interactions between the Project and the Firm as a Whole

One assumption embedded in the quantitative model is that firm profit will be maximized if project profit is maximized. However, development decisions must be made in the context of the firm as a whole. The two key interactions between the project and the firm are *externalities* and *strategic fit.*

EXHIBIT 15-15
The broader context of a development project.

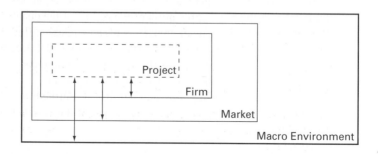

- *Externalities:* An externality is an "unpriced" cost or benefit imposed on one part of the firm by the actions of a second; costs are known as negative externalities and benefits as positive externalities. As an example of a positive externality, development learning on one project may benefit other current or future projects but is paid for by the first project. How should the other projects account for such benefits gained at no additional cost? How should the first project account for resources spent which benefit not only itself, but also other current or future projects?

- *Strategic fit:* Decisions of the development team must not only benefit the project, but also be consistent with the firm's overall product plan and technology strategy. For example, how well does a proposed new product, technology, or feature fit with the firm's resources and objectives? Is it compatible with the firm's emphasis on technical excellence? Is it compatible with the firm's emphasis on uniqueness?

Because of their complexity and uncertainty, externalities and strategic fit are very difficult to quantify. This does not mean these issues should not be considered; rather, they must be considered qualitatively. See Chapter 3, Product Planning, for discussion of some strategic planning issues which cut across multiple projects.

Interactions between the Project and the Market

We have modeled explicitly only price and volume as the key externally driven factors. In effect, we have held the actions and reactions of the market constant. To model project value accurately, we must relax the *ceteris paribus* assumption to recognize that a development team's decisions impact the market and that market events impact the development project. The market environment is impacted by the actions of not only the development team but also three other groups:

- *Competitors:* Competitors may provide products in direct competition or products which compete indirectly as substitutes.
- *Customers:* Customers' expectations, incomes, or tastes may change. Changes may be independent or may be driven by new conditions in markets for complementary or substitute products.
- *Suppliers:* Suppliers of inputs to the new product are subject to their own markets' competitive pressures. These pressures may, indirectly through the value chain, impact the new product.

Actions and reactions of these groups most often impact expected price and volume, but they can have second-order effects as well. For example, consider a new competitor that has rapid product development cycles and that seems to value market share rather than short-term profitability. Clearly, the entrance of such a new competitor would change our expected price and volume. Further, we may attempt to accelerate our own development efforts in response. Thus, the competitor's actions may impact not only our sales volume forecasts but also our planned development schedule.

Interactions between the Project and the Macro Environment

We must relax the *ceteris paribus* assumption to take into account key macro factors:

- *Major economic shifts:* Examples of typical major economic shifts which impact the value of development projects are changes in foreign exchange rates, materials prices, or labor costs.

- *Government regulations:* New regulations can destroy a product development opportunity. On the other hand, a shift in the regulatory structure of an industry can also spawn entire new industries.
- *Social trends:* As with government regulations, new social concerns such as increased environmental awareness can also destroy existing industries or create new ones.

Macro factors can have important impacts on development project value. However, these effects are difficult to model quantitatively because of inherent complexity and uncertainty.

The product development team for the CI-700 faced many qualitative issues during development of the product. We present three of the key qualitative issues that the Polaroid team encountered and describe the impacts of these issues on the project. The examples illustrate not only the limitations of quantitative analysis but also the importance of qualitative analysis.

Carrying Out Qualitative Analysis

For most project teams, the most appropriate qualitative analysis method is simply to consider and discuss the interactions between the project and the firm, the project and the market, and the project and the macro environment. Then the team considers these interactions in concert with the results of the quantitative analysis in order to determine the most appropriate relative emphasis on development speed, development expense, manufacturing cost, and product performance. We provide three examples below of the qualitative analysis for the CI-700.

While we believe this informal approach is most appropriate for decisions made at the level of the project team, more structured techniques are available, including strategic analysis, game theory, and scenario analysis techniques. References for each of these techniques are included in the bibliography.

Example 1: Decrease in the Price of a Substitute Product

Color laser printers are a substitute product for the CI-700. While color laser printers are more costly than the CI-700, they limit the potential of the CI-700's market by placing a ceiling on the price Polaroid can charge. During the CI-700 development, manufacturers of color laser printers achieved several important technological breakthroughs. The advances allowed the manufacturers to offer color laser printers at significantly lower prices. The Polaroid team was faced with a change in the competitive environment which was caused by others and which invalidated fundamental assumptions of the financial model for the CI-700.

In this example, it is clear that the CI-700 project cannot be considered in isolation. The original sales volume forecasts incorporated implicit assumptions about the color laser printer market, but the assumptions were invalidated by advances in color laser printer technology. While the color laser printer breakthrough would have been difficult to predict, quantitative analysis helped the CI-700 development team understand the sensitivity of the value of the project to this development. By using the model to estimate the sensitivity of NPV to changes in sales volume, the team was able to quickly grasp the magnitude of the change in project value. The combination of quantitative and qualitative analysis convinced the team to move more quickly and to further reduce the cost of the product for the project to remain viable.

Example 2: Increased Competition in a Complementary Product Market

The demand for the CI-700 is closely linked to the prices of personal computers (PCs). PCs are said to be complementary products to the CI-700 because a decrease in the price of PCs increases demand for the CI-700. Cheaper PCs both increase PC sales and allow buyers to afford additional peripheral products such as the CI-700. Thus, a decrease in the price of PCs would increase the value of the CI-700 project.

During the CI-700 development project, intense competition in the PC market further increased with new entrants and rapid technological development. The Polaroid team was faced with a change in the competitive environment for their product. Quantitative analysis helped the CI-700 development team understand how project value was impacted. By using the model to estimate the sensitivity of NPV to changes in sales volume, the team was able to quickly understand the magnitude of the change in project value, which they hoped would partially offset the pressure from the falling prices for color printers.

Example 3: The "Option" Value of Creating a Good Platform Product

The CI-700 was the first product of its kind produced by Polaroid, and the development team recognized that many of their development decisions would impact potential future generations of the product line. For example, the ease with which future generation products could be built around the same basic technological platform depended upon design decisions made in the CI-700 project. The team could choose to increase development spending and development time in order to facilitate the development of potential future generation products, even if doing so would not make economic sense in the context of a single product. They chose, however, to push ahead quickly with the CI-700 without extreme efforts to accommodate future models. The argument was that the future of the market was so uncertain that the risk of not getting to market in time outweighed the potential usefulness of the product as a platform for future products.

Summary

Product development teams must make many decisions in the course of a development project. Economic analysis is a useful tool for supporting this decision making.

- The method consists of four steps:
  1. Build a base-case financial model.
  2. Perform a sensitivity analysis to understand the relationships between financial success and the key assumptions and variables of the model.
  3. Use the sensitivity analysis to understand project trade-offs.
  4. Consider the influence of qualitative factors on project success.

- Quantitative analysis using NPV techniques is practiced widely in business. The technique forces product development teams to look objectively at their projects and their decisions. At the very least, they must go through the process of creating realistic project schedules and budgets. Financial modeling provides a method for quantitatively understanding the key profit drivers of the project.

- Quantitative techniques such as financial modeling and analysis rest upon assumptions about the external environment. This environment is constantly changing and may be influenced by a development team's decisions or by other uncontrollable factors. Further, quantitative analysis, by its very nature, considers only that which is measurable, yet many key factors influencing the project are highly complex or uncertain and are thus difficult to quantify.

- Qualitative analysis emphasizes the importance of such difficult-to-quantify issues by asking specifically what the interactions are between the project and the rest of the firm, the market, and the macro environment.

- Together, quantitative and qualitative techniques can help ensure that the team makes economically sound development decisions.

References and Bibliography

Many current resources are available on the Internet via
www.ulrich-eppinger.net

Bayus provides an interesting analysis of trade-offs involving development time and product performance factors.

> Bayus, Barry L., "Speed-to-Market and New Product Performance Trade-offs," *Journal of Product Innovation Management,* Vol. 14, 1997, pp. 485–497.

For a thorough review of discounted cash flow techniques as well as option theory from a general financial perspective, see the classic text on corporate finance.

> Brealey, Richard A., and Stewart C. Myers, *Principles of Corporate Finance,* seventh edition, McGraw-Hill, New York, 2002.

Smith and Reinertsen include a thorough discussion of how to model the economics of development time.

> Smith, Preston G., and Donald G. Reinertsen, *Developing Products in Half the Time,* Van Nostrand Reinhold, New York, 1991.

Management literature has criticized business schools and MBA programs for their lack of relevance. In particular, MBAs are criticized for their reliance on quantitative techniques such as discounted cash flow. This is an interesting case study reviewing the issues.

> Linder, Jane C., and H. Jeff Smith, "The Complex Case of Management Education," *Harvard Business Review,* September–October 1992, pp. 16–33.

Michael Porter's strategic analysis techniques have become standard fare for business school students. His 1980 text on strategic analysis has been very influential. In his 1985 book, Porter presents a general structured approach to scenario analysis, a technique originally developed by Royal Dutch Shell for planning under uncertainty.

> Porter, Michael E., *Competitive Strategy: Techniques for Analyzing Industries and Competitors,* The Free Press, New York, 1980.
> Porter, Michael E., *Competitive Advantage: Creating and Sustaining Superior Performance,* The Free Press, New York, 1985.

Game theory can be used to analyze competitive interactions. Oster provides a view of strategic analysis and game theory from a microeconomic perspective:

Oster, Sharon M., *Modern Competitive Analysis,* Oxford University Press, New York, 1990.

Copeland and Antikarov provide a detailed treatment of *real options* and analysis of projects with uncertainty and decision points.

Copeland, Tom, and Vladimir Antikarov, *Real Options: A Practitioner's Guide,* Texere, New York, 2001.

Exercises

1. List five reasons firms may choose to pursue a product even if the quantitative analysis reveals a negative NPV.
2. Build a quantitative model to analyze the development and sale of a bicycle light. Assume that you could sell 20,000 units per year for five years at a sales price (wholesale) of $20 per unit and a manufacturing cost of $10 per unit. Assume that production ramp-up expenses would be $20,000, ongoing marketing and support costs would be $2,000 per month, and development would take another 12 months. How much development spending could such a project justify?
3. Compute the trade-off rules for the case described in Exercise 2.

Thought Questions

1. Can you think of successful products that never would have been developed if their creators had relied exclusively on a quantitative financial model to justify their efforts? Do these products share any characteristics?
2. One model of the impact of a delay in product introduction is that sales are simply shifted later in time. Another model is that some of the sales are pushed beyond the "window of opportunity" and are lost forever. Can you suggest other models for the implications of an extension of product development time? Is such an extension ever beneficial?
3. How would you use the quantitative analysis method to capture the economic performance of an entire line of products to be developed and introduced over several years?

Appendix A

Time Value of Money and the Net Present Value Technique

This appendix provides a very basic tutorial on net present value for those who are unfamiliar with this concept.

Net present value (NPV) is an intuitive and powerful concept. In essence, NPV is simply a recognition of the fact that a dollar today is worth more than a dollar tomorrow. NPV calculations evaluate the value today *(present value)* of some future income or expense.

Say that a bank will give an *interest rate* of 8 percent per time period (the time period could be a month, a quarter, or a year long). If we invest $100 today for one time period at an interest rate of 8 percent, how much will the bank pay out after one time period? If we let r be the interest rate and C be the amount invested, then the amount received after one time period is

$$(1 + r) \times C = (1 + 0.08) \times 100 = (1.08) \times 100 = \$108$$

Thus, if we invest $100 for one time period at an interest rate of 8 percent, we will receive $108 at the end of the time period. In other words, $100 today is worth $108 received in the next time period.

Now, let's say that we have invested some amount, C', for one time period at an interest rate r. Let's also say that after the one time period, the amount received back is $100, and the interest rate is 8 percent. How much, then, was invested originally? We can find C', the original investment, by doing the reverse of what we did in the previous example:

$$(1 + r) \times C' = \$100$$

$$C = \frac{\$100}{1 + r} = \frac{\$100}{1 + 0.08} = \frac{\$100}{1.08} = \$92.59$$

Thus, if we invest $92.59 for one time period at an interest rate of 8 percent, we will receive $100 at the end of the time period. In other words, $92.59 today is worth $100 received in the next time period.

We have just shown how a dollar today is worth more than a dollar tomorrow. Of course, $100 today is worth $100. But what is $100 received *next* time period worth in *today's* dollars? The answer is $92.59 as we showed in the last example. Stated another way, the *present value* of $100 received in the next time period is $92.59 at a *discount rate* of 8 percent. So, present value is the value in today's dollars of some income received or expense paid out in a future period.

Now let's look at the result of investing $100 at 8 percent for longer periods:

One time period: $(1 + r) \times C = (1 \times 0.08) \times \$100 = \$108$

Two time periods: $(1 + r) \times (1 + r) \times \$100 = (1 + 0.08)^2 \times \$100 = \$116.64$

Three time periods: $(1 + r) \times (1 + r) \times (1 + r) \times \$100 = (1 + 0.08)^3 \times \$100 = \$125.97$

As we did earlier, let's find the present value of three separate investments of $100 received after one, two, and three time periods:

One time period: $(1 + r) \times C' = \$100$

$$C' = \frac{\$100}{1 + 0.08} = \$92.59$$

The present value of $100 received next time period is $92.59.

Two time periods: $(1 + r) \times (1 + r) \times C' = \$100$

$$C' = \frac{\$100}{(1 + 0.08)^2} = \$85.7$$

The present value of $100 received after two time periods is $85.73.

Three time periods: $(1 + r) \times (1 + r) \times (1 + r) \times C' = \$100$

$$C' = \frac{\$100}{(1 + 0.08)^3} = \$79.38$$

The present value of $100 received after three time periods is $79.38.

We found the present value of these three separate investments. Let's say instead that we had one investment that paid out $100 in each of time periods one, two, and three. What would that investment be worth today? The answer is simply the sum of the individual present values, or $257.70. The sum of the present values is called the *net present value,* or NPV. NPV is the present value of all cash inflows and all cash outflows. The present value of a cash *outflow* is just the negative of a cash *inflow* of the same amount.

We can summarize the present value calculation into a convenient formula. The present value (PV) of an amount C received (or paid out) t time periods from now is

$$PV = \frac{C}{(1 + r)^t}$$

Some calculators have a special present value function on them which can do the calculations quickly. Most computer spreadsheet programs have special financial functions which automatically do the present value calculations. The information required for these special functions is the future amount paid out, the interest rate, and the number of time periods of the investment.

What Interest Rate Should We Use?

The interest rate (also called the discount rate, discount factor, or hurdle rate) to use is our own or our company's "opportunity cost of capital." It is called the opportunity cost of capital because it is the return forgone by investing in the project rather than in other investments. Stated another way, the discount factor is the reward that investors demand for accepting delayed payment. A project which has a positive NPV must be earning more than the opportunity cost of capital and is thus a good investment. Note that many firms apply a constant hurdle rate to all their investment decisions. In recent years most firms have been using discount factors of 10 to 20 percent.

Sunk Costs Are Irrelevant for Net Present Value Calculations

In the context of product development decision making, costs that have already been incurred are termed *sunk costs.* Because sunk costs are past and irreversible outflows, they cannot be affected by present or future decisions, so they should be ignored for NPV calculations. To clarify this point, let's consider an example of the familiar "cut our losses" argument: "We've already spent over $600 million and nine years with no product to show for it, and you want me to approve another $90 million? That's crazy!" While this type of argument might sound logical, in fact the amount of money already spent is not important for the decision of whether or not to spend $90 million more. What is important is how much extra profit will be gained from investing the additional $90 million.

Say that the expected profit from product sales is $350 million. Let's look at the NPV of the two options (assume all numbers given are present values):

"Cut Our Losses"		"Invest $90 Million More"	
Additional amount invested:	$0	Additional amount invested:	–$90
Profits from product sales:	0	Profits from product sales:	350
NPV of "cut losses" decision:	$0	NPV of "invest" decision:	$260
Total invested:	–$600	Total invested:	–$690
Total project return:	–$600	Total project return:	–$340

Because the "invest" decision has a positive NPV, the firm should proceed. While it is clear that the firm will lose money on the project in either case, the $600 million already spent is a sunk cost and should not impact the invest-or-cut-losses decision. Of course, the sunk cost argument is a cold analytical perspective; there is a saying that "sunk costs are only relevant to the manager who sunk them." Project managers with a long record of negative total project returns may find that sunk costs are extremely relevant to their ability to get support for future projects.

Appendix B

Modeling Uncertain Cash Flows Using Net Present Value Analysis

Product development projects face many perils. For example, the team may think that the manufacturing cost for a particular new product will be $40 per unit. However, the cost could be much higher or it might even be lower. The team does not know for sure until the product is actually built. The team may forecast sales for the new product, but the forecasts depend on (among other things) when competitors get their versions to the market, and this information will not be available until their products are actually introduced. These uncertainties that are particular to a project are called *project-specific risks*. How should project-specific risks be accounted for? Some development teams increase the discount rate to offset uncertainty about the outcomes. However, such an arbitrary increment in the discount rate would be applied uniformly to both certain and uncertain cash flows. Fortunately, better approaches are available if the team is able to estimate the probabilities of uncertain cash flows.

Instead of using arbitrary adjustments to the discount rate, development teams should strive for realistic forecasting of cash flows. These forecasts can be supplemented with sensitivity analysis to understand the impact of the full range of possible outcomes for the uncertain factors. Project-specific risks should be considered only in the expected cash flows and not in the discount rate.

Sensitivity analysis can be performed by systematically varying the model parameters, such as product price or manufacturing cost, to understand how critically the net present value depends on specific values for these parameters. A basic analysis can be performed one variable at a time, as explained in the body of this chapter, or combinations of variables can be adjusted to form realistic scenarios. A more sophisticated analysis can be performed using Monte Carlo simulation based on assumed probability distributions for the parameters in the model.

Note that there is a second type of risk, *general market risk,* which is not specific to the project. General market risk stems from the fact that there are economywide perils, which threaten all businesses and projects. Although entire books on calculating market risk have been written, for our purposes it suffices to say that market risk is typically accounted for by inflating the discount rate.

Analyzing Scenarios

Sometimes project teams face discrete scenarios that are clearly foreseeable and that will have a direct and significant influence on the project outcome. For example, a team may have filed a patent application on a novel and distinctive product concept. If the patent is allowed, then the team expects to face much less of a competitive threat than if the patent is not allowed. These two scenarios can be modeled as a *decision tree,* as shown in Exhibit 15-16. (In this case, there is not an explicit decision, but rather an outcome of an uncertain process. These diagrams are nevertheless called decision trees by convention.) The two branches of this tree represent the two scenarios the team envisions. The present value of the project can be analyzed for each scenario taken independently. The team can also assign a probability to each scenario. Given these inputs, the team can now calculate the expected net present value for the project accounting for the two possible scenarios:

$$NPV = P_a \times PV_a + P_b \times PV_b \qquad \text{where } P_a + P_b = 1$$

For the situation depicted by the decision tree shown in Exhibit 15-16,

$$NPV = 0.60 \times \$6,500,000 + 0.40 \times \$1,500,000 = \$4,500,000$$

This kind of analysis is appropriate when discrete and distinct scenarios can be envisioned and when these scenarios have substantially different cash flows.

Analyzing Scenarios with Decision Points

When analyzing product development projects, the team should recognize that most development projects can be discontinued or redirected based on the latest information available. Such decision points may occur at the time of major milestones or reviews. This flexibility to expand or contract a project is financially valuable. The notion of decision points with the ability to change an investment is the subject of an entire field of analysis called

EXHIBIT 15-16

A situation in which two discrete scenarios can be envisioned.

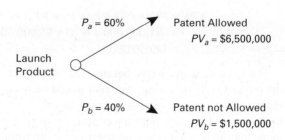

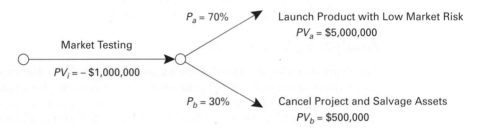

EXHIBIT 15-17 In this situation, the team can either launch the product immediately and face a great deal of market risk or it can test the market and then decide whether to launch the product.

real options. Copeland and Antikarov (2001) provide a detailed treatment of this subject. Here we provide a way to think about scenarios containing decision points.

Consider the scenario depicted in Exhibit 15-17. A team is contemplating the launch of a product in an entirely new category, which is an inherently risky type of project. The team could just launch the project and hope for success, or it could spend time and money testing the product in the marketplace. If it invests in market testing, the team may discover that the product is not viable, in which case it has the option to cancel the project. Alternatively, it may discover that the market is highly responsive to the new product, in which case it can launch with confidence and a much higher associated expected value of the future cash flows.

As a base case, the team analyzes the value of just launching the product without investigation. Given the team's assessment of the likelihood of success, the present value is $2 million for this plan. The value of market testing followed by a decision to proceed or not can be analyzed as follows. In this case, the team spends an additional $1 million for investigation. After investigation there is a 70 percent chance that the team will launch the product and reap positive cash flow of $5 million. There is a 30 percent chance that the team will decide to cancel the project, reaping only $0.5 million in salvage value. Thus, the net present value of the project is

$$NPV = PV_i + P_a \times PV_a + P_b \times PV_b$$
$$= -\$1,000,000 + 0.70 \times \$5,000,000 + 0.30 \times \$500,000$$
$$= \$2,650,000$$

Based on these estimates, because the net present value exceeds that of just launching the product without testing, the team would be better off spending the $1 million to test the market. There are, of course, many factors that influence a decision about whether to launch a product with high uncertainty or to perform further investigation. Economic modeling can be used as one perspective for informing this kind of decision.

CHAPTER SIXTEEN

Managing Projects

Courtesy of Eastman Kodak Company

EXHIBIT 16-1
The Cheetah microfilm cartridge.

This chapter was developed in collaboration with Stephen Raab.

A manufacturer of microfilm imaging equipment approached the Eastman Kodak Company to design and supply the microfilm cartridges for use with a new machine under development (Exhibit 16-1). The target specifications were similar to previous products developed by the cartridge group at Kodak. However, in contrast to the usual 24-month development time, the customer needed prototype cartridges for demonstration at a trade show in just 8 months, and production was to begin 4 months later. Kodak accepted this challenge of cutting its normal development time in half and called its efforts the Cheetah project. Effective project management was crucial to the successful completion of the project.

For all but the simplest products, product development involves many people completing many different tasks. Successful product development projects result in high-quality, low-cost products while making efficient use of time, money, and other resources. *Project management* is the activity of planning and coordinating resources and tasks to achieve these goals.

Project management activities occur during *project planning* and *project execution.* Project planning involves scheduling the project tasks and determining resource requirements. The project plan is first laid out during the concept development phase, although it is a dynamic entity and continues to evolve throughout the development process.

Project execution, sometimes called *project control,* involves coordinating and facilitating the myriad tasks required to complete the project in the face of inevitable unanticipated events and the arrival of new information. Execution is just as important as planning; many teams fail because they do not remain focused on their goals for the duration of the project.

This chapter contains five remaining sections. We first present the fundamentals of task dependencies and timing, along with three tools for representing relationships among project tasks. In the second section we show how these principles are used to develop an effective product development plan. In the third section we provide a set of guidelines for completing projects more quickly. After that, we discuss project execution, and finally we present a process for project evaluation and continuous improvement.

Understanding and Representing Tasks

Product development projects involve the completion of hundreds or even thousands of tasks. This section discusses some of the fundamental characteristics of interacting tasks—the "basic physics" of projects. We also present three ways to represent the tasks in a project.

Sequential, Parallel, and Coupled Tasks

Exhibit 16-2 displays the tasks for three portions of the Cheetah project. The tasks are represented by boxes, and the information (data) dependencies among the tasks are represented by arrows. We refer to this representation as an *information-processing view* or a *data-driven perspective* of product development because most of the dependencies involve transfer of information (data) between the tasks. We say that task B is *dependent* on task A if an output of task A is required to complete task B. This dependency is denoted by an arrow from task A to task B.

EXHIBIT 16-2

The three basic
types of task
dependencies:
(a) sequential,
(b) parallel, and
(c) coupled.

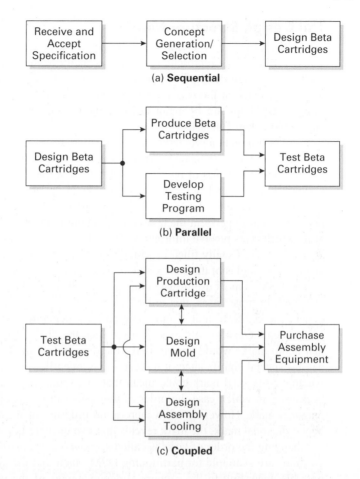

(a) **Sequential**

(b) **Parallel**

(c) **Coupled**

Exhibit 16-2(a) shows three tasks, two of which are dependent on the output of another task. These tasks are *sequential* because the dependencies impose a sequential order in which the tasks must be completed. (Note that when we refer to tasks being "completed" sequentially, we do not necessarily mean that the later task cannot be started before the earlier one has been completed. Generally the later task can begin with partial information but cannot finish until the earlier task has been completed.) Exhibit 16-2(b) shows four development tasks. The middle two tasks depend only on the task on the left, but not on each other. The task on the right depends on the middle two tasks. We call the middle two tasks *parallel* because they are both dependent on the same task but are independent of each other. Exhibit 16-2(c) shows five development tasks, three of which are *coupled*. Coupled tasks are mutually dependent; each task requires the result of the other tasks in order to be completed. Coupled tasks either must be executed simultaneously with continual exchanges of information or must be carried out in an iterative fashion. When coupled tasks are completed iteratively, the tasks are performed either sequentially or simultaneously with the understanding that the results are tentative and that each task will most likely be repeated one or more times until the team converges on a solution.

The Design Structure Matrix

A useful tool for representing and analyzing task dependencies is the *design structure matrix* (DSM). This representation was originally developed by Steward (1981) for the analysis of design descriptions and has more recently been used to analyze development projects modeled at the task level (Eppinger et al., 1994, Eppinger, 2001). Exhibit 16-3 shows a DSM for the 14 major tasks of the Cheetah project. (Kodak's actual plan included more than 100 tasks.)

In a DSM model, a project task is assigned to a row and a corresponding column. The rows and columns are named and ordered identically, although generally only the rows list the complete names of the tasks. Each task is defined by a row of the matrix. We represent a task's dependencies by placing marks in the columns to indicate the other tasks (columns) on which it depends. Reading across a row reveals all of the tasks whose output is required to perform the task corresponding to the row. Reading down a column reveals which tasks receive information from the task corresponding to the column. The diagonal cells are usually filled in with dots or the task labels, simply to separate the upper and lower triangles of the matrix and to facilitate tracing dependencies.

The DSM is most useful when the tasks are listed in the order in which they are to be executed. In most cases, this order will correspond to the order imposed by sequential dependencies. Note that if only sequentially dependent tasks were contained in the DSM, then the tasks could be sequenced such that the matrix would be lower triangular; that is, no marks would appear above the diagonal. A mark appearing above the diagonal has special significance; it indicates that an earlier task is dependent on a later task. An above-diagonal mark could mean that two sequentially dependent tasks are ordered backward, in which case the order of the tasks can be changed to eliminate the above-diagonal mark. However, when there is no ordering of the tasks that will eliminate an above-diagonal mark, the mark reveals that two or more tasks are coupled.

Changing the order of tasks is called *sequencing* or *partitioning* the DSM. Simple algorithms are available for partitioning DSMs such that the tasks are ordered as much as possible according to the sequential dependencies of the tasks. Inspection of a partitioned DSM reveals which tasks are sequential, which are parallel, and which are coupled and will require simultaneous solution or iteration. In a partitioned DSM, a task is part of a sequential group if its row contains a mark just below the diagonal. Two or

EXHIBIT 16-3

Simplified design structure matrix for the Kodak Cheetah project.

Task		
Receive and accept specification	A	
Concept generation/selection	B	
Design beta cartridges	C	
Produce beta cartridges	D	
Develop testing program	E	
Test beta cartridges	F	
Design production cartridge	G	
Design mold	H	
Design assembly tooling	I	
Purchase assembly equipment	J	
Fabricate molds	K	
Debug molds	L	
Certify cartridge	M	
Initial production run	N	

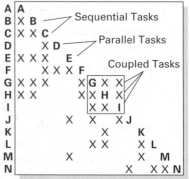

more tasks are parallel if there are no marks linking them. As noted, coupled tasks are identified by above-diagonal marks. Exhibit 16-3 shows how the DSM reveals all three types of relationships.

More sophisticated use of the DSM method has been a subject of research at MIT since in the 1990s. Much of this work has applied the method to larger projects and to the development of complex systems such as automobiles and airplanes. Analytical methods have been developed to help understand the effects of complex task coupling (Smith and Eppinger, 1997); to predict the distribution of possible project completion times and costs (Cho and Eppinger, 2001); and to help plan organization designs based on product architectures (Eppinger, 1997).

DSM practitioners have found that creative uses of the DSM's graphical display of project task relationships can be highly insightful for project managers in both the planning and execution phases. The chapter appendix presents a larger DSM model in which several overlapping phases of coupled development activities are represented.

Gantt Charts

The traditional tool for representing the timing of tasks is the Gantt chart. Exhibit 16-4 shows a Gantt chart for the Cheetah project. The chart contains a horizontal time line created by drawing a horizontal bar representing the start and end of each task. The filled-in portion of each bar represents the fraction of the task that is complete. The vertical line in Exhibit 16-4 shows the current date, so we can observe directly that task D is behind schedule, while task E is ahead of schedule.

A Gantt chart does not explicitly display the dependencies among tasks. Task dependencies constrain, but do not fully determine, the timing of the tasks. The dependencies dictate which tasks must be completed before others can begin (or finish, depending on

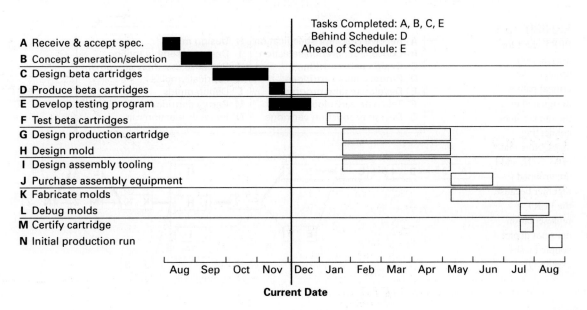

EXHIBIT 16-4   Gantt chart for the Cheetah project.

the nature of the dependency) and which tasks can be completed in parallel. When two tasks overlap in time on a Gantt chart, they may be parallel, sequential, or iteratively coupled. Parallel tasks can be overlapped in time for convenience in project scheduling because they do not depend on one another. Sequential tasks might be overlapped in time, depending on the exact nature of the information dependency, as described below in the section on accelerating projects. Coupled tasks must be overlapped in time because they need to be addressed simultaneously or in an iterative fashion.

PERT Charts

PERT (program evaluation and review technique) charts explicitly represent both dependencies and timing, in effect combining some of the information contained in the DSM and Gantt chart. While there are many forms of PERT charts, we prefer the "activities on nodes" form of the chart, which corresponds to the process diagrams that most people are familiar with. The PERT chart for the Cheetah project is shown in Exhibit 16-5. The blocks in the PERT chart are labeled with both the task and its expected duration. Note that the PERT representation does not allow for loops or feedback and so cannot explicitly show iterative coupling. As a result, the coupled tasks G, H, and I are grouped together into one task. The graphical convention of PERT charts is that all links between tasks must proceed from left to right, indicating the temporal sequence in which tasks can be completed. When the blocks are sized to represent the duration of tasks, as in a Gantt chart, then a PERT diagram can also be used to represent a project schedule.

The Critical Path

The dependencies among the tasks in a PERT chart, some of which may be arranged sequentially and some of which may be arranged in parallel, lead to the concept of a *critical path*. The critical path is the longest chain of dependent events. This is the single se-

EXHIBIT 16-5
PERT chart for the Cheetah project. The critical path is designated by the thicker lines connecting tasks. Note that tasks G, H, and I are grouped together because the PERT representation does not depict coupled tasks explicitly.

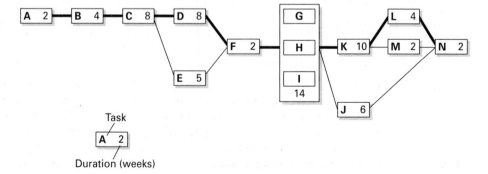

quence of tasks whose combined required times define the minimum possible completion time for the entire set of tasks. Consider for example the Cheetah project represented in Exhibit 16-5. Either the sequence C-D-F or the sequence C-E-F defines how much time is required to complete the four tasks C, D, E, and F. In this case, the path C-D-F requires 18 weeks and the path C-E-F requires 15 weeks, so the critical path for the whole project includes C-D-F. The critical path for the project is denoted by the thick lines in Exhibit 16-5. Identifying the critical path is important because a delay in any of these *critical tasks* would result in an increase in project duration. All other paths contain some *slack,* meaning that a delay in one of the noncritical tasks does not necessarily create a delay for the entire project. Exhibit 16-4 shows that task D is behind schedule. Because task D is on the critical path, this delay, if not corrected, will result in a delay of the completion of the entire project.

Several software packages are available for producing Gantt charts and PERT charts; these programs can also compute the critical path.

Baseline Project Planning

The project plan is the roadmap for the remaining development effort. The plan is important in coordinating the remaining tasks and in estimating the required development resources and development time. Some measure of project planning occurs at the earliest stages of product development, but the importance of the plan is highest at the end of the concept development phase, just before significant development resources are committed. This section presents a method for creating a *baseline project plan.* After establishing this baseline, the team considers whether it should modify the plan to change the planned development time, budget, or project scope. The results of the concept development phase plus the project plan make up the *contract book.*

The Contract Book

We recommend that a contract book be used to document the project plan and the results of the concept development phase of the development process. The concept of a contract book is detailed by Wheelwright and Clark (1992). The word *contract* is used to emphasize that the document represents an agreement between the development team and the senior management of the company about project goals, direction, and resource requirements. The book is sometimes actually signed by the key team members and the senior managers of the company. A table of contents for a contract book is shown in Exhibit 16-6, along with references to the chapters in this book where some of these contents are discussed.

Below we discuss elements of the project plan: the project task list, team staffing and organization, the project schedule, the project budget, and the project risk areas.

Project Task List

We have already introduced the idea that a project consists of a collection of tasks. The first step in planning a project is to list the tasks which make up the project. For most product development projects the team will not be able to list every task in great detail; too much uncertainty remains in the subsequent development activities. However, the team will be able to list its best estimate of the remaining tasks at a general level of detail. To be most useful during project planning, the task list should contain from 50 to 200

EXHIBIT 16-6
Table of contents of a contract book for a project of moderate complexity.

Item	Approximate Pages	See Chapter(s)
Mission Statement	1	3
Customer Needs List	1–2	4
Competitive Analysis	1–2	3, 4, 5, 7, 8
Product Specifications	1–3	5
Sketches of Product Concept	1–2	6, 10
Concept Test Report	1–2	8
Sales Forecast	1–3	8, 15
Economic Analysis/Business Case	1–3	15
Manufacturing Plan	1–5	11
Project Plan		
Task List	1–5	2, 16
Design Structure Matrix	2–3	16
Team Staffing and Organization	1	2, 16
Schedule (Gantt and/or PERT)	1–2	16
Budget	1	16
Risk Plan	1	16
Project Performance Measurement Plan	1	16
Incentives	1	16
	Total 18–38 Pages	

items. For small projects, such as the development of a hand tool, each task may correspond, on average, to a day or two of work for a single individual. For medium-sized projects, such as the development of a computer printer, each task may correspond to a week of work for a small group of people. For a large project, such as the development of an automobile, each task may correspond to one or more months of efforts for an entire division or subteam. For large projects, each of the tasks identified at this level may be treated as its own development project with its own project plan.

An effective way to tackle the generation of the task list is to consider the tasks in each of the remaining phases of development. For our generic development process, the phases remaining after concept development are system-level design, detail design, testing and refinement, and production ramp-up. (See Chapter 2, Development Processes and Organizations.) In some cases, the current effort will be very similar to a previous project. In these cases, the list of tasks from the previous project is a good starting point for the new task list. The Cheetah project was very similar to dozens of previous efforts. For this reason, the team had no trouble identifying the project tasks. (Its challenge was to complete them quickly.)

After listing all of the tasks, the team estimates the effort required to complete each task. Effort is usually expressed in units of person-hours, person-days, or person-weeks, depending on the size of the project. Note that these estimates reflect the "actual working time" that members of the development team would have to apply to the task and not the "elapsed calendar time" the team expects the task to require. Because the speed with which a task is completed has some influence on the total amount of effort that must be applied to the task, the estimates embody preliminary assumptions about the overall project schedule and how quickly the team will attempt to complete tasks. These estimates are typically derived from past experience or the judgment of experienced members of the development team. A task list for the Cheetah project is shown in Exhibit 16-7.

EXHIBIT 16-7
Task list for the
Cheetah project.
(This task list is
abbreviated for
clarity; the
actual list
contained over
100 tasks.)

Task	Estimated Person-Weeks
Concept Development	
Receive and accept specification	8
Concept generation/selection	16
Detail Design	
Design beta cartridges	62
Produce beta cartridges	24
Develop testing program	24
Testing and Refinement	
Test beta cartridges	20
Design production cartridge	56
Design mold	36
Design assembly tooling	24
Purchase assembly equipment	16
Fabricate molds	16
Debug molds	24
Certify cartridge	12
Production Ramp-up	
Initial production run	16
Total	**354**

Team Staffing and Organization

The project team is the collection of individuals who complete project tasks. Whether or not this team is effective depends on a wide variety of individual and organizational factors. Smith and Reinertsen (1991) propose seven criteria as determinants of the speed with which a team will complete product development; in our experience these criteria predict many of the other dimensions of team performance as well:

1. There are 10 or fewer members of the team.
2. Members volunteer to serve on the team.
3. Members serve on the team from the time of concept development until product launch.
4. Members are assigned to the team full-time.
5. Members report directly to the team leader.
6. The key functions, including at least marketing, design, and manufacturing, are on the team.
7. Members are located within conversational distance of each other.

While few teams are staffed and organized ideally, these criteria raise several key issues: How big should the team be? How should the team be organized relative to the larger enterprise? Which functions should be represented on the team? How can the development team of a very large project exhibit some of the agility of a small team? Here we address the issues related to team size. Chapter 1, Introduction, and Chapter 2, Development Processes and Organizations, address some of the other team and organizational issues.

The minimum number of people required on the project team can be estimated by dividing the total estimated time to complete the project tasks by the planned project duration. For example, the estimated task time for the Cheetah project was 354 person-weeks. The team hoped to complete the project in 12 months (or about 50 weeks), so the minimum possible team size would be seven people. All other things being equal, small teams seem to be more efficient than large teams, so the ideal situation would be to have a team made up of the minimum number of people, each dedicated 100 percent to the project.

Three factors make realizing this ideal difficult. First, specialized skills are often required to complete the project. For example, one of the Cheetah tasks was to design molds. Mold designers are highly specialized, and the team could not use a mold designer for a full year. Second, one or more key team members may have other unavoidable responsibilities. For example, one of the engineers on the Cheetah project was responsible for assisting in the production ramp-up of a previous project. As a result, she was only able to commit half of her time to the Cheetah project initially. Third, the work required to complete tasks on the project is not constant over time. In general, the work requirement increases steadily until the beginning of production ramp-up and then begins to taper off. As a result, the team will generally have to grow in size as the project progresses in order to complete the project as quickly as possible.

After considering the need for specialized skills, the reality of other commitments of the team members, and the need to accommodate an increase and subsequent decrease in workload, the project leader, in consultation with his or her management, identifies the full project staff and approximately when each person will join the team. When possible, team members are identified by name, although in some cases they will be identified only by area of expertise (e.g., mold designer, industrial designer). The project staffing for the Cheetah project is shown in Exhibit 16-8.

Project Schedule

The project schedule is the merger of the project tasks and the project timeline. The schedule identifies when major project milestones are expected to occur and when each project task is expected to begin and end. The team uses this schedule to track progress

Person	Month:	1	2	3	4	5	6	7	8	9	10	11	12
Team Leader		100	100	100	100	100	100	100	100	100	100	100	100
Schedule Coordinator		25	25	25	25	25	25	25	25	25	25	25	25
Customer Liaison		50	50	50	50	25	25	25	25	25	25	25	25
Mechanical Designer 1		100	100	100	100	100	100	100	100	50	50	50	50
Mechanical Designer 2			50	100	100	100	100	100	100	50			
CAD Technician 1			50	100	100	100	100	100	100	100	50	50	50
CAD Technician 2					50	100	100	100	100	100	50		
Mold Designer 1		25	25	25	25	100	100	100	100	25	25	25	
Mold Designer 2						100	100	100	100				
Assembly Tool Designer		25	25	25	25	100	100	100	100	100	100	50	50
Manufacturing Engineer		50	50	100	100	100	100	100	100	100	100	100	100
Purchasing Engineer			50	50	100	100	100	100	100	100	100	100	100

EXHIBIT 16-8   Project staffing for the Cheetah project. Numbers shown are approximate percentages of full time.

and to orchestrate the exchange of materials and information between individuals. It is therefore important that the schedule is viewed as credible by the entire project team.

We recommend the following steps to create a baseline project schedule:

1. Use the DSM or PERT chart to identify the dependencies among tasks.
2. Position the key project milestones along a timeline in a Gantt chart.
3. Schedule the tasks, considering the project staffing and other critical resources.
4. Adjust the timing of the milestones to be consistent with the time required for the tasks.

Project milestones are useful as anchor points for the scheduling activity. Common milestones include design reviews (also called phase reviews or design gates), comprehensive prototypes (e.g., alpha prototype, beta prototype), and trade shows. Because these events typically require input from almost everyone on the development team, they serve as powerful forces for integration and act as anchor points on the schedule. Once the milestones are laid out on the schedule, the tasks can be arranged between these milestones.

The Cheetah schedule was developed by expanding the typical project phases into a set of approximately 100 tasks. The major milestones were the concept approval, the testing of beta prototype cartridges, the trade show demonstration, and production ramp-up. Relationships among these activities and the critical path were documented using a combined PERT/Gantt chart.

Project Budget

Budgets are customarily represented with a simple spreadsheet, although many companies have standard budgeting forms for requests and approvals. The major budget items are staff, materials and services, project-specific facilities, and spending on outside development resources.

For most projects the largest budget item is the cost of staff. For the Cheetah project, personnel charges made up 80 percent of the total budget. The personnel costs can be derived directly from the staffing plan by applying the *loaded* salary rates to the estimated time commitments of the staff on the project. Loaded salaries include employee benefits and overhead and are typically between two and three times the actual salary of the team member. Many companies use only one or two different rates to represent the cost of the people on a project. Average staff costs for product development projects range from $2,000 to $5,000 per person-week. For the Cheetah project, assuming an average cost of $3,000 per person-week, the total cost for the 354 person-weeks of effort would be $1,062,000.

Early in the development project, uncertainty of both timing and costs are high, and the forecasts may only be accurate within 30 to 50 percent. In the later stages of the project the program uncertainty is reduced to perhaps 5 percent to 10 percent. For this reason some margin should be added to the budget as a contingency. A summary of the Cheetah project budget is shown in Exhibit 16-9.

Project Risk Plan

Projects rarely proceed exactly according to plan. Some of the deviations from the plan are minor and can be accommodated with little or no impact on project performance. Other deviations can cause major delays, budget overruns, poor product performance, or high manufacturing costs. Often the team can assemble, in advance, a list of what might go wrong, that is, the areas of risk for the project.

Item	Amount
Staff salaries	
354 person-weeks @ $3,000/week	$1,062,000
Materials and Services	125,000
Prototype Molds	75,000
Outside Resources, Consultants	25,000
Travel	50,000
Subtotal	$1,337,000
Contingency (20%)	$267,400
Total	$1,604,400

EXHIBIT 16-9   Summary budget for the Cheetah project. The production tooling and equipment are accounted for as manufacturing costs rather than as part of the development project budget. (Kodak figures are disguised and listed here only for illustration.)

EXHIBIT 16-10

Risk plan for the Cheetah project.

Risk	Risk Level	Actions to Minimize Risk
Change in customer specifications	Moderate	• Involve the customer in process of refining specifications. • Work with the customer to estimate time and cost penalties of changes.
Poor feeding characteristics of cartridge design	Low	• Build early functional prototype from machined parts. • Test prototype in microfilm machine.
Delays in mold-making shop	Moderate	• Reserve 25% of shop capacity for May–July.
Molding problems require rework of mold	High	• Involve mold maker and mold designer in the part design. • Perform mold filling computer analysis. • Establish design rules for part design. • Choose materials at end of concept development phase.

After identifying each risk, the team can prioritize the risks. To do so, some teams use a scale combining severity and likelihood of each risk. A complete risk plan also includes a list of actions the team will take to minimize the risk. In addition to pushing the team to work to minimize risk, the explicit prioritization of risk during the project planning activity helps to minimize the number of surprises the team will have to communicate to its senior management later in the project. The risk plan for the Cheetah project is shown in Exhibit 16-10.

Modifying the Baseline Plan

The baseline project plan embodies assumptions about how quickly the project should be completed, about the performance and cost goals for the product, and about the resources to be applied to the project. After completing a baseline plan, the team should consider whether some of these assumptions should be revisited. In particular, the team can usually choose to trade off development time, development cost, product manufacturing cost, product performance, and risk. For example, a project can sometimes be completed more quickly by spending more money. Some of these trade-offs can be explored quantitatively using the economic analysis techniques described in Chapter 15, Product Development Economics. The team may also develop contingency plans in case certain risks cannot be overcome. The most common desired modification to the baseline plan is to compress the schedule. For this reason, we devote the next section to ways the team can accelerate the project.

Accelerating Projects

Product development time is often the dominant concern in project planning and execution. This section provides a set of guidelines for accelerating product development projects. Most of these guidelines are applicable at the project planning stage, although a few can be applied throughout a development project. Accelerating a project before it has begun is much easier than trying to expedite a project that is already underway.

The first set of guidelines applies to the project as a whole.

- *Start the project early.* Saving a month at the beginning of a project is just as helpful as saving a month at the end of a project, yet teams often work with little urgency before development formally begins. For example, the meeting to approve a project plan and review a contract book is often delayed for weeks because of difficulty in scheduling a meeting with senior managers. This delay at the beginning of a project costs exactly as much time as the same delay during production ramp-up. The easiest way to complete a project sooner is to start it early.

- *Manage the project scope.* There is a natural tendency to add additional features and capabilities to the product as development progresses. Some companies call this phenomenon "feature creep" or "creeping elegance," and in time-sensitive contexts it may result in an elegant product without a market. Disciplined teams and organizations are able to "freeze the design" and leave incremental improvements for the next generation of the product.

- *Facilitate the exchange of essential information.* As shown in the DSM representation, a tremendous amount of information must be transferred within the product development team. Every task has one or more internal customers for the information it produces. For small teams, frequent exchange of information is quite natural and is facilitated by team meetings and colocation of team members. Larger teams may require more structure to promote rapid and frequent information exchange. Blocks of coupled tasks revealed by the DSM identify the specific needs for intensive information exchange. Computer networks and collaboration software tools can facilitate regular information transfer within large and dispersed product development teams.

The second set of guidelines is aimed at decreasing the time required to complete the tasks on the critical path. These guidelines arise from the fact that the only way to reduce the time required to complete a project is to shorten the critical path. Note that a decision to allocate additional resources to shortening the critical path should be based on the value of accelerating the entire project. For some projects, time reductions on the critical path can be worth hundreds of thousands, or even millions, of dollars per week.

- *Complete individual tasks on the critical path more quickly.* The benefit of recognizing the critical path is that the team can focus its efforts on this vital sequence of tasks. The critical path generally represents only a small fraction of the total project effort, and so additional spending on completing a critical task more quickly can usually quite easily be justified. Sometimes completing critical tasks more quickly can be achieved simply by identifying a task as critical so that it gets special attention, starts earlier, and is not interrupted. Note that the accelerated completion of a critical task may cause the critical path to shift to include previously noncritical tasks.

- *Aggregate safety times.* The estimated duration of each task in the project generally includes some amount of "safety time." This time accounts for the many normal but unpredictable delays which occur during the execution of each task. Common delays include: waiting for information and approvals, interruptions from other tasks or projects, and tasks being more difficult than anticipated. Goldratt (1997) estimates that built-in safety doubles the nominal duration of tasks. Although safety time is added to the expected task duration to account for random delays, these estimates become targets during execution of the tasks, which means that tasks are rarely completed early and many tasks overrun. Goldratt recommends removing the safety time from each task along the critical path and aggregating all of the safety time from the critical path into a single *project buffer* placed at the end of the project schedule. Because the need to extend task duration occurs somewhat randomly, only some of the tasks will actually need to utilize time from the project buffer. Therefore, a single project buffer can be smaller than the sum of the safety times that would be included in each estimate of task duration, and the critical path may be completed sooner. In practice, the project buffer may only need to start with time equal to half of the shortened critical path duration. Goldratt has developed these ideas into a project management method called *Critical Chain.* In addition to the project buffer, the method uses *feeder buffers* to protect the critical path from delays where noncritical tasks feed into the critical path. Each feeder buffer aggregates the safety times of the tasks on a noncritical path. Exhibit 16-11 illustrates the use of project and feeder buffers.

- *Eliminate some critical path tasks entirely.* Scrutinize each and every task on the critical path and ask whether it can be removed or accomplished in another way.

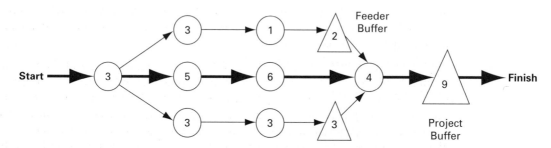

EXHIBIT 16-11 The Critical Chain method aggregates the safety time along the critical path into the project buffer. Feeder buffers protect the critical path from delays. In this illustration, nominal task durations (in days) are given for each task, and the critical path is shown with thicker arrows linking the critical tasks.

- *Eliminate waiting delays for critical path resources.* Tasks on the critical path are sometimes delayed by waiting for a busy resource. The waiting time is frequently longer than the actual time required to complete the task. Delays due to waiting are particularly prominent when procuring special components from suppliers. Sometimes such delays can be avoided by ordering an assortment of materials and components in order to be sure to have the right items on hand, or by purchasing a fraction of the capacity of a vendor's production system in order to expedite the fabrication of prototype parts. These expenses may make perfect economic sense in the context of the overall development project, even though the expenditure may seem extravagant when viewed in isolation. In other cases, administrative tasks such as purchase order approvals may become bottlenecks. Because in past cartridge development projects periodic budget approvals had caused delays, the Cheetah project leader began early to pursue aggressively the necessary signatures so as not to hold up the activities of the entire team.

- *Overlap selected critical tasks.* By scrutinizing the relationships between sequentially dependent tasks on the critical path, the tasks can sometimes be overlapped or executed in parallel. In some cases, this may require a significant redefinition of the tasks or even changes to the architecture of the product. (See Chapter 9, Product Architecture, for more details on dependencies arising from the architecture of the product.) In other cases, overlapping entails simply transferring partial information earlier and/or more frequently between nominally sequential tasks or freezing the critical upstream information earlier. Krishnan (1996) provides a framework for choosing various overlapping strategies.

- *Pipeline large tasks.* The strategy of *pipelining* is applied by breaking up a single large task into smaller tasks whose results can be passed along as soon as they are completed. For example, the process of finding and qualifying the many vendors which supply the components of a product can be time-consuming and can even delay the production ramp-up if not completed early enough. Instead of waiting until the entire bill of materials is complete before the purchasing department begins qualifying vendors, purchasing could qualify vendors as soon as each component is identified. Pipelining in effect allows nominally sequential tasks to be overlapped.

- *Outsource some tasks.* Project resource constraints are common. When a project is constrained by available resources, assigning tasks to an outside firm or to another group within the company may prove effective in accelerating the overall project.

The final set of guidelines is aimed at completing coupled tasks more quickly. Recall that coupled tasks are those which must be completed simultaneously or iteratively because they are mutually dependent.

- *Perform more iterations quickly.* Much of the delay in completing coupled tasks is in passing information from one person to another and in waiting for a response. If the iteration cycles can be completed at a higher frequency, then the coupled tasks can sometimes be competed more quickly. Faster iterations can be achieved through faster and more frequent information exchanges. In the Cheetah project, the mechanical engineer would work closely with the mold designer, who would in turn work closely with the mold maker. In many cases, these three would share a single computer display for the purpose of exchanging ideas about how the design was evolving from their three different perspectives.

- *Decouple tasks to avoid iterations.* Iterations can often be reduced or eliminated by taking actions to decouple tasks. For example, by clearly defining an interface between two interacting components early in the design process, the subsequent design of the two components can proceed independently and in parallel. The definition of the interface may take some time in advance, but the avoidance of time-consuming iterations may result in net time savings. (See Chapter 9, Product Architecture, for a discussion of establishing interfaces in order to allow the independent development of components.)

- *Consider sets of solutions.* Iterations involve the exchange of information about the evolving product design. Rather than exchanging point-value estimates of design parameters, in some cases the use of ranges or sets of values may facilitate faster convergence of coupled tasks. Researchers have recently described the application of such set-based approaches to concurrent engineering at Toyota (Sobek et al., 1999).

Project Execution

Smooth execution of even a well-planned project requires careful attention. Three problems of project execution are particularly important: (1) What mechanisms can be used to coordinate tasks? (2) How can project status be assessed? and (3) What actions can the team take to correct for undesirable deviations from the project plan? We devote this section to these issues.

Coordination Mechanisms

Coordination among the activities of the different members of the team is required throughout a product development project. The need for coordination is a natural outgrowth of dependencies among tasks. Coordination needs also arise from the inevitable changes in the project plan caused by unanticipated events and new information. Difficulties in coordination can arise from inadequate exchanges of information and from organizational barriers to cross-functional cooperation. Here are several mechanisms used by teams to address these difficulties and facilitate coordination.

- *Informal communication:* A team member engaged in a product development project may communicate with other team members dozens of times per day. Many of these communications are informal; they involve a spontaneous stop by someone's desk or a telephone call to solicit a piece of information. Good informal communication is one of the mechanisms most useful in breaking down individual and organizational barriers to cross-functional cooperation. Informal communication is dramatically enhanced by locating the core members of the development team in the same work space. Allen (1977) has shown that communication frequency is inversely related to physical separation and falls off rapidly when people are located more than a few meters from one another (Exhibit 16-12). In our experience, electronic mail and, to a lesser extent, voice mail also provide effective means of fostering informal communication among people who are already well acquainted with one another.

- *Meetings:* The primary formal communication mechanism for project teams is meetings. Most teams meet formally at least once each week. Many teams meet twice each week, and some teams meet every day. Teams located in the same work space need fewer formal meetings than those whose members are geographically separated. Time spent exchanging information in meetings is time not spent completing other project

EXHIBIT 16-12

Communication frequency versus separation distance. This relationship shown is for individuals with an organizational bond, such as belonging to the same product development team.

Source: Allen, 1977

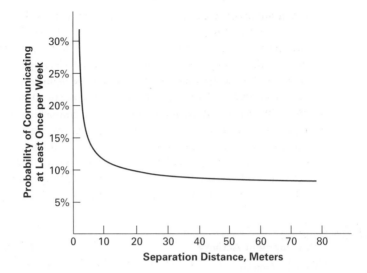

tasks. In order to minimize the amount of time wasted in meetings, some teams that hold frequent meetings meet standing up to emphasize that the meeting is intended to be quick. Other techniques for controlling the length of meetings include preparing a written agenda, appointing someone to run the meeting, and holding the meeting just before lunchtime or near the end of the day when people are anxious to leave. We recommend that team meetings be held at a regular time and place so that no extra effort is expended in scheduling the meeting and in informing the team of its time and location.

- *Schedule display:* The most important information system in project execution is the project schedule, usually in the form of a PERT or Gantt chart. Most successful projects have a single person who is responsible for monitoring the schedule. On small projects, this is usually the team leader. Larger projects generally have a designated person other than the project leader who watches and updates the schedule regularly. On the Cheetah project, Kodak provided a part-time project analyst who kept the schedule current on a weekly basis and reported to the project team leader. The team members understood the importance of accurate schedule projections and were very cooperative in this effort. Schedule updates are usually displayed in Gantt chart form (Exhibit 16-4).

- *Weekly updates:* The *weekly status memo* is written by the project leader and is distributed on paper, in electronic form, or even by voice mail to the entire extended project team, usually on Friday or over the weekend. The memo is usually one or two pages long and lists the key accomplishments, decisions, and events of the past week. It also lists the key events of the coming week. It is sometimes accompanied by an updated schedule.

- *Incentives:* Some of the most basic organizational forms, such as functional organizations which use functional performance reviews, may inhibit the productive collaboration of team members across functions. The implementation of project-based performance measures creates incentives for team members to contribute more fully to the project. Having both a project manager and a functional manager contribute to individual performance reviews leading to promotions, merit increases, and bonuses sends a strong message that project results are highly valued. (See Chapter 2, Development Processes and Organizations, for a discussion of various organizational forms, including project, functional, and matrix organizations.)

- *Process documents:* Each of the methods presented in this book also has an associated information system which assists the project team in making decisions and provides documentation. (By information systems we mean all of the structured means the team uses to exchange information, not only the computer systems used by the team.) For example, the concept selection method uses two concept selection matrices to both document and facilitate the selection process. Similarly, each of the other information systems serves both to facilitate the logical execution of the process step and to document its results. Exhibit 16-13 lists some of the important information systems used at the various stages of the development process.

EXHIBIT 16-13

Information systems which facilitate product development decision making, team consensus, and the exchange of information.

Development Activity	Information Systems Used
Product planning	Product segment map Technology roadmap Product-process change matrix Aggregate resource plan Product plan Mission statement
Customer needs identification	Customer needs lists
Concept generation	Function diagrams Concept classification tree Concept combination table Concept descriptions and sketches
Concept selection	Concept screening matrix Concept scoring matrix
Product specifications	Needs-metrics matrix Competitive benchmarking charts Specifications lists
System-level design	Schematic diagram Geometric layout Differentiation plan Commonality plan
Detailed design	Bill of materials Prototyping plan
Industrial design	Aesthetic/ergonomic importance survey
Product development economics	NPV analysis spreadsheet
Project management	Contract book Task list Design structure matrix Gantt chart PERT chart Staffing matrix Risk analysis Weekly status memo Buffer report Postmortem project report

Assessing Project Status

Project leaders and senior managers need to be able to assess project status to know whether corrective actions are warranted. In projects of modest size (say, fewer than 50 people) project leaders are fairly easily able to assess the status of the project. The project leader assesses project status during formal team meetings, by reviewing the project schedule, and by gathering information in informal ways. The leader constantly interacts with the project team, meets regularly with individuals to work through difficult problems, and is able to observe all of the information systems of the project. A team may also engage an expert from outside the core team to review the status of the project. The goal of these reviews is to highlight areas of risk and to generate ideas for addressing these risk areas.

Project reviews, conducted by senior managers, are another common method of assessing progress. These reviews tend to correspond to the end of each phase of development and are key project milestones. These events serve not only to inform senior managers of the status of a project but also to bring closure to a wide variety of development tasks. While these reviews can be useful milestones and can enhance project performance, they can also hinder performance. Detrimental results arise from devoting too much time to preparing formal presentations, from delays in scheduling reviews with busy managers, and from excessive meddling in the details of the project by those reviewing the project.

The Critical Chain method uses a novel approach to monitoring the project schedule. By simply monitoring the project buffer and the feeder buffers of the project (described briefly above), the project manager can quickly assess the criticality of each path and the estimated project completion time. If tasks consume the project buffer faster than the critical path is being completed, the project runs the risk of slipping the end date. A buffer report therefore provides a concise update on the project status in terms of progress of the critical path and its feeder paths.

Corrective Actions

After discovering an undesirable deviation from the project plan, the team attempts to take corrective action. Problems almost always manifest themselves as potential schedule slippage, and so most of these corrective actions relate to arresting potential delays. Some of the possible actions include:

- *Changing the timing or frequency of meetings:* Sometimes a simple change from weekly to daily meetings increases the "driving frequency" of the information flow among team members and enables more rapid completion of tasks. This is particularly true of teams that are not already colocated (although if the team is highly dispersed geographically, meetings can consume a great deal of travel time). Sometimes simply moving a weekly meeting from a Tuesday morning to a Friday afternoon increases the urgency felt by the team to "get it done this week."

- *Changing the project staff:* The skills, capabilities, and commitment of the members of the project team in large measure determine project performance. When the project team is grossly understaffed, performance can sometimes be increased by adding the necessary staff. When the project team is overstaffed, performance can sometimes be increased by removing staff. Note that adding staff in a panic at the end of a project can lead to delays in project completion because the increased coordination requirements may outweigh the increase in human resources.

- *Locating the team together physically:* If the team is geographically dispersed, one way to increase project performance is to locate the team in the same work space. This action invariably increases communication among the team members. Some benefit of "virtual colocation" is possible with electronic mail, video conferencing, and other network-based collaboration tools.

- *Soliciting more time and effort from the team:* If some team members are distributing their efforts among several projects, project performance may be increased by relieving them of other responsibilities. Needless to say, high-performance project teams include team members who regularly deliver more than a 40-hour work week to the project. If a few critical tasks demand extraordinary effort, most committed teams are willing to devote a few weeks of 14-hour days to get the job done. However, 60- or 70-hour weeks cannot reasonably be expected from most team members for more than a few weeks without causing fatigue and burnout.

- *Focusing more effort on the critical tasks:* By definition, only one sequence of tasks forms the critical path. When the path can be usefully attacked by additional people, the team may choose to temporarily drop some or all other noncritical tasks in order to ensure timely completion of the critical tasks.

- *Engaging outside resources:* The team may be able to retain an outside resource such as a consulting firm or a supplier to perform some of the development tasks. Outside firms are typically fast and relatively economical when a set of tasks can be clearly defined and when coordination requirements are not severe.

- *Changing the project scope or schedule:* If all other efforts fail to correct undesirable deviations from the project plan, then the team must narrow the scope of the project, identify an alternative project goal, or extend (slip) the project schedule. These changes are necessary to maintain a credible and useful project plan.

Postmortem Project Evaluation

An evaluation of the project's performance after it has been completed is useful for both personal and organizational improvement. This review is often called a *postmortem project evaluation,* although more friendly names are appropriate (Smith, 1996). The postmortem evaluation is usually an open-ended discussion of the strengths and weaknesses of the project plan and execution. This discussion is sometimes facilitated by an outside consultant or by someone within the company who was not involved in the project. Several questions help to guide the discussion:

- Did the team achieve the mission articulated in the mission statement?
- Which aspects of project performance (development time, development cost, product quality, manufacturing cost) were most positive?
- Which aspects of project performance were most negative?
- Which tools, methods, and practices contributed to the positive aspects of performance?
- Which tools, methods, and practices detracted from project success?
- What problems did the team encounter?
- What specific actions can the organization take to improve project performance?

- What specific technical lessons were learned? How can they be shared with the rest of the organization?

A postmortem report is then prepared as part of the formal closing of the project. These reports are used in the project planning stage of future projects to help team members know what to expect and to help identify what pitfalls to avoid. The reports are also a valuable source of historical data for studies of the firm's product development practices. Together with the project documentation, and particularly the contract book, they provide "before and after" views of each project.

For the Cheetah project, the postmortem discussion involved six members of the core team and lasted two hours. The discussion was facilitated by a consultant. The project was completed on time, and despite the aggressive schedule, so much of the discussion focused on what the team had done to contribute to project success. The team agreed that the most important contributors to project success were:

- Empowerment of a team leader.
- Effective team problem solving.
- Emphasis on adherence to schedule.
- Effective communication links.
- Full participation from multiple functions.
- Building on prior experience in cartridge development.
- Use of computer-aided design (CAD) tools for communication and analysis.
- Early understanding of manufacturing capabilities.

The Cheetah team also identified a few opportunities for improvement:

- Use of three-dimensional CAD tools and plastic molding analysis tools.
- Earlier participation by the customer in the design decisions.
- Improved integration of tooling design and production system design.

Summary

Successful product development requires effective project management. Some of the key ideas in this chapter are:

- Projects consist of tasks linked to each other by dependencies. Tasks can be sequential, parallel, or coupled.
- The longest chain of dependent tasks defines the critical path, which dictates the minimum possible completion time of the project.
- The design structure matrix (DSM) can be used to represent dependencies. Gantt charts are used to represent the timing of tasks. PERT charts represent both dependencies and timing and are frequently used to compute the critical path.
- Project planning results in a task list, a project schedule, staffing requirements, a project budget, and a risk plan. These items are key elements of the contract book.

- Most opportunities for accelerating projects arise during the project planning phase. There are many ways to complete development projects more quickly.
- Project execution involves coordination, assessment of progress, and taking action to address deviations from the plan.
- Evaluating the performance of a project encourages and facilitates personal and organizational improvement.

References and Bibliography

Many current resources are available on the Internet via
www.ulrich-eppinger.net

There are many basic texts on project management, although most do not focus on product development projects. PERT, critical path, and Gantt techniques are described in most project management books, including Kerzner's classic text. Kerzner also discusses project staffing, planning, budgeting, risk management, and control.

Kerzner, Harold, *Project Management: A Systems Approach to Planning, Scheduling, and Controlling,* seventh edition, Wiley, New York, 2001.

Several authors have written specifically about the management of product development. Wheelwright and Clark discuss team leadership and other project management issues in depth.

Wheelwright, Stephen C., and Kim B. Clark, *Revolutionizing Product Development: Quantum Leaps in Speed, Efficiency, and Quality,* The Free Press, New York, 1992.

The design structure matrix (DSM) was originally developed by Steward in the 1970s. More recently, this method has been applied to industrial project planning and improvement by Eppinger and his research group at MIT.

Steward, Donald V., *Systems Analysis and Management: Structure, Strategy, and Design,* Petrocelli Books, New York, 1981.

Eppinger, Steven D., et al., "A Model-Based Method for Organizing Tasks in Product Development," *Research in Engineering Design,* Vol. 6, No. 1, 1994, pp. 1–13.

Smith, Robert P., and Steven D. Eppinger, "Identifying Controlling Features of Engineering Design Iteration," *Management Science,* Vol. 43, No. 3, March 1997, pp. 276–293.

Eppinger, Steven D., "A Planning Method for Integration of Large-Scale Engineering Systems," *International Conference on Engineering Design,* Tampere, Finland, August 1997, pp. 199–204.

Eppinger, Steven D., Murthy V. Nukala, and Daniel E. Whitney, "Generalized Models of Design Iteration Using Signal Flow Graphs," *Research in Engineering Design,* Vol. 9, No. 2, 1997, pp. 112–123.

Cho, Soo-Haeng, and Steven D. Eppinger, "Product Development Process Modeling Using Advanced Simulation," *ASME Conference on Design Theory and Methodology,* Pittsburgh, PA, no. DETC-21691, September 2001.

Eppinger, Steven D., "Innovation at the Speed of Information," *Harvard Business Review,* Vol. 79, No. 1, January 2001, pp. 149–158.

Krishnan provides a framework for overlapping nominally sequential tasks, explaining under what conditions it is better to transfer preliminary information from upstream to downstream and when it may be better to freeze the upstream task early.

> Krishnan, Viswanathan, "Managing the Simultaneous Execution of Coupled Phases in Concurrent Product Development," *IEEE Transactions on Engineering Management,* Vol. 43, No. 2, May 1996, pp. 210–217.

Goldratt developed the Critical Chain method of project management. This approach aggregates safety times from each task into project and feeder buffers, allowing the project to be tracked by monitoring these buffers.

> Goldratt, Eliyahu M., *Critical Chain,* North River Press, Great Barrington, MA, 1997.

Smith and Reinertsen provide many ideas for accelerating product development projects, along with interesting insights on team staffing and organization.

> Smith, Preston G., and Donald G. Reinertsen, *Developing Products in Half the Time,* Van Nostrand Reinhold, New York, 1991.

Sobek, Ward, and Liker present the principles of set-based concurrent engineering, in which product development teams reason about sets of possible design solutions rather than using only point-based values to describe the evolving design.

> Sobek II, Durward K., Allen C. Ward, and Jeffrey K. Liker, "Toyota's Principles of Set-Based Concurrent Engineering," *Sloan Management Review,* Vol. 40, No. 2, Winter 1999, pp. 67–83.

Allen has extensively studied communication in R&D organizations. This text includes the results of his seminal empirical studies of the influence of physical layout on communication.

> Allen, Thomas, J., *Managing the Flow of Technology: Technology Transfer and the Dissemination of Technological Information within the R&D Organization,* MIT Press, Cambridge, MA, 1977.

Kostner offers guidance for leaders of geographically dispersed teams.

> Kostner, Jaclyn, *Virtual Leadership: Secrets from the Round Table for the Multi-Site Manager,* Warner Books, New York, 1994.

Markus explains that electronic mail can facilitate rich interactions between project team members, in addition to traditional rich media such as face-to-face meetings.

> Markus, M. Lynne, "Electronic Mail as the Medium of Managerial Choice," *Organization Science,* Vol. 5, No. 4, November 1994, pp. 502–527.

Hall presents a structured process for risk identification, analysis, and management, with application examples in software and systems engineering. (See also Kerzner, 2001.)

> Hall, Elaine M., *Methods for Software Systems Development,* Addison-Wesley, Reading, MA, 1998.

Smith presents a 12-step process for project review and evaluation, leading to ongoing improvement of the product development process.

> Smith, Preston G., "Your Product Development Process Demands Ongoing Improvement," *Research-Technology Management,* Vol. 39, No. 2, March–April 1996, pp. 37–44.

Exercises

1. The tasks for preparing a dinner (along with the normal completion times) might include:

   a. Wash and cut vegetables for the salad (15 minutes).
   b. Toss the salad (2 minutes).
   c. Set the table (8 minutes).
   d. Start the rice cooking (2 minutes).
   e. Cook rice (25 minutes).
   f. Place the rice in a serving dish (1 minute).
   g. Mix casserole ingredients (10 minutes).
   h. Bake the casserole (25 minutes).

   Prepare a DSM for these tasks.

2. Prepare a PERT chart for the tasks in Exercise 1. How fast can one person prepare this dinner? What if there were two people?

3. What strategies could you employ to prepare dinner more quickly? If you thought about dinner 24 hours in advance, are there any steps you could take to reduce the time between arriving home the next day and serving dinner?

4. Interview a project manager (not necessarily from product development). Ask him or her to describe the major obstacles to project success.

Thought Questions

1. When a task on the critical path (e.g., the fabrication of a mold) is delayed, the completion of the entire project is delayed, even though the total amount of work required to complete the project may remain the same. How would you expect such a delay to impact the total cost of the project?

2. This chapter has focused on the "hard" issues in project management related to tasks, dependencies, and schedules. What are some of the "soft," or behavioral, issues related to project management?

3. What would you expect to be some of the characteristics of individuals who successfully lead project teams?

4. Under what conditions might efforts to accelerate a product development project also lead to increased product quality and/or decreased product manufacturing costs? Under what conditions might these attributes of the product deteriorate when the project is accelerated?

Appendix

Design Structure Matrix Example

One of the most useful applications of the design structure matrix (DSM) method is to represent well-established, but complex, engineering design processes. This rich process modeling approach facilitates:

- Understanding of the existing development process.
- Communication of the process to the people involved.
- Process improvement.
- Visualization of progress during the project.

Exhibit 16-14 shows a DSM model of a critical portion of the development process at a major automobile manufacturer. The model includes 50 tasks involved in the digital-mock-up (DMU) process for the layout of all of the many components in the engine compartment of the vehicle. The process takes place in six phases, depicted by the blocks of activities along the diagonal. The first two of these phases (project planning and CAD data collection) occur in parallel, followed by the development of the digital assembly model (DMU preparation). Each of the last three phases involves successively more accurate analytical verification that components represented by the digital assembly model actually fit properly within the engine compartment area of the vehicle.

In contrast to the simpler DSM model shown in Exhibit 16-3, where the squares on the diagonal identify sets of coupled activities, the DSM in Exhibit 16-14 uses such blocks to show which activities are executed together (in parallel, sequentially, and/or iteratively) within each phase. Arrows and dashed lines represent the major iterations between sets of activities within each phase.

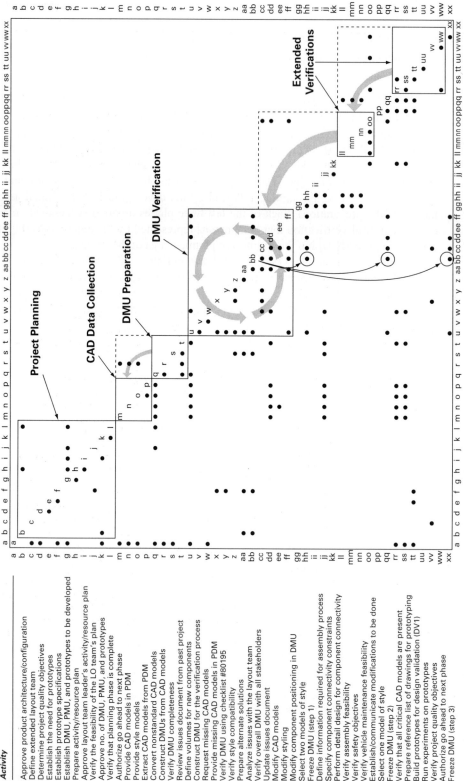

EXHIBIT 16-14 Design structure matrix model of the digital mock-up (DMU) process used to validate layout of the automobile's engine compartment.